VISION, SCIENCE AND LITERATURE, 1870–1920: OCULAR HORIZONS

AF213444

Science and Culture in the Nineteenth Century

Series Editor: Bernard Lightman

Titles in this Series

1 Styles of Reasoning in the British Life Sciences: Shared Assumptions, 1820–1858
James Elwick

2 Recreating Newton: Newtonian Biography and the Making of Nineteenth-Century History of Science
Rebekah Higgitt

3 The Transit of Venus Enterprise in Victorian Britain
Jessica Ratcliff

4 Science and Eccentricity: Collecting, Writing and Performing Science for Early Nineteenth-Century Audiences
Victoria Carroll

5 Typhoid in Uppingham: Analysis of a Victorian Town and School in Crisis, 1875–1877
Nigel Richardson

6 Medicine and Modernism: A Biography of Sir Henry Head
L. S. Jacyna

7 Domesticating Electricity: Expertise, Uncertainty and Gender, 1880–1914
Graeme Gooday

8 James Watt, Chemist: Understanding the Origins of the Steam Age
David Philip Miller

9 Natural History Societies and Civic Culture in Victorian Scotland
Diarmid A. Finnegan

10 Communities of Science in Nineteenth-Century Ireland
Juliana Adelman

11 Regionalizing Science: Placing Knowledges in Victorian England
Simon Naylor

12 The Science of History in Victorian Britain: Making the Past Speak
Ian Hesketh

13 Communicating Physics: The Production, Circulation and Appropriation of Ganot's Textbooks in France and England, 1851–1887
Josep Simon

14 The British Arboretum: Trees, Science and Culture in the Nineteenth Century
Paul A. Elliott, Charles Watkins and Stephen Daniels

Forthcoming Titles

Popular Exhibitions, Science and Showmanship, 1840–1910
Joe Kember, John Plunkett and Jill Sullivan (eds)

VISION, SCIENCE AND LITERATURE, 1870–1920: OCULAR HORIZONS

BY

Martin Willis

Published by the University of Pittsburgh Press, Pittsburgh, Pa., 15260
Copyright © 2016, University of Pittsburgh Press
All rights reserved

10 9 8 7 6 5 4 3 2 1

Cataloging-in-Publication is available from the British Library

ISBN 13: 978-0-8229-8190-9 Hardback: 978-1-84893-234-0
ISBN 10: 0-8229-8190-4

CONTENTS

For Ruth

ACKNOWLEDGEMENTS

I have been lucky enough to enjoy the support of many individuals and institutions in undertaking the research for this book. The University of Glamorgan's Department of Humanities, and in particular the English Research Unit, supported my writing of several chapters with the award of a sabbatical in 2007, as well as with the funding to participate in numerous conferences and seminars since my appointment in 2003. My colleagues in English have been a source of continual inspiration and intellectual acuity. I wish particularly to thank Jane Aaron, Alice Entwistle, Kevin Mills, Andy Smith, Diana Wallace and Jeff Wallace, who have encouraged and critiqued, and have been enthusiastic about the book and its aims throughout its reasonably long gestation. I also owe thanks to my former colleague, Rachel Hewitt, a recent research fellow with the Glamorgan Research Centre for Literature, Science and the Arts.

I have also been extremely fortunate to work closely with Keir Waddington, of the University of Cardiff, on a number of cross-university research projects, and have benefited enormously from his detailed, thoughtful (and quick!) responses to several draft chapters. Keir has, I think, read more of this book at each of its stages than anyone else. I owe him a particular debt of gratitude for doing so.

In the wider research community of the history of science and literature I am grateful to many scholars and friends who have also commented on drafts, asked important questions of seminar and conference papers, and suggested additions and amendments. I wish to thank particularly Dan Cordle, John Holmes, Alice Jenkins, Laura Otis, Vike Plock, Sharon Ruston, Charlotte Sleigh, Alex Warwick and Michael Whitworth for their encouragement throughout the project, and the members of the BSLS and BSHS for various discussions which have sharpened my thinking and improved every chapter. I would also like to extend my gratitude to the organizers and institutions who invited me to speak at seminars and conferences, and the comments of their audiences and delegates. I benefited a great deal from sharing my ideas in such circumstances at the universities of Birmingham, Cambridge, Cardiff, Edinburgh, Glasgow, Keele, Kent, Oxford, Northumbria, Nottingham Trent and Westminster. I also owe considerable thanks to my editor, Bernie Lightman, and three anonymous reviewers for Pickering & Chatto for their invaluable advice and suggestions. The final book is a better one for their interventions.

A number of archivists and librarians gave valuable support to my research. I am thankful for the help of archivists at the Wellcome Library and at Senate House, in London; at the Bodleian Library, Oxford; several Cambridge libraries; the British Library; Cardiff University Library; and the Library of Congress in Washington DC. I also owe considerable thanks to the knowledgeable archivists at the Petrie Museum and in the archive of the Egypt Exploration Fund, London. Special thanks also go to Antoinette Beiser and Lauren Amundson of the Hendricks Center for Planetary Studies at the Lowell Observatory in Flagstaff, Arizona, for their superb support as I worked through the archive of the observatory's founder, Percival Lowell.

I have received financial support from the Wellcome Trust, the British Academy, HEFCW's Strategic Insight Programme and the Scientific Instruments Society. I could not have conducted much of my archival research without their funding and certainly would not have been able to travel so extensively to assess and examine such a range of sources.

I owe a small group of friends my gratitude in these acknowledgements, too. For several years now they have had to suffer my spontaneous and often lengthy monologues on vision and the Victorians. It's too late for an apology, but I can at least thank them for being both patient and, when I needed it, enthusiastic. Thanks, then, to Anthony, Eryl, James, John, Lewis, Liz, Melanie, Rich, Rose and Simon. I owe them numerous dinners and a truckle of cheese.

I would not be thanking anyone were it not for Ruth. Without her energy and commitment to our life together and all we do in it, I would never have finished this book, and perhaps never even have started it. I hate at any moment to disappoint her, but I must do so here. Despite her belief that I have failed to deal with a key theme in visual culture – which she calls 'man-eyes' and which apparently affects man's ability to see things directly in front of him – I have never found any evidence of it in the historical literature. But, of course, I might have missed it.

LIST OF FIGURES

INTRODUCTION: OCULAR HORIZONS: VISION, SCIENCE AND LITERATURE

> I've taken to the eye, my boy. There's a fortune in the eye. A man grudges a half-crown to cure his chest or his throat, but he'd spend his last dollar over his eye. There's money in ears, but the eye is a gold mine!
>
> Arthur Conan Doyle[1]

This is a book about vision and its historical fragility.[2] It deals particularly with vision in the nineteenth and early twentieth centuries, or what might be called Victorian and modernist ways of seeing. If this were its only organizing principles it would be a vast book, indeed it would probably be a small library of books, such is the extent of scholarly interest in vision, visuality and perception. However, it has a particular focus (I recognize this pun and want to return to it in a moment). In it, I consider the role of vision across science and literature: how instruments, objects, places, people, eyes, ideologies, discourses and imaginations together make the many ways of seeing that characterize the second half of the nineteenth century and the opening decades of the twentieth. This is a project, largely, of cultural phenomenology. I am not interested in the things scientists or writers (and their fictional characters) saw, but what they did and thought when they looked at them, and what they said about that looking. I am interested, therefore, in the phenomenon of seeing and the cultures of observation. Perhaps more than anything else, I attend to the discourses of seeing, looking and observing: the language and vocabulary employed in discussions of vision, both consciously and unconsciously. Back to my pun: I used that cliché of academic scholarship – 'the particular focus' – to highlight how vocabularies of vision permeate language and are often foundational in the writing of exploratory disciplines (in which category I would place both scientific and literary discourse). Commonly such vocabulary offers access to knowledge: it shines a light upon, makes clear, opens one's eyes to, illuminates, highlights, emerges from the dark and, of course, brings into focus. Despite this imperial linguistic quality vision did not, in the period I will be discussing, offer easy access to anything. Indeed, vision was fragile: characterized as illusory as often as it was penetrative, or found to be opaque as readily as it was perspicuous.

This sense of vision as subject to myriad stresses that enhanced or limited its power, strengthened or weakened it, is registered in the book's subtitle: ocular horizons. The horizons of ocularity, viewed from the science and literature of the late Victorian and Edwardian periods, show vision as a merging of the biological functions of the eye and optic nerve with the myriad cultural functions carried by the viewer. It is in this period that vision became, and was continually reinforced as, a negotiation between the actual and the metaphoric, often articulated as the real and the imagined. However, vision did not function simply as a series of binaries. Instead, acts of seeing were performances of extraordinary variation that occurred at each individual site of visual exchange. It is in the particulars of who is looking, at what, where, when and with whom that an appropriate understanding of the vision of this period can be found. What I shall claim, however, is that these ways of seeing are still in place now. There is no radical shift in vision from the nineteenth century into the twentieth century, or indeed from the early to the later twentieth century. How we see in the present is rooted in how the Victorians came to see, and while we might all now look at different things, and at things the Victorians could not see, our visual epistemology still owes something to theirs.

Historiography of Victorian and Modern Vision

By making such a claim I am arguing against one of the abiding paradigms of visual culture: that there was a recognizable change in visuality in the late nineteenth century. Either by slow evolution or sudden revolution it has been generally accepted that vision moved from objectivity to subjectivity, and in doing so became 'modern'. One of the most influential of these analyses of vision, especially in relation to scientific technologies, is Jonathan Crary's articulation of the change from objective to subjective vision.[3] Crary's work of the 1990s remains a touchstone not only for history and philosophy of science scholarship on nineteenth-century vision but also for cultural interrogations of vision and visuality. Crary argues that the classical vision of the eighteenth century, marked by a determined objectivity, gave way in the first half of the nineteenth century to a modern, subjective vision which remains in place in modernist visuality in the early twentieth century. Writing at the same time as Crary, James Krasner's work on visual perception in Darwin's evolutionary science and post-Darwinian narrative, supports Crary's reading of a shift from objectivity to subjectivity.[4] For Krasner, however, this change occurs only after Darwin's publication of his evolutionary theory in 1859, and is characterized specifically by a turn also from exterior to interior vision. Krasner argues that new understandings of the physiology of the eye had a particular influence upon vision in the late nineteenth and early twentieth century, leading to an understanding that vision was an interior process which organized and ordered sight rather than a way of seeing the exterior world that remained disorganized and incomplete. Krasner's and Crary's powerful reading of vision as moving from objectivity to subjectivity as the nineteenth century gave way to the twentieth century became (and perhaps still remains) the accepted paradigm of contemporary scholarship. When Carol T. Christ and John O. Jordan introduced their important collection of essays on the Victorian visual imagination in 1995 they identified the turn to subjectivity leading to modernity as one of the most persuasive accounts of nineteenth-century vision.[5] Similarly, Kevin Z. Moore, in his account of research on Victorian visuality in 1997, saw a difference between the external vision of the Victorians and the internalized subjective vision of the moderns.[6] Kate Flint, too, in her wide-ranging work of 2000 on the visual imagination accepted that vision became increasingly subjective in the second half of the nineteenth century, and even Srdjan Smajić, writing in 2010, saw relativism and subjectivism as the dominant visual mode after the 1840s, as well as arguing that this was becoming increasingly focused inwardly from the last decades of the nineteenth century.[7]

Nevertheless, this construction of vision's evolution has not gone without criticism. Indeed Christ and Jordan, while identifying its importance, noted that there was a plausible alternative account of vision as increasingly realist, moving from photography into modern cinema. Ultimately, they claim, the Victorians were interested in the contested middle ground; the conflict between subjectivity and objectivity. Jennifer Green-Lewis, in her excellent account of photography (1996), says something similar in arguing that characteristic of photographic visual culture is an oscillation between truth and fiction.[8] Chris Otter is also sceptical of the objective to subjective paradigm; in his fascinating 2008 account of light and vision in nineteenth-century liberal politics he argues that this evolution is not at all clear, nor the first stage (objectivity) even likely to be so clearly demarcated.[9]

In the sciences and imaginative literatures that I interrogate in the ensuing chapters, I have found no clear evidence of the existence of a shift from the objective to the subjective in vision. Nor have I found that Victorian vision differs to any significant extent from the vision of the modernist period. It seems to me that the conflict between subjectivity and objectivity identified by Christ and Jordan (and also in a different context by Green-Lewis) is decidedly closer to the complex visual culture of the nineteenth century than those accounts of Crary and Krasner. Yet nineteenth- and early twentieth-century vision does not always function as an easy binary of these two modes. The Victorians and Edwardians were themselves very conscious of these labels (which Moore wrongly denies) and caught them up or cast them aside depending upon their position with regard to vision at specific moments, in distinct locations and in relation to particular individuals or institutions. They did not trust the simple categorization of vision as either objective or subjective, and we should not therefore attempt to flatten the contours of their sophisticated visuality by applying them retrospectively and simplistically.

A further defining feature of the scholarship on vision, and specifically in the history of scientific technologies, has been the effort to fit the differing modalities of vision into a Foucaldian structure of discipline and control. This is apparent in Green-Lewis's argument that photography did not only show or reveal images of Britain but also controlled its representation. It is also central to Carla Yanni's understanding of visual displays of science and their relationship to a defining and controlling museum architecture (2005).[10] Kate Flint, too, in her discussion of scientific instruments argues that they made the natural world visible and therefore controllable. Audrey Jaffe's analysis of representation and spectatorship in 2000 is perhaps the most obviously Foucaldian of all these critical readings; claiming that nineteenth-century culture was entirely defined by spectatorial encounters and that these should be read as self-scrutinizing conflicts of power.[11] Chris Otter has been the most vociferous opponent of such Foucaldian analysis. He argues that visual experience was caught somewhere between freedom and restraint, and cannot therefore be fitted neatly into Foucault's principle that the control of visual authority determined unidirectional relationships of power.[12] Otter is, I believe, quite right to question the validity of such approaches to nineteenth-century vision. As I shall show, scientific instruments that extended the visual capacity of the eye (the microscope and the telescope, say) did not exert control over vision, nor did those observers who used the technologies. Indeed visual technologies just as often undermined visual authority, or came to stand as metaphors for vision's fragility. More broadly, scientific observation (even without the use of instrumental aid) was rarely able to proclaim itself as an ultimate authority, and never for very long.

In part this was because seeing was actually very difficult. Nineteenth- and twentieth-century observers (everyone, that is) had lots to see, and could very often only see a little of it. While there was a proliferation, as Flint argues, of image and display in nineteenth-century society, there were also many new ways of seeing that were inaccessible to large numbers of people. Not everyone could look through a telescope at Mars or the transit of Venus, not everyone could put their eye to the microscope and see infusoria or later, bacteria. Nor could everyone see the world of spirits that the séance revealed, as Smajić points out, or the many other invisible or inward worlds that appeared to, or were imagined to, exist. Moore claims that by the mid-nineteenth century the image had replaced the imagination in visual culture, but Flint's argument that the relationship between the visible and invisible, and therefore between sight and imagination, was more complex and slippery than this, is nearer the truth. Smajić, too, agrees with Flint, arguing that sight was both corporeal (residing in the eye) and perceptual (occurring only in the mind). What they do not say, however, is that the porous boundaries of seen and unseen, and of image and imagination, led to a new epistemology of vision that held the actual and the imagined in fragile suspension. Each visual encounter had the potential to push the observer and the observed into either category, or both, or to shift between them. Nor could the observer ever be certain of the outcome of such encounters because the control they had over their own vision was limited, and their control of the object of their observation even less so.

Certainly those who looked, and especially those who looked at new things, as scientists and fiction writers did, had an opportunity to try to define their visual encounters (and thereby their vision *in toto*) but it was never the case that they looked alone, or that they were aware of all the pressures exerted on their vision. Inevitably, their discourses reveal both these efforts at definition as well as their recognition of their vision's fragility. As I shall show in Part III, the archaeologist looked at the newly-discovered artefact, dated and defined it, but then wondered whether it was anything more than a fantasy of a foreign land. And as I argue in Part II, the astronomer saw canals on Mars, but others claimed he had poor eyesight and was being hoodwinked by an optical illusion. I will reveal, in Part I, that while one microscopist claimed he had found the cause of disease in a tiny microbe, a second thought it was the product of his imagination. Each of these is a different visual encounter (and each of them I examine in detail in the chapters to follow) which gives rise to variable understandings of vision. Each, though, is characterized to some extent by a recognition of vision's fragile status and of the potential for vision to be manipulated as it is employed in the search for new knowledge.

Ultimately, however, it is only by attending to the individual visual encounter that any organizing principles of Victorian and modern vision can be identified. Nevertheless, each of these encounters is not so limited as to be without significance. Since they traverse several areas of Victorian knowledge, in order to consider science's relationship to literary culture and social politics, they do offer some indication of how certain characteristics of vision become components of shared ways of seeing that articulate, even if only partially, how Victorian and modern society observed.

Organizing Principles

What individual visual encounters have I chosen to consider, and why? Part I investigates the function of microscopy in medicine, and particularly in the context of disease transmission. It offers an analysis of scientific writing on the microscope and disease allied with a series of Gothic fictions that represent them imaginatively. It concludes with a specific case study of the British Institute of Preventive Medicine. Part II turns from the microscope to the telescope by considering astronomical debates, popular science writing and fiction that focus on the planet Mars. The work of the astronomer Percival Lowell situates the analysis temporally and culturally across the late Victorian and Edwardian periods. In Part III I consider the archaeological work of Egyptologist William Matthew Flinders Petrie and the Egypt Exploration Fund. Here again elite science in placed in dialogue with popular archaeology and with literature – both fiction and travel writing. Finally, Part IV centres on ophthalmology and optics to interrogate their emergence in creative arts (fiction and stage magic). In the productions of Harry Houdini and Arthur Conan Doyle vision itself is under investigation as a cultural phenomenon, and it is in their important conclusions that the book is brought to an end.

These varying explorations may make the whole appear rather kaleidoscopic. Yet they have been chosen particularly to allow for both an extended reach and a tightened interplay. First, I wished to consider visual technologies of differing scale; hence the opening two parts on the small, via the microscope, and the large, through the telescope. But I wanted, too, to consider vision in a scientific field that did not rely upon instrumentation. Archaeology offered the opportunity to investigate the discovery of things previously unseen but to do so without the reliance on visual technologies. It stands, in this respect, as an important corrective or alternative to microscopy and telescopy. Finally, I saw it as important to deal directly with sciences closely connected to the eye itself. Part IV's analysis of ophthalmology and optics allows space for close attention to vision in its key bodily site.

The organizing titles of the four parts – small, large, past and future – is not intended to suggest potential oppositions but rather complementarity. I do not address the large scale vision of the telescope by recourse, say, to its difference from the smaller visual scale of the microscope. Rather, the decision to look at both in turn was taken to allow me to explore the kinds of visual culture they reveal on their own terms and in the parallels that might (or might not) exist between them. In turn, parts III and IV are not oppositional in focusing on the past and the future. Indeed, these are conceived rather similarly. The past represents archaeology's attempts to visualize ancient historical sites and objects from the perspective of the late Victorian and Edwardian present. The future also situates the present in the late Victorian and Edwardian period but highlights how the creative and cultural use of optical and ophthalmological research looked forward to the future of visuality in the twentieth century.

Each of the four parts of this volume is not limited therefore to consideration in terms of any other single part, but can be thought of in connection to any of them. For example, the role of optical devices in the final chapter are as interesting in the context of microscopes in the first as they are by comparison with the ophthalmological instruments discussed alongside them. In fact I have specifically sought to draw out connections across each of the parts. The microscopist Jabez Hogg, whom I discuss in the first chapter, makes a reappearance in Part IV writing about the ophthalmoscope. Florence Nightingale's work on disease prevention, again in the first part, may be reflected on through her travel writing about Egypt in Part III. H. G. Wells's work on popular science, considered in detail in Part II, uses Arthur Conan Doyle's fiction as its exemplar, which I then go on to talk about in Part IV. And Harry Houdini's one attempt at fiction, which I discuss in Part IV, situates the narrative action in the Egyptian tombs which Flinders Petrie worked with in Part III. Such cross-correspondences are a conscious tactic employed to engender a freer interplay between each part of the book, and to suggest that a cohesive understanding of vision can only emerge from multiple perspectives.

Inevitably there are aspects of visual culture absent here. I made a choice not to include two interesting areas of investigation that have been given prominence in recent critical work: photography and art. Both can be very valuably discussed in the context of the history of science, or in literary scholarship, and have been. Nor do I attend to the other senses with which vision is inextricably connected. In his book on vision, Srdjan Smajić very modestly acknowledges how he feels the loss of the other senses.[13] I do not share this sense of loss with him. In fact, I would argue that vision's overriding importance within the sensorium necessitates a focus only on that premier sense. Most nineteenth-century and early twentieth-century commentators would, I think, agree with me. When Charles Darwin wanted to illustrate evolution he used the eye as his example, just as William Paley had done to explain God's existence and greatness and to refute atheism some half a century earlier. At the opening of the twentieth century, Georg Simmel's sociology of the senses began, predictably, with the visual sense. Sociology, science and religion all give primacy to vision. And, of course, as Conan Doyle's Stark Munro reminds us, in this introduction's epigraph, there may be money in the ear, but 'the eye is a gold mine!'

Reading Science and Literature

There is one remaining organizing principle that I have left unmentioned. The book as a whole is designed as a refutation of an influential view (in the philosophy of science in particular) that by the nineteenth century the sciences and imaginative disciplines, nay the imagination itself, had parted company never again to meet. This thesis has been disseminated very successfully by Lorraine Daston in several of her engaging analyses of scientific history from the 1990s and 2000s. Daston argues that the imagination's role in science was devalued (and thereafter disappeared) during the late eighteenth-century Enlightenment and was thereafter replaced by nineteenth-century objectivity.[14] Imagination, I shall argue, neither disappeared nor was devalued, but continued throughout the nineteenth century to give impetus to the making of knowledge. Indeed both the literary text that deals with scientific themes and scientific discourses and practices employ the imagination in similar ways, and it is in their uses of the imagination that the convergence between science and literature is at its strongest.

Since Gillian Beer (1983) and George Levine (1988) suggested that the connections between literary and scientific texts were far closer than we had previously imagined, work in the field of literature and science (as Beer would call it) or science and literature (as Levine prefers) has focused its attention on both the use of literary devices in science writing and the influence of scientific ideas on imaginative literature.[15] In tandem with this, however, the paradigm of science and literature as two separate cultures has continued its strong influence. Recently, however, fresh analyses of the relationships between literature and science have set aside the two cultures and also offered new ways of thinking about the interrelationships between scientific and literary narrative. Most interesting has been Ralph O'Connor's extensively evidenced argument of 2007 that science should be examined as literature.[16] In fact, O'Connor's use of this phrase is slightly misleading, for he does not wish to suggest that scientific writing is actually imaginative fiction, but rather that science writing is not only valuable because of its factual content, but because it is a narrative form (and employs narrative genres) akin, and sometimes exactly equivalent, to literary form (and its genres). This is a compelling argument, and in O'Connor's example of popular geological writing, it is articulated through a tremendous range of sources and with great deftness.

O'Connor's analysis of science as literature has been an important influence on my efforts here to reclaim the imagination's role in scientific knowledge-making. Important, first, because it revealed the value in considering form and structure as well as metaphor and meaning in linking science with literature. Yet it is also important because it highlights what is, in my view, one side of a more radical integration of science with literature. Imaginative writing about science in fiction mirrors some of the methods of the scientific text.[17] Variously, literary texts, like their scientific counterparts, employ diagrams and illustrations to construct and enhance meaning; they draw connections between their own investigations and the investigative work of other writers and communicators; they reflect upon their own practice, especially in relation to its objectivity; they act as virtual witnesses for scientific experiments and events; and, like the *tableaux vivants* of seventeenth- and eighteenth-century works of popular science, they offer above all else an examination, and understanding, of science's effects within the human social world. Each of these are methods found in scientific discourse, but also in the material practices of science, such as the work of microscopy, where meaning is made by illustrating, contextualizing and witnessing. To read Amelia Edwards's travel book on Egypt, *A Thousand Miles Up The Nile* (1876), or H. G. Wells's *The War of the Worlds* (1898), is to discover how imaginative writing on science (here, on archaeology and astronomy) is abundant with illustration and pictorial evidence, cross-references to contemporary scientific knowledge, the self-discipline required for appropriate negotiation of subjective and objective analysis, and a consistent examination of the role science plays in the world. Literary writers do, then, make efforts in form, structure and content to produce fictional narratives that parallel scientific texts. They do this not simply as a form of homage but to highlight their own contribution to disciplinary knowledge.

Yet where scientific narrative and literary narrative most obviously overlap is in the employment of the imagination. Through the evocation of wonder, and the engagement with science as a visual spectacle, the imagination was part of scientific analysis from the very beginning, providing the inspiration for investigation and then taking those investigations further when rational inductive methods were no longer suitable or available. The imagination was not, therefore, a last resort for the scientist but implicated in all scientific work. In creating a literary work the role of the imagination is usually accepted as incalculably important, and in existence from the very first creative moments. While this is true, it is important to recognize that in fictional texts that aimed to investigate scientific ideas, writers channelled their imagination in the same ways as scientists. Literary narratives built their fictional science on rational foundations, drawing on existing knowledge to speculate about science's meanings and influence. To this extent, the imagination employed by the writer of fictional texts is employed in the same functional capacity as that used by the astronomer or archaeologist. For instance, Wells's *The War of the Worlds* is based around the same generic principles Wells used in an essay to describe the very best works of popular science. Similarly, Edwards's travel book on Egypt was only completed after she had returned to Britain to undertake two years of study in contemporary archaeology. In each case their imagination is used in combination with rational knowledge. The literary text is clearly, in these examples, parallel to the scientific text. The continued existence and vital importance of the imagination undermines the case for its subjugation by Enlightenment objectivity. And as I hope to show in this book the imagination not only played a key role in science in the Victorian and Edwardian periods, it was also the thread that stitched together its science and literature. There is, I will argue, not two different cultural forms, but one cultural continuum where the making of public scientific knowledge is juggled between elite scientific writing and practice, popular science, fiction and creative performance. All have an investment and a role in building and moulding science's role in society; and it is through attention to their visual cultures that this comes into focus.

PART I: SMALL

1 MICROSCOPY AND DISEASE: SCIENCE, IMAGINATION AND THE PHANTASMAGORIA

The microscope was one of the most profound signifiers of new modes of scientific vision that came to maturity in the nineteenth century. As Graeme Gooday noted in 2008, the microscope became 'the iconic instrument of laboratory epistemology', representing access to new knowledge, objectivity and trust in the procedures of scientific experimentation.[1] Yet before the microscope attained its position as one of the pre-eminent scientific instruments, which it did only in the last two decades of the nineteenth century, the new ocular horizons it presented to those who put their eye to its lens were unstable and ambiguous. Indeed in one of the key areas of biological research, the study of infectious diseases, the microscope both offered increased hope for the advancement of knowledge and made knowledge unpredictably haphazard. William Henry, the writer of one of the first extended reports on disease for the new British Association for the Advancement of Science in 1834, said in his discussion of infection that while 'our senses give us no insight into the properties of this subtile [*sic*] agent' it is more unfortunate that we do not 'derive any assistance from the most refined instruments'.[2] It was not that Henry's microscope did not add to his visual capacity, but that it failed to provide either the necessary 'magnifying power' or the steady 'clearness of definition' needed for thorough investigation.[3] The microscope tantalized: it hinted at the existence of objects that might lie just beyond its scopic capacity but was unable to bring them into focus. Imaginative writers of fiction with an interest in the instruments used by savants deftly represented this problem of visual identification. Fitz-James O'Brien, for example, in his fantasy of the perfect microscopic lens, wrote that the microscope of the present (he was writing in the 1850s) may 'lay bare a world of enchantments' but that this was as likely to be 'the result of an optical illusion' as it was a true representation of the natural world.[4] Such epistemological uncertainty was the defining feature of what this chapter will call 'microscopic vision', and it continued to be one of the most influential modalities of seeing across the nineteenth century.

This opening chapter traces microscopic vision via its most consequential outlet: investigations of infectious disease. It draws on the practices and writing of experimenters whose sympathies lay with contagionist laws of disease transmission, as well as those who defined themselves as miasmatists, or were involved in the sanitary science movement. The chapter considers how the objects of research (Henry's subtle agent, the microbe, or the germ) were given definition and representation by the ways in which they were encountered through the microscope. These representations were not restricted to the work of epidemiologists, medical practitioners or microscopists. Writers of Gothic fiction, given impetus by microscopic research on infection, evoked the tropes of the Gothic to reimagine objects of infection as active, changeable organisms, just as microscopists described them. In the public sphere, too, the microbe or germ was similarly cast as an active agent. For example, the protests against the proposed creation of a new British Institute for Preventive Medicine in London in the early 1890s very clearly builds a narrative of infection that is founded upon microscopic vision. Indeed the study of microscopic vision illuminates how tightly experimental practice, public understandings of science, and imaginative responses to scientific knowledge were bound together. They all focus their attention on the problematics of vision, its instability and potential for trickery. Each also finds representational material in the language and register of the Gothic. Finally, all define microscopic vision as productive of the phantasmagoric; as a complex interplay between science, technology and the imagination. Exploring these parallels also highlights the interpenetration of science, society and literary culture. This is not the kind of hierarchical interpenetration in which scientific influence gives fiction a patina of rational objectivity, or the lay public ventriloquizes elite knowledge. Rather it is an entanglement of different arenas of knowledge-making, where, as Ken Alder argues, a fictional or public narrative might be 'a kind of truth, a work of verisimilitude' just as readily as science (or the history of science).[5]

Microscopic Vision, Imagination and Objectivity

What changes in vision did looking through a microscope engender? In the first instance the power of the microscope to reveal new and ever smaller worlds provoked observers to marvel at the spectacle of the minuscule. In the many popular accounts of microscopy it is this marvelling that dominates, and it is a sense of the marvellous as fascinating and charismatic, as Lorraine Daston and Katherine Park define it, that microscopists evoke when they examine the ways of seeing that the microscope offered.[6] William Carpenter, for example, noted in 1856 that:

> We cannot long scrutinize 'the world of the small' to which we thus find access, without having the conviction forced upon us, that all size is but relative, and that mass has nothing to do with real grandeur. There is something in the extreme of minuteness, which is no less wonderful, – might it not almost be said, no less majestic? – than the extremes of vastness. If the mind loses itself in the contemplation of the immeasurable depths of space, and of the innumerable multitudes of stars and systems by which they are peopled, it is equally lost in wonder and admiration, when the eye is turned to those countless multitudes of living beings which a single drop of water may contain.[7]

For Carpenter, and others who wrote with similar effusion of the microscope, the new worlds that it revealed were also marvellous because they were set apart 'from more mundane phenomena'.[8] Yet in being outside normal visual understanding, microscopic vistas were also beyond perception, caught between the real and the imagined. Indeed, for one microscopic enthusiast writing in 1806 it was a complex interaction between these two ontologically different spaces:

> The microscope displays to us, in each object, a thousand others which escaped our knowledge. Yet, in every object discovered by this instrument, others yet remain unseen, which the microscope itself can never bring to notice. What wonders should we see, if we could continually improve those glasses, which are invented for the assistance of our sight. Imagination may, in some measure, supply the defect of our eyes, and make it seem as a mental microscope to represent in each atom thousands of new and invisible worlds.[9]

The role of the imagination is to construct what the microscope might reveal when the eye and the lens have reached their limit. Clearly, this departs from what might be verifiable observational fact and enters the realm of fiction. The medical polymath and eye surgeon, Jabez Hogg, describes the experience of looking through the microscope in 1856 as 'astounding [in] its revelations' but notes also that it leads to a point where 'our facts become stranger than fiction, and far beyond the imaginings of the most poetic brain'.[10] In similar fashion in 1846 Gideon Mantell compares the discoveries of 'that noblest instrument of natural philosophy, the Microscope' to 'the realisation of the imaginings of Shakespear [*sic*] and of Milton, or of the speculations of Locke and of Bacon'.[11] Microscopic vision, then, was represented as both the actualization of what had previously only been imagined and the starting point for a new process of imagining what might still be out of visual reach. To look through the microscope was to engage, therefore, in a visual phantasmagoria: where, as Terry Castle explains, the technological production of visual phenomena is entangled with the phantasms of imaginary experience.[12]

For the study of infectious disease developments in the power of the microscope promised to give new impetus to research on the causes of infection. Most especially in histology, the microscope's technological extension of the capabilities of the human eye suggested that any presently unknown organic constituents of blood would soon be made visible, including those that played a part in infecting the body. As Figure 1.1 shows, bacteria invisible to the naked eye (in the top two images) were discovered by the microscope (below) and revealed as active agents of disease (bottom right and left). In this Figure the 'schimmelpilze' (fungus or mould) and 'Eiterbakterien' (bacterial matter) are transformed by microscopic vision into bearers ('träger') and exciters ('erreger') of infection and disease.

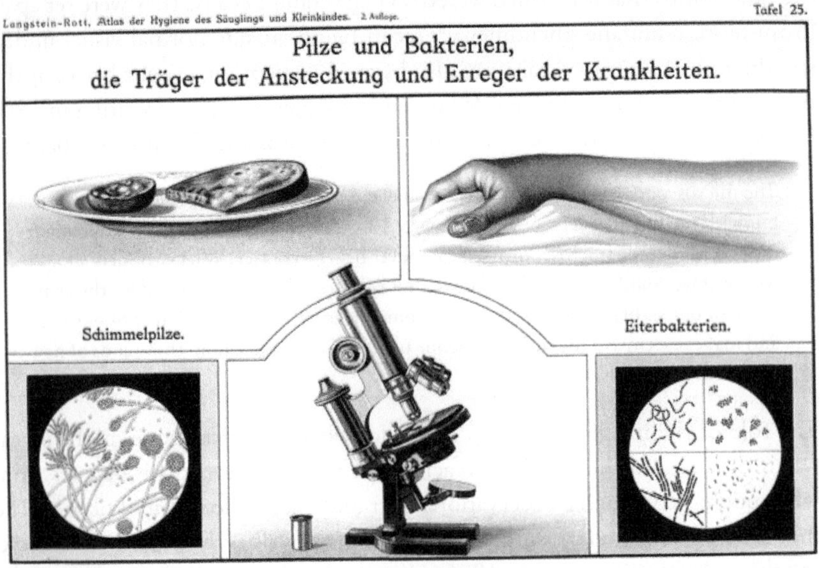

Figure 1.1: A German pictorial representation of the potential risks of infection from bacteria. The invisibility of the bacteria in the top two images are countered by them being made visible in the two microscopic slides in the lower half of the picture, made possible, by the microscope. Source: Wellcome Historical Images; reproduced with permission of the Wellcome Trust.

However, until the advent (and acceptance) of bacteriology in the 1890s, microscopic observation – or rather the microscopic vision that emerged from observation – did not bring to an end debates about infection and its causes but rather created the conditions from which a range of different understandings of infection were articulated, often by recourse to the interplay between fact and fiction.[13] Contagionism was the dominant disease theory of the early nineteenth century and, certainly in Britain, was the position adopted by the state in defending the nation and its citizens from infection.[14] Yet one state-appointed expert on contagion, Edward Bancroft, in his governmental report on contagion from 1806, wrote of it in terms very similar to those microscopists uncertain whether their observations were real or the product of the imagination: 'our belief frequently depends upon supposed causes and effects, whose existence and relations are not capable of being either seen, heard or felt; and yet men will frequently imagine they have seen, heard or felt all that is necessary to warrant their belief'.[15] As Bancroft's discomforting shift around the issues of seeing infection show, new microscopic studies only served to interweave apparent truths with an imagined infective agent that remains beyond ocular (and other sensual) limits.

Anti-contagionists (those who believed that infection was environmental in origin and arose primarily from poisonous miasmas proceeding from stagnating organic matter) also employed the imagination in an effort to 'see' the agents of disease. As microscopic lenses continued to improve, some of the most influential figures of the sanitary science movement (who were avowed anti-contagionists) brought the evidence of microscopic investigations to the fore in promoting their desire to see improvements in sanitary conditions across Britain. Thomas Southwood Smith, the founder of the London Fever Hospital and the physician who was invited to conduct the public anatomical examination of Jeremy Bentham's body, employed microscopical evidence to support his anti-contagionist belief. In an early work on epidemics (1866), Southwood Smith argued that it is the dust that accumulates in unclean domestic spaces that is likely to cause illness: 'the moisture in the air of a crowded room may be condensed by ice', he explains, and then examined under a microscope. The microscope will reveal that the moisture 'is indeed pernicious, for it is an animal poison ... it collects in large quantities on the furniture and walls of dirty houses'.[16] Another leading figure in the sanitary science movement, Florence Nightingale, agreed wholeheartedly with Southwood Smith's conclusions, arguing in her hugely popular *Notes on Nursing for the Labouring Classes* (1860), that nurses should 'keep the air [the patient] breathes as pure as the external air' and 'to rid the walls, carpets, furniture, ledges, etc., of the organic matter and dust, dust consisting greatly of organic matter, with which they become saturated, and which really makes the room musty'.[17]

Yet for all the certainty that these views appear to express, they were destabilized by sanitary scientists' recognition that the imagination was required in order to accept that disease resided in household dust. As Nightingale argued, to come to terms with the existence of disease in the surrounding environment required 'the want of observation *simple*, and the want of observation *compound*, compounded, that is, with the imaginative faculty'.[18] Nightingale is here giving definition to the microscopic vision that dominated the study of infectious disease; that phantasmagoric entanglement of what the technological instrument allows an observer to see, combined, or 'compounded' as she puts it, with what can only be called up by the imagination. John Tyndall, too, in his investigations into the constituent parts of the dust of his own London drawing-room, which he published in *Nature* in 1870, recognizes the powerful interplay between observation and imagination. He first notes that 'the air of our London rooms is loaded with this organic dust, nor is the country free from its pollution'. He goes on to describe, in a synecdoche of microscopic observation, that a 'sufficiently powerful beam [of light] causes the air in which the dust is suspended to appear as a semi-solid rather than a gas' and that the 'disgust' he feels at the sight of this is not 'abolished by the reflection that, although we do not see the nastiness, we are churning it in our lungs every hour and minute of our lives'. However, his paper concludes with a recognition that it may still be either 'a truth or falsehood' whether this dust carries disease, and he cites as a caution 'such papers as those of Dr. Budd, of Bristol, on cholera, scarlet-fever and small-pox ... who is a man of strong imagination, and may occasionally take a flight beyond his facts'.[19]

Tyndall's paper is often credited as coming close to a nascent germ theory that was not fully articulated until Pasteur's work a decade later. Yet even when the germ theory of disease came to be debated by chemists and biologists in the 1880s, concerns about the use of the microscope were still commonly discussed, and microscopic vision utilized to suggest that the germ might still be a product of the imagination. Pasteur's rival Félix-Archimède Pouchet, for example, hinted at the microscope's complex visuality in his efforts to deny the existence of the germ: 'microscopy says, "These [germs] are tangible things; one can generally feel and see them. Whoever speaks of them is bound to show them." But this is what no one has yet done.'[20] Anti-contagionists were also strong opponents of germ theory, which threatened, they felt, to undermine their project of sanitary improvement. For Florence Nightingale the germ was nothing more than an imagined truth, a phantasmagoria of the senses produced by microscopic visions of microbial life:

> The insanity of our doctrines about 'germs' and proceedings about cholera is so virulent. Our whole Indian experience tends to, nay actually proves, that cholera is not communicable from person to person, that it is a local disease, depending on pollution of buildings, earth, air and water, the quarantines, cordons, medical inspection and the like are all fatal aggravations of the disease ... Is this not so?, that attendants do not 'catch' the disease from the sick, anymore than they do from poisoned cases. Is not all this so? You look at Europe and Egypt now – it is this doctrine of 'germs' which has 'poisoned' us.[21]

Nightingale's articulation of germ *theory* as a form of infection illuminates the ease with which scientific knowledge can itself be rejected as a false epistemology emerging from the imagination.

Nightingale also suggests, however, that drawing on 'experience' might allow the observer to make rational decisions about the visual evidence provided by

Gothic Infection

Before considering microscopic vision specifically it is important to establish that Stoker and Le Fanu did employ the figure of the vampire as a representation of infection. In fact, neither was particularly innovative in doing so. The earliest vampire fiction, John Polidori's 'The Vampyre' (1819), imagined the vampire as a metaphor for disease.[33] The first vampiric attack takes place in the kind of domestic space which sanitarians would have believed to be indicative of disease-causing miasmas: 'a hovel that hardly lifted itself up from the masses of dead leaves and brushwood which surrounded it'.[34] From his encounter with the vampire in such a place, the narrative's central character Aubrey 'was seized with a most violent fever'.[35] The story clearly connects Aubrey's illness with the attack of the vampire, suggesting that the vampire itself is being imagined as either a disease or a disease-causing miasma. This is further reinforced when the vampire Lord Ruthven – whose ruthless romantic pursuit of young women of wealth and status hint at his potential for moral infection – acknowledges the link between himself and Aubrey's final illness: 'When he heard of Aubrey's ill-health, he readily understood himself to be the cause of it'.[36] Here, then, is one of the earliest articulations of fiction's connection of the vampire with disease: where the vampire is imagined as the causative agent of disease. Polidori's credentials for writing about disease were impressive. He was a qualified physician, completing his medical training at the Edinburgh Medical School in 1818, where one of his fellow students was the sanitarian Southwood Smith.[37] His first professional position, as personal physician to Lord Byron, was at the instigation of Dr Henry Halford, who later led a commission on epidemic diseases in the aftermath of the cholera outbreaks in Britain.

Le Fanu's 'Carmilla' represents disease even more assuredly than Polidori. First, the narrative's vocabulary invites a medical reading: the evolution of the story's vampirism is littered with both medical terminology and specific medical images. The narrator and central protagonist, Laura, situates her victimization by the vampiric Carmilla within prevailing discourses on infection. She describes her declining health as similar to the 'mysterious complaint'[38] suffered by the local working classes, and imagines that Carmilla has 'been stricken with the strange epidemic that ... had invaded the country'.[39] Laura's father, dismissive of the existence of vampires, regards local superstition as an 'infect[ion]' and employs medical expertise – two doctors – to restore Laura's health.[40] By the conclusion of the story Carmilla's death is likened to ridding the land of a plague.[41] More specifically, it is clear that Le Fanu infuses his narrative with imaginative reconfigurations of miasmatist discourse. Carmilla's unexpected arrival at the home of Laura and her father is prefigured by a change in the surrounding environmental conditions, where 'over the sward and grounds a thin film of mist was stealing, like smoke'.[42] Carmilla arrives from out of this mist, a portentous symbol of miasma from which the disease of vampirism emerges. Even in 1834 William Henry's report on the laws of contagion had conceded that there were 'several points of analogy between the operation of marsh miasms, and that of contagious poisons'.[43]

Dracula also draws considerably on miasmatic conceptions of infection. The female vampires that attack Jonathan Harker first appear as 'quaint little specks floating in the rays of the moonlight'[44] and Count Dracula's tomb gives off 'a deathly, sickly odour', both typical exemplars of the exciting causes due to an unhealthy environment that dominated anti-contagionist rhetoric: the 'vapours and exhalations ... possessing a specific power of exciting fever' that sanitary scientists so feared.[45] The specks of matter in the air described in Harker's journal are proven later to herald the arrival of the vampire: Lucy Westenra suffers a similar attack from 'a whole myriad of little specks' and Mina Murray's infection begins with 'a thin streak of white mist'.[46] Foul smells also consistently indicate the presence of potential infection; Dr Seward finds a putrid atmosphere in Lucy's tomb, to contrast with the 'fresh and pure' night air outside.[47] More potently, Dracula's tomb at Carfax Abbey gives off 'an earthy smell, as of some dry miasma, which came through the fouler air'.[48] At the same time several of these representations, especially Harker's and Lucy's descriptions of the vampire, employ the same descriptive vocabulary as microscopists. Jabez Hogg, in his mid-Victorian history of the microscope describes how looking through the lens of the microscope reveals 'the minutest fragments of matter'.[49] Gideon Mantell describes the same view as illuminating 'numberless myriads of creatures'[50] which may well 'swarm in the air' around us, too small to be seen with the naked eye. Similarly again, Thomas Dick describes *animalculae* in 1851 as those 'myriads of living creatures' and asks 'how many of these invisible tribes there may be throughout the air'?[51] Stoker's deployment of the language of microscopy reinforces the analogy between the vampire and the *animalculae* of miasmas.

Scholarly work on vampire fictions has tended to focus on sexuality rather than disease, although there have been one or two key readings that at least approach their medical and scientific contributions.[52] A number of critics have been alert enough to highlight this deficiency, and to point towards metaphors of disease as important areas for further consideration. William Hughes, a leading critic of the work of Bram Stoker, has argued that 'modern criticism's preoccupation with sexuality dominates – and indeed inhibits the development of – the debate on vampirism'.[53] Alex Warwick, one of the most perceptive readers of the contextual histories of vampire fiction, provided an important reminder of the central concerns of vampire fiction in her essay on their evolution. 'The trope of infection', claims Warwick, is 'one of the vital components of the [vampire] myth'.[54]

It is not only that Le Fanu and Stoker offer an imaginative representation of the etiology of disease. They also interrogate the fragile balance between the real and the imagined, emerging from microscopic vision, that so influenced scientific discourses on infection across the nineteenth century. In 'Carmilla', for example, the guardians of the two young women who fall prey to the vampire, both believe that the vampire (representing infection itself) is nothing but superstition; the product of the imagination. One of them scorns the local doctor for suggesting that a vampire has caused the fatal illnesses in the neighbourhood, as the narrator Laura describes: 'He [the doctor] and papa emerged from the room together, and I heard papa laugh, and say as they came out: "well, I do wonder at a wise man like you. What do you say to hippogriffs and dragons?"'.[55] Likewise, the other guardian, General Spielsdorf, admits that his was as disbelieving an attitude when he talked with his physician: 'being myself wholly sceptical as to the existence of any such portent as the vampire, the supernatural theory of the good doctor furnished, in my opinion, but another instance of learning and intelligence oddly associated with some one hallucination'.[56] Both men regard the vampire as a creature of superstition and believe that the expert opinion they have been given is an example of knowledge being fooled into falsehood by the imagination.

Stoker's depiction of the conflicting evidence of microscopic vision is more complete than Le Fanu's. In *Dracula* he reveals not only the phantasmagoric character of such vision, but also interrogates the observational expertise that claimed greater objectivity. This is achieved through a series of conflicts between different medical characters that is very clearly a reimagining of the germ theory debate of the 1880s. The novel's central medical practitioner, Dr Seward, is an exemplary figure in Stoker's analysis: he is the doctor who fails to find evidence of the vampire, or the microbe, and therefore distrusts its very existence. The first example of this comes in his analysis of Lucy's blood, observed through a microscope. Seward can see that Lucy is 'bloodless' but his 'qualitative analysis gives quite a normal condition'.[57] Just as Pasteur's rival, Pouchet, used the fact of the microscope's ambiguous status to challenge the existence of the germ, so too does Seward find that Van Helsing's story of the vampire lacks credibility because he has been unable to recognize its presence for himself. However, it is his paucity of visual expertise that is soon shown to be the real cause of his misdiagnosis. Seward's lack of visionary adeptness contrasts markedly with Van Helsing's 'quick look', leading him to admit that 'I have not seen' the mark on Lucy's throat that Van Helsing points to as the cause of her disease.[58] Seward's inability to see the aetiology of Lucy's disease leads him to dismiss the real cause, even when he comes closest to recognizing it: 'It at once occurred to me that this wound, or whatever it was, might be the means of that manifest loss of blood; but I abandoned the idea as soon as formed, for such a thing could not be'.[59]

Van Helsing, as Stoker's proponent of germ theory, is finally drawn into an argument with Seward over the cause of Lucy's illness and its apparent recurrence in the children of Hampstead Heath. Their discussion is one of the most important of the novel, for it goes to the heart of the germ theory debate of the 1880s and 1890s. Van Helsing pushes Seward to account for Lucy's illness, which he cannot without 'data on which to found a conjecture'.[60] Van Helsing's apparently oblique response is to tell Seward, 'You do not let your eyes see nor your ears hear, and that which is outside your daily life is not of account to you'.[61] He goes on: 'Do you not think that there are things which you cannot understand, and yet which are; that some people see things that others cannot?'[62] Forced into a retreat, Seward admits: 'I feel like a novice blundering through a bog in a mist, jumping from one cassock to another in the mere blind effort to move on without knowing where I am going'.[63] But even in the midst of his blindness Seward understands that Van Helsing wishes him 'not to let some previous conviction injure the receptivity of my mind with regard to some strange matter'.[64]

This confrontation follows the contours of the sceptical opposition to germ theory that were common in Britain in the last decade and a half of the nineteenth century.[65] Opposition to the germ theory of disease often rested upon the vagaries of microscopic vision. There were two reasons for this: first, microscopic vision's challenge to the observer's ability to differentiate between true vision and imaginatively altered vision meant that even where microbes were observed, their role in disease was often disregarded as the result of an over-productive imagination. Second, germs were elusive, notoriously difficult to capture under the microscope and impossible to observe with the naked eye, unlike the filth of a disease-ridden area of Britain's cities or the effervescent secretions of a fermenting liquid. John Hughes Bennett stated this position succinctly in an 1868 article for the *Edinburgh Medical Journal*:

> It is perhaps unnecessary to remark, that our modern microscopes have reached such perfection that we can examine with the greatest accuracy particles of only one fifty-thousandth of an inch in diameter, which are much more minute than the smallest recognizable ova or seeds. As all efforts, therefore, to discover the supposed germs in the atmosphere with our best instruments have failed, many scientific men who had personally investigated the subject were once more led into the belief of, at least, an equivocal or doubtful generation of the lowest forms of animal and vegetable life.[66]

The germ could be easily dismissed, therefore, as completely without foundation in reality and, as Nancy Tomes has shown, persuading sceptics 'that they co-existed with an invisible world of microorganisms ... [was] daunting'.[67] This did not stop germ theorists maintaining that their vision of disease, based in part on their microscopic expertise, was closer to the truth than the previous paradigms of contagion and anti-contagion. As Louis Pasteur claimed in 1888, 'a new science has been born ... it has caused a veritable revolution in our knowledge of virulent and contagious disease'.[68]

Stoker cleverly constructs the divergent views of Seward and Van Helsing as an investigation of microscopic vision: he asks questions about expertise and objectivity, about the role of the eye of the observer and about the influence of specific ways of seeing that emerge from the practice of microscopy. Ultimately, Van Helsing's visual expertise wins out over Seward's fragmentary and occluded vision; the objectivity of the acute and experienced observer trumps the microscopist caught up in phantasmagoric seeing. Yet there is an irony here, which is also present in Le Fanu's 'Carmilla', in the fact that it is the existence of the mythological vampire that becomes the most objective and rational position in each of the fictions. If this appears to be a reversal of expectations, it is because such a reversal precisely mirrors the most striking effect of discourses on disease: that the specific object of investigation (the *animalculae*, dust mote, microbe or germ) is always depicted as Gothic. These disease objects become Gothic both because they are always anthropomorphized as an active agent of disease transmission, and because they are very difficult to see, even with the technological addition of the microscope's lens. This combination of activity and opacity makes them increasingly spectral; much like the phantasm or ghost of popular superstition.

Characterizing infection as a Gothic spectre was not restricted to disease theorists and microscopists, or to fiction writers interested in reimagining their debates in new contexts. In popular discourse, too, disease was imaginatively remade as a Gothic phantasm. A very good example of such discourse can be seen in the case of the public response to the British Institute for Preventive Medicine, which was founded in the early 1890s as a British version of Pasteur's Parisian institute, and which bought land in Chelsea, central London, with a view to building its first premises. From the beginning the British Institute encountered considerable public opposition, primarily owing to the common perception that, like Pasteur's institute, the British Institute would practice vivisection.[69] There was also particular attention paid to the British Institute's experimental role in the study of disease. Inevitably such attention came from outside the scientific community, for instance, in questions posed by local residents or investigations made by members of Parliament.

One key moment in the several-year-long debate about the British Institute's decision to build in Chelsea was a deputation made to the home secretary, Herbert Asquith, by several Chelsea residents, led by the local MP, Charles Whitmore. The discussion, transcribed verbatim for Asquith by one of his civil servants, turns quickly to the role of germs, and the position of the British Institute as 'a possible source of infection and danger.'[70] Asquith posed the question to the deputation of whether they had any reason to believe that the British Institute would have dangerous infectious potential. Replying for the deputation, the Chelsea Vestry member, Mr Irons, said, 'we only judge from what has occurred in the papers and from the things which have been done in the Pasteur Institute in Paris. I have read Grant Allen's "Devil's Dyke" which gives a graphic account of the dangers of careless handling of the persons engaged in such places'.[71] Irons's response is a fascinating articulation of the movement of scientific knowledge from one sphere of discourse to another. The basis for Irons's fears originates not in scientific writing but in journalism and in an essay by a writer of fantastic fiction. These fears are, as he later acknowledges, that 'very dangerous germs and death dealing poisons and gases would be let loose and might poison the vicinity'.[72] By the conclusion of the discussion with Asquith the deputation felt they had been able clearly to identify their displeasure with the building of the British Institute in Chelsea. Whitmore concluded by saying that 'he thought the germs of the objections of the deputation had been recognised', to which Asquith replied, 'Germs, Germs! Those are very dangerous things aren't they Mr Whitmore?'[73] There is extraordinary linguistic diversity in the use of the word 'germ' here. Yet its variant meanings each attest to the definition constructed by microscopic vision of the germ as an active spectre or phantasm. The word is used by Whitmore to describe an invisible essence that must be sought out, by Irons, to articulate an active and unruly miasma that must be secured, and finally by Asquith (comically) as an anthropomorphized agent of danger.

Far from signifying any shift in meaning that takes place as scientific knowledge enters the public sphere, this episode has its mirror in a later discussion of germs by some of Britain's leading scientists. The British Institute's scientific investigators described the germ in the same Gothic mode when they too made a deputation, this time to the Board of Trade, some six weeks after Whitmore's visit to the Home Secretary. The scientific deputation was a heavyweight one: its list of participants included G. G. Stokes, president of the Royal Society, J. Norman Lockyer, editor of *Nature*, Sir Edwin Saunders, surgeon to Queen Victoria, and the biologists Joseph Lister and Sir John Lubbock. The chemist Sir Henry Roscoe, also present, noted that 'the deputation ... is one I think of exceptional importance. It represents not only the whole body of medical opinion in this country ... but also the whole of the scientific element.'[74] While the major portion of the deputation's meeting is taken up with discussions of governmental support for scientific activity, a defence of the practice of vivisection in research, and the British Institute in particular, when the subject of the conversation shifted to infectious diseases, it also shifted register into a more Gothic language. Lubbock captured this with his closing comments:

> Vivisection has no doubt been referred to today but it is after all but a very small part of the whole work before us, unless indeed vivisection is to be understood as applying to the bacteria; but I would venture to remind you that although Acts of Parliament may prevent us from destroying the bacteria, they cannot prevent the bacteria from destroying us.[75]

As was the case with the previous deputation, the material object of infection, on this occasion bacteria, is charged with an active life. Indeed, Lubbock imagines the bacteria as an animal in its own right, subject to the same processes of vivisection. Lubbock also conjures the bacteria's spectral status; in particular its ability to resist being controlled by the regulating structures of human society, in which context it remains invisible, uncontrollable and constantly dangerous.

Phantasmagorias

Lubbock's partly satirical comments about bacteria and Acts of Parliament seem to be at some distance from the microscopist who looks through the lens of his microscope and finds an image which is difficult to identify as either real or imagined. Yet Lubbock's identification of bacteria as the dangerous spectre of infectious disease is only available to him because of the discourses that emerged from microscopic vision. While he might be certain of bacteria's existence – thereby identifying himself as an expert, objective observer – he is also able to define it imaginatively as an active yet invisible creature of infection. As Stephen Jacyna has argued, it was relatively common for microscopists to be uncertain whether the evidence of their microscopic investigations revealed 'a structure ... fundamental to all living things' or an entirely 'fictitious entity', and such epistemological indecision 'reveals how it was possible to swing from one to the other pole in this controversy in a remarkably short time'.[76] Even so, the appearance of the same characterization of infection in such diverse arenas – from microscopy to bacteriology, through sanitary science, fiction and the socio-political sphere – suggests that there is a more foundational principle at work that is elastic enough to sustain such interconnections despite the pressure put upon them by being stretched across numerous boundaries.

Common to each incarnation of the object of infection is its phantasmagoric properties; a product of its visual relationship with new technologies of vision and the perception of the observer. The phantasmagoria is a construct of science and as both Jonathan Crary and Terry Castle show, it was a late eighteenth-century technology employed in scientific education and public entertainment.[77] Phantasmagoria was defined in very particular ways, which resonate with the defining features of microscopic vision. Phantasmagoria emerged in the complex interplay between the technological object and the observer. Castle argues that the original phantasmagoric effect was produced by 'the use of a magic lantern' that illuminated spectral figures which were 'made to increase and decrease in size, to advance and retreat, dissolve, vanish and pass into each other'.[78] It was not only that the phantasmagoria produced phantasms, however, but that the technologies that did so hid the means by which they achieved their visual spectacle. For Crary, this was an 'occultation of production', a 'mystification' which made more phantasmagoric the images seen by the observer.[79] Such technological secrecy additionally gave rise to the possibility that the phantasms were not the product of technology but instead a trick of the observer's imagination. As Castle argues, such ambiguity gave the phantasmagoria great metaphoric possibility, as it constructed an epistemology that 'mediated oddly between rational and irrational imperatives'.[80]

Despite Terry's convincing history of the evolution of the phantasmagoria as a form of visuality that continued to be influential across the nineteenth century, historians of science have generally followed one line of Crary's inquiry which argues that the optical technologies of the nineteenth century 'were insufficiently "phantasmagoric"' because they revealed their operational structures.[81] In part this reading of phantasmagoria's decline is based upon an acceptance of the view, most forcefully put forward by Daston and Park, that the scientific Enlightenment of the late eighteenth century created a context for a more deliberate and objective science in the nineteenth century, a science that rejected the marvellous or the wonderful as too subjective and imaginative.[82] Yet Crary also notes that efforts to undermine scientific phantasmagoria did not fully succeed. Instead, they 'simply collapsed' the technological phantasmagoria 'onto a single human subject, transforming each observer simultaneously into the magician and the deceived'.[83] However, the evidence of microscopy indicates that phantasmagoric technology did not decline, nor did it collapse onto the observer. The visual experience afforded by the microscope was very precisely phantasmagoric in both its mystification of the means by which the extraordinary images were produced, and in the ways in which it gave rise to questions of accurate or imaginative seeing. Microscopy, then, suggests that Crary's analysis was too limited but also, and more importantly, that Daston's reading of the loss of the imagination from the sciences is, at least, too generalized or even mistaken.

Although microscopic commentators would not have been able to articulate this historical sense of their own vision it is the case that their belief in the microscope's own occult power does reveal itself in their writing. Jabez Hogg, for example, talks of discovering things 'with' the microscope rather than by employing it as an aid to vision, and an anonymous writer of a book of popular microscopy discusses what the 'microscope discovers' rather than what the observer discovers by using it.[84] These are unconscious acknowledgments of the active involvement of the microscope in the visions which reach the eye of the observer. The microscope becomes a participant in the creation of vision suggestive of a hidden power that is part of the defining feature of the phantasmagoria. When investigators of disease were confronted, then, with a microscopic image apparently showing the organic material of infection, it was in this context that they responded to them. Unsurprisingly, many reflected as the physician and physiologist John Hughes Bennett did, in asking 'what are these little beasts?' or rejected them, as did Florence Nightingale and Southwood Smith, as the product of the phantasmagoric imagination.[85] The visual similarity of the technological phantasmagoria (magic lantern) and the microscope is particularly evident when a microscope slide and a magic lantern plate are placed alongside one another. As figures 1.2 and 1.3 show, the images seen by observers of a magic lantern show and microscopists were similarly staged, appeared to be produced by a somehow super-scopic machinery, and were developed on the observer's retinal canvas through unseen (at the moment of viewing) technological apparatus. Of course as both the microscope and the magic lantern used light in combination with lenses and mirrors organized within a tube, the images they produced were also similar in shape and shading.

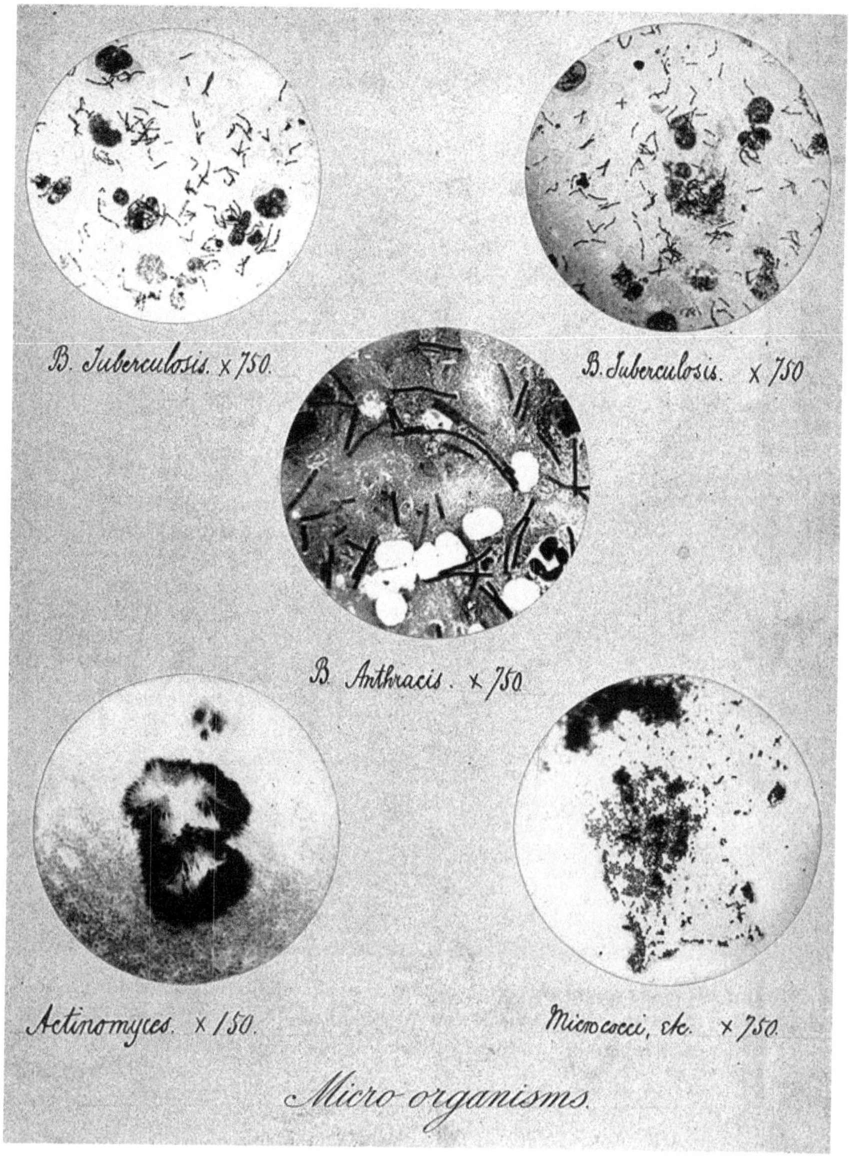

Figure 1.2: Microscopic images of anthrax and tuberculosis bacilli, similar in their visual representation to the slides of the magic lantern. Source: Wellcome Historical Images; reproduced with permission of the Wellcome Trust.

Figure 1.3: The phantasmagoric images produced by the magic lantern, striking in their similarity to the images of bacteria as seen through the microscope. Source: Wellcome Historical Images; reproduced with permission of the Wellcome Trust.

While the phantasmagoric has been rather under-investigated in history of science scholarship this is not the case in other disciplines. Particularly within social science the phantasmagoria has been central to defining the emergence of consumer capitalism. This is primarily as a result of Walter Benjamin's important work on consumption in his case study of the Paris arcades in the 1920s.[86] For Benjamin, phantasmagoria accurately described the 'spectralization of subject-object relationships' in the modern, that is later nineteenth and early twentieth century, marketplace.[87] While Benjamin's analysis impacts directly on visual culture it appears rather disconnected to the role of science in the modern world. However, Benjamin's employment of phantasmagoria is a direct response to, and rejection of, Max Weber's thesis that scientific knowledge had led to an entirely rational and bureaucratic modern society, which Weber had described as 'disenchanted' in his 1918 lecture 'Science as a Vocation'.[88] Benjamin argues that, contrary to Weber's view, 'irrationality did not disappear in the enlightenment but continued in modernity'.[89] For Benjamin, then, the phantasmagoric remained an important part of nineteenth-century knowledge production even within the apparently objective sciences. Historians of science are beginning to return to Weber's work on disenchantment in order to shine fresh light on nineteenth-century scientific epistemologies. George Levine, for example, has begun the task of reconsidering 'from the perspective of Darwinian naturalism, the narrative of disenchantment, fashioned so powerfully and convincingly by Max Weber'.[90] In this context it is clearly also valuable to reconsider Benjamin's employment of the phantasmagoria as one way of offering a critical reappraisal of Weber's view. Indeed since Benjamin's own argument about the continuing influence of phantasmagoria is in striking parallel to discourses on the microscopic vision of disease, reading the characterization of disease through the lens of his work on consumption can only further illuminate the role of phantasmagoria in the former.

Benjamin's work immediately opens the possibility of considering infection in its socio-political context, and in particular its relationship to consumer capitalism. The social and political consequences of infectious disease have of course been considered at length by historians of medicine, most profitably in the present context by Anne Hardy, Frank Mort and Keir Waddington.[91] In particular Waddington shows, in his extensive analysis of bovine tuberculosis, not only how medicine and social politics interact, but also how 'the public and commercial spheres', specifically 'trade interests', are drawn into debates about infection.[92] Using features of the phantasmagoria this analysis might be extended to consider how the organic materials of disease, brought under the purview of vision with the aid of the microscope, are themselves commodities within the networks of consumer capitalism.

As Benjamin explains, through his reading of Marx, objects become commodities by being made phantasmagoric.[93] Their mode of production is hidden and their imagined or spectral value takes its place. Objects of infectious disease, already defined as phantasmagoric by microscopic vision, therefore had the potential to be defined in relation to other commodities when they were employed in debates about the significance of infection within networks of capital. The case of the British Institute of Preventive Medicine reveals this process of signification at work. One of the most active groups to protest at the building of the British Institute in Chelsea were the local residents who owned or rented property within the vicinity of the proposed site. As the *Westminster News* reported in July 1896, 'Chelsea ratepayers were against the licensing [of the British Institute since] they must regard the Institute, if licensed, as a standing menace to Chelsea'.[94] The reason for their opposition was a fear that the Institute would be 'insanitary', both from those who were carrying infection and entered the neighbourhood seeking inoculation at the Institute and from the 'poisons and microbes' which would be collected and stored there.[95] As Charles Whitmore argued, 'the erection of such a building ... must have the effect of depreciating property in the neighbourhood' and it would therefore be a far better idea to site the Institute at some distance from Chelsea where the 'value of property [is] not increasing'.[96] Property owners in Chelsea were particularly vehement in their opposition, organizing a petition which they claimed was the 'most extensively signed ... since the days of the Chartists'.[97] They spoke particularly about how the 'diseases flying out of the windows of the Institute' would 'affect the "status" of the residents' and 'depreciate the value' of their homes.[98]

An interesting exchange takes place in this depiction of bacteriology's conflict with an energized public. It is not only the British Institute but disease itself that is made an active participant in an argument about the effects of science on the economics of a specific area. The object of disease, the microbe or poison, becomes the key opponent to the commodity of personal property. The residents of Chelsea imagine disease as particularly phantasmagoric: it is hidden from their view along with its mode of production but still holds the supernatural power of the spectre that might haunt them. It is also the very possibility of disease acting phantasmagorically that leads them to be concerned about their reduced status as owners of capital. Only in their belief that the phantasm of disease might escape the confines of the Institute and enter the residential areas around it does it have any particular power over their property. Disease is defined here as the disruptive element of a productive capitalist network: it is an intruder that challenges the value of existing commodities.

Dangerous and threatening, intruding into the spaces of consumption, phantasmagoric disease is an (unwanted) Gothic commodity. It is not only in the popular imagination that infection holds onto its 'mysterious residues' when it confronts the Victorian market economy.[99] Stoker and Le Fanu also consider disease as a gothic commodity, especially within the context of imperial trade. 'Carmilla' and *Dracula* are narratives of imperial consumption, in which infection (represented by the vampire, of course) is exchanged for money. That is, disease is made one of the commodities of trade with foreign nations. Le Fanu's short story deals with this subtly, but it is nevertheless at the heart of the narrative. His key protagonists, Laura and her father, are part of a colonising civil service in Styria. Their success in the imperial project is represented by the castle in which they live and the servants they maintain. In return for the imperial progress Laura's father had made, he had married a local woman, of the ancient Karnstein family. It is after the death of this woman, Laura's mother, that the narrative commences, and the reader discovers that the vampire is herself a daughter of the Karnstein family, and therefore related both to Laura's mother and to Laura. The story is, then, a microcosm of imperial relations, in which the castle represents England (it is where Laura and her father take tea and read Shakespeare) and Styria the foreign trading partner.[100] The vampire, Carmilla, moves between these two geographic sites as a representative of imperial trade. But she is also (as a Karnstein) the commodity gained by Laura's father in the capitalist exchanges that have taken place. Again unwanted and unrecognized, and indeed 'beautiful' to look at, the vampire is for Le Fanu a Gothic commodity produced by the economics of colonialism.[101] In both microscopy and chemical-biological studies of infection, disease was likewise seen in the context of an emerging global capitalism. At the Exhibition of Works of Industry of all Nations, for example, the work of microscopists took a leading role, and the manufacture of microscopes by British microscopists was honoured as an example of the trading excellence of Britain by the Great Exhibition Committee.[102] The British state also recognized that an increase in global trade meant a potential increase in infectious disease. The key strategy for containing the spread of disease was to install quarantine procedures in areas felt to be most susceptible to infection. These areas were identified as the port cities where the exchange of commodities between Britain and other nations was most common. As Krista Maglen has shown, the prevention of infection was most rigorously enforced at places of economic importance, and was generally to the detriment of individuals rather than market systems.[103] What such measures recognize, if not explicitly, is that infection itself was an unwanted commodity always potentially hidden amongst other commodities arriving in Britain to be distributed in the capitalist market. Quarantine was, therefore, another way of putting in place the kind of surveillance of disease objects undertaken by microscopic observation.

Stoker, like Le Fanu, also recognizes the close conceptual connection between the phantasmagoric object of infection and the phantasmagoric commodity. *Dracula* has been read as an imperialist fiction by several critics, most profitably by Stephen D. Arata, who sees in Dracula's attacks on British subjects and on British soil Stoker attempting to articulate the anxieties of reverse colonization.[104] The novel also, however, aligns disease with imperialism, by narrating the story of Jonathan Harker's efforts to provide Dracula with British commodities (in particular Carfax Abbey) in return for a share of his wealth. In doing this, Harker brings back to Britain not only the accumulated capital of his trade with Dracula, but also infection (represented by the vampire). Only too late does Harker recognize that the phantasm of infection has been the invisible commodity that he has unwittingly traded, leading him to spend the remainder of the narrative seeking out Dracula. When Harker does eventually confront Dracula, Stoker takes the opportunity to reinforce the association he has constructed between the commodity and disease. In the central action of this scene, Harker slashes at Dracula with his knife, 'making a wide gap whence a bundle of bank-notes and a stream of gold fell out'.[105] This symbolic cutting open of Dracula signals the fact that Harker recognizes his phantasmagoric status as the spectral image of disease who hides his infectious potential beneath the respectable attire of the gentleman. But this should also be read as Harker revealing Dracula's status as a commodity in an imperial exchange. The fact that he bleeds money should signal that his essence is capital, despite his phantasmagoric appearance as something far more complex and valuable (at least as Harker had initially believed).[106]

Fictions like these provide access to some of the ways in which the construction of scientific knowledge is transformed in other cultural spaces. The phantasmagoria may have begun as a term used to define a specific type of scientific technology but it quickly became conceptual, defining an epistemology of vision that microscopists found useful in capturing the observational experience of microscopical research. However, the concept of the phantasmagoria was not the exclusive property of microscopists or bacteriologists. It had significance also in the genre of the Gothic and was employed in Gothic fiction to articulate the spectrality and supernaturalism of the imagination. When used in this context the phantasmagoria was also redefined as Gothic. Further, the phantasmagoria became a useful concept for political commentators to give definition to an emerging market economy, and in particular to give substance to the commodity. The study of infectious disease, and specifically the microscopic investigations that gave it such impetus, was only one of several cultural practices in which the phantasmagoria was employed conceptually as a way of expressing new knowledge. It is certainly true that scientific work on infection, and the parallel developments in microscopy, provided inspiration for the Gothic fictions of Le Fanu and Stoker. What the case of the British Institute and the fictions of Le Fanu and Stoker tell us, then, is that the ways of producing knowledge within science are rarely exclusive to science. Just as importantly, they also reveal that when new scientific knowledge is defined in society it is not likely to be defined on its own terms. Rather it is defined by its accretion to other forms of knowledge-making that may be accepted as offering as many truths as science about how we should see the world.

2 MICROSCOPY AND DISEASE: PLACE AND IDENTITY IN LABORATORY SCIENCE AND FICTION

By 1894, some five years after leading Victorian scientists had called upon the state to support the creation of a 'Pasteur Institute in Britain', the building of the British Institute of Preventive Medicine on the Chelsea Embankment was nearing completion.[1] This year, however, was not to be characterized by celebrations of British achievements in bacteriology, so long the preserve of French and German science, but by fierce public opposition, both to the Institute and the location of its laboratories. One particularly public protest took place on 28 April 1894, when protestors – drawn largely from anti-vivisection and labour groups – conducted a parade and mass meeting in Pimlico, which passed by the site of the British Institute laboratories on the way to its rallying point on the Old Pimlico Pier. The handbill for this event very specifically set out its grounds for opposition: 'to protest against the Erection of the proposed Institute of Preventive Medicine (so called) on the CHELSEA EMBANKMENT (Near Chelsea Bridge)'.[2] Clearly, the geographical placement of the laboratory was one key aspect of public opposition, with 'public' in this instance consisting of a miscellany of anti-vivisectionists, suffragists, radical club activists, working-class and friendly societies members, and local residents. The other key areas of opposition to the British Institute were its desire to undertake vivisection, and the concern that the diseases studied at the Institute would be 'disseminat[ed] ... by the germs flying about in the air' throughout Chelsea.[3] The rally and meeting was significant enough to receive press attention, and reports gave the numbers attending at around 300.[4]

However it is not only the characterization of the British Institute as a 'poison den and a murder house', as C. A. Smart, president of the Chelsea Labour League, described it, that is interesting about this episode of public opposition to laboratory science. It is also that the scientists of the British Institute, led by the Institute's secretary, the experimental pathologist Marc Armand Ruffer, appointed a private detective to attend the rally, introduce the British Institute's view wherever possible and complete a written report. As Figure 2.1 reveals, the report completed on the Institute's behalf placed the opposition's advertisements and claims within the parentheses of an official communication, and presented the activities and personnel from a particular perspective.

29th. APRIL 1894.

REPORT OF PARADE AND MASS MEETING AT PIMLICO.

TO PROTEST AGAINST THE ERECTION OF THE BRITISH INSTITUTE OF PREVENTIVE MEDICINE.

The meeting was called apparently under the auspices of the Pimlico Radical Club the principal members of the procession being members of tnat Association, assisted by a small contingen of the Nelson (Borough of Strand) Radical Club.

Attached is the announcement handbill.

THE PROPOSED PASTEUR INSTITUTE.

A PARADE AND MASS MEETING
will be held on
SUNDAY, APRIL 29TH.,

To Protest against the Erection
of the proposed Institute of
Preventive Medicine (so called)

on the
CHELSEA EMBANKMENT
(Near Chelsea Bridge),

The Procession will start from
the Pimlico Radical Club, Sussex
Street, at 1 o'clock sharp, and
will meet on the Old Pimlico
Pier at 3.30 p.m. when addresses
will be given by the following
Ladies and Gentlemen:

W. EVERITT (CHAIRMAN).
C. A. SMART (PRESIDENT CHELSEA LABOUR LEAGUE.)
MISS JESSIE CRAIGEN.
F. LONGMAN (WEST LONDON ANTI VACCINATION LEAGUE.)
H. COREY. C. FITZPATRICK.
H. LITSTEY. ARTHUR WESTCOTT.

WORKING MEN ATTEND AND PROTEST AGAINST THE ERECTION OF THIS INSTITUTE.

Though announced for one o'clock it was 2 p.m. before the assembly was sounded by the bandmaster of the Operative Bricklayer's Society, and the procession formed. First came the banner of the Pimlico Radical Club the front side of which

Figure 2.1: The opening page of the British Institute's report on the protest against their Chelsea laboratory in April 1894. Source: Archive of the Lister Institute, Wellcome Library; reproduced with permission of the Wellcome Trust.

This report – written, of course, from a position of sympathy with the British Institute's practices and objectives – reveals how the individuals involved in protesting against the Institute were described in ways designed to discredit their arguments and undermine their character. For example, the anonymous private detective writes of the suffragist Jessie Craigen, who took a leading role in the parade and later spoke at the meeting, that she took her place on a wagon 'plastered with extracts from the [anti-vivisectionist] opinions of eminent judges and lawyers, &c' and which, symbolically, 'fell down soon after starting'.[5] The detective's description of Craigen is more striking. She was, he writes, 'a stout elderly lady of dark complexion with a stubby beard and a strong moustache, who was sucking oranges the greater portion of the journey and looked very happy and cheerful. A palefaced man kept her company together with several little boys and girls, who distributed pamphlets indiscriminately.'[6] More commonly Craigen was described in rather awe-struck terms: her statuesque (sometimes called massive) frame and deep baritone voice made a significant impression and she was one of the most sought after of public speakers from the 1870s through the 1890s.[7] Yet, here, the private detective depicts her as a degenerate, masculine figure, dark-skinned, manic and even child-like, and accompanied by a suspiciously quiet assistant and a number of children with no understanding of the contents of their propaganda sheets. Craigen is simultaneously the gypsy, the manly woman, the foreigner, the deviant and the child. Such an attenuation of derogatory characteristics emerge from her association with the protest movement and her particular position on scientific experimentation. Craigen's identity comes under pressure, then, in the specifics of her relationship with bacteriology and laboratory science.

This chapter will explore the relationship between identity and disease. In particular it will investigate how the ways of seeing that emerged in microscopy – which the previous chapter called microscopic vision – bled into other cultural arenas to become a more interrogative method of observing and imagining individual identity. One focus of the interconnectedness of scientific research on disease and socio-cultural identity will be fictional accounts of disease's effects in the world, fictions that conduct what might be called thought experiments on the impact of infection in the social, cultural and political spheres. Another will be laboratory science, and in particular the British Institute of Preventive Medicine, an example from which opened the chapter. There is already significant critical work that supports this approach to the history of disease. In the history of science both Frank Mort and Nancy Tomes, among others, have revealed the important connections between disease, politics and gender, and disease theories and everyday domestic practices.[8] In the field of literature and science, Laura Otis has argued for the cultural significance and influence of disease research and microscopy and has shown that influence at work in fictional accounts of infection.[9] The study of laboratories has a long history, which has recently turned towards a more holistic approach to laboratory culture, by placing, as Graeme Gooday argues, 'the laboratory more carefully into a wider social canvas'.[10] Rarely, though, are these complexly overlapping areas of scientific and cultural significance considered together. Yet it is important to do so. If, as Robert E. Kohler suggests, 'lab history is of necessity also social history', then it is vital to ask not only how laboratories 'shape public meanings of knowledge', as Kohler would rightly wish to do, but also how public knowledge shapes the laboratory.[11] That public knowledge can be identified in imaginative fictions, the press and other public discourses such as organized meetings, published pamphlets, guides and handbills.[12] Indeed, Kohler, who has reflected on these issues very usefully in the context of laboratory science, notes that one 'premier issue' for further investigation is 'authority, both epistemic and social'.[13] But from where does authority emerge? Where is it challenged?

One way (among many) to address these questions is the route taken in the remainder of this chapter: to consider the role of visual authority in microscopy and disease and to trace its influence across several spheres, from microscopy to bacteriology, in fiction and the laboratory, via the press and public commentary. These are all cultural spaces where infection and its meanings are debated, altered or reinforced. These are also spaces where the visual complexity of microscopic seeing combines with the invisibility of disease to disrupt perception. Infection becomes something that is not simply observed but also imagined: reconstructed in language and made into metaphor.

Place and Personal Identity

An early nineteenth-century example of the availability of microscopic vision as a metaphor for different ways of seeing is the anonymously authored travel guide to London, *How To Live in London: or, The Metropolitan Microscope and Stranger's Guide*, produced by the West London publisher Joseph Smith in 1828. This satirical guide drew on the microscope as a metaphor of vision to explain its method of describing London to the uninitiated. The introduction set out the aims of the guide:

> [To] Elucidat[e] the manner and means by which thousands exist in apparent respectability without friends, profession, trade, or fortune! Explaining also how this may be effected honestly: And, on the other hand, containing hints to the unwary, to avoid the stratagems of swindlers, tricks of thieves, gamblers, Cyprians, and all those who live by plundering those they appear to befriend; with a few cautionary and instructive remarks on Lawyers, Pawnbrokers, and Auctioneers![14]

The metropolitan microscope reveals the hidden talent of some of its population for criminal and immoral behaviour. The trope of microscopic vision allows the author to articulate his ability to penetrate the superficial and fake surface of the city and delve into the inner truths of its methods and motives. Yet there are none too subtle hints of the ideology of the observer in his categorizing of 'Cyprians' alongside swindlers, thieves and gamblers. Microscopic vision, in this example, is constructed as a defender of truth and usurper of the power of the trickster but its powerful new way of seeing is also a way of stigmatizing those whose identity the observer might believe to be corrupt. Indeed, this new way of seeing suggests new ways of constructing identity: through discovering inner truths to character and moral constitution that cannot be deduced from outward appearance. Moreover, it also suggests how microscopic ways of seeing might be reconfigured in specific locations. Joseph Smith was a jobbing publisher working in Seven Dials, an area of considerable poverty in 1820, although one with a thriving publishing and printing trade based around the popular ballad.[15] Smith's travel guide reflects what uses the metaphor of microscopic vision can be put to in areas of London where criminal activity was high and immigrant populations numerous (although these are not necessarily connected). In Smith's guide the microscope is used as a vehicle to connect one immigrant group to specific types of crime.

It was not uncommon for other scientific and cultural spaces to draw a very similar connection between microscopic vision and character. In the medical profession such observational praxis was taken up with some enthusiasm, especially by those microscopists who already held that the microscope was a tool for the discovery of nature's truths, rather than the more circumspect who were concerned to know what kind of truth the microscope might lead the observer to imagine. Since the microscope could give access to the body's interior in ways that were not previously possible, it appeared sensible to make use of its penetrative vision in making decisions between individual bodies whose outward appearances appeared identical. As the *Lancet* showed, medical decision-making of this kind, based upon microscopic vision, could easily shift from a biological to an ideological imperative. In an article on the microscopical character of milk from 1844 the *Lancet*'s contributor noted that:

> Recent inquiries have shown that human milk, examined by the microscope, presents different character ... The microscope, then, will enable us to determine, in doubtful cases, whether a given milk be of a strong or weak class, and will guide the physician in the choice of a nurse whenever the question turns on the advisability of one or the other of these kinds.[16]

Here the choice between two or more nurses is made on the basis of microscopical surveillance. Whatever the medical pragmatism of such a decision it is apparent that the quality of the milk identified by the microscope is very quickly deflected into a series of decisions about 'character' and 'class'. Social identity is constructed from a biological signifier. The repercussions of microscopic ways of seeing are significant: they allow for increasingly intimate surveillance of the individual body; they lead to new and invasive methods for constructing identity; and they align biological phenomena with social and moral categories.

Microscopic vision also had an effect on the many different, but interconnected, disease hypotheses proposed during the nineteenth century. As Mort argues, contagionist strategies of quarantine and containment took on an extra dimension as the power of the microscope began to reveal that the existence of disease, as well as its potential for infecting others, could only properly be identified by looking within the body.[17] If it were the case that disease was largely invisible to the naked eye then those either infected with disease, or having the potential to be infected, became a greater danger to those around them. Microscopic vision, therefore, gave enough knowledge to those coming into contact with potential disease-carriers for them to be increasingly fearful of contamination by an invisible source. Containment of individuals presumed to be infected was therefore expanded and methods used subsequent to quarantine – such as disinfecting or burning – were used much more liberally. Southwood Smith captured the concern of many at the extent to which contagionist strategies led to social unrest and the destruction of community:

> If indeed the emanation thrown off from the living body formed permanent and powerful poisons ... we could never meet in society, for we should poison each other; the first symptom of illness would be the signal for the abandonment of the sick, and we should be compelled by a due regard to self-preservation to withhold from persons afflicted with disease every kind and degree of assistance that required personal attendance.[18]

Microscopic vision led directly, then, to a greater fear of the invisible and, as a result, to a more robust separation from society of the contaminated individual. This should be seen as an example of the opportunity that microscope vision offers for increased surveillance. Although disguised as a safe national medical practice it actually becomes a practice of containment which allows the potentially contaminated to be looked at.

Edgar Allan Poe had anticipated the potential effects of microscopic vision in human communities in his tale, 'The Masque of Red Death' (1842). When an unknown stranger, costumed as a victim of the red death, appears at a masked ball, he is feared, reviled, constantly watched and eventually attacked. As the attack begins the gathered 'revellers' 'gasped in unutterable horror at finding the grave-cerements and corpse-like mask which they had handled with so violent a rudeness, untenanted by any tangible form'.[19] Poe's Gothic fantasy imagines a phantasmagoric scene of surveillance, hatred and invisibility; with the potentially contaminated depicted as a spectral vampire ostracized yet continually surveyed by the community. In Poe's diseased community, contagionist practice is a precursor to violent internecine conflict, a Gothic melodrama of the dangers of unfettered microscopic vision allied to specific medical theory.

Regardless of Poe's and Southwood Smith's protestations at contagionist strategies, sanitary science was not immune to the effects of microscopic vision. Indeed sanitary science took microscopic vision further in the direction of surveillance and individual morality than any other disease theory. Southwood Smith was as guilty of this as other anti-contagionists. In a revealing example in *The Common Nature of Epidemics*, Smith discusses overcrowding as a predisposing cause of disease in urban homes. His example of how this functions, however, is to cite the introduction of lodgers into the home. 'I see every day in the Fever Hospital', he said in evidence, 'the consequence of taking in such lodgers'.[20] Smith's argument is a subtle, and perhaps unconscious, sleight of hand, but it highlights the influence of microscopic vision's alteration of individual identity. Overcrowding may be the cause, but blame is placed upon the stranger, the unknown individual who has not been under surveillance and may therefore be a carrier of invisible disease-causing agents. This is less a fear of increased conditions for disease and more a concern with how individual identity might be hidden by the unscrupulous or unknowing stranger.

As we have seen in the case of the wet nurse, it was relatively easy to slip from scientific questions of interior health to questions of social status and moral well-being. Smith's warning against the unknown lodger betrays the beginning of such a move in sanitary science, but it is left to Florence Nightingale to complete it. In a letter she wrote to Henry Parker in 1867 she stated:

> There are sick streets, like sick people, which one can recognise at once ... and sick streets produce sick people and bad people ... the sick streets will always bear the same fruits – each in its degree – we shall always reap the same harvest of sickness (and consequent pauperism) and of vice from them.[21]

The collapsing together of disease and morality is clear in Nightingale's conclusion that sickness inevitably leads to vice. It was the microscope that allowed sanitary science to see the sick street and to draw moral conclusions from its vision. Southwood Smith revealed the importance of microscopic analysis of the interiors of houses in one of his early works: 'The moisture in the air of a crowded room may be condensed by ice', he explains, and then examined under a microscope. The microscope will reveal that the moisture 'is indeed pernicious, for it is an animal poison ... it collects in large quantities on the furniture and walls of dirty houses'.[22] So microscopic vision leads to a discovery of sickness within the home, sickness which is expanded to take in first the street and thereafter the street's inhabitants. From the revealing vision of the microscope the sanitarian deduces the immoral identity of the community.

Sheridan Le Fanu's 'Carmilla' offers an imaginative consideration of the connections made between geographical location, identity and infection. His short story uses his own understanding of the Irish experience of disease to reveal how (national) identity, place and infection were consistently intertwined. The narrative's location in Styria appears at first to locate it far from Le Fanu's Ireland. Yet the placing of the events in Styria is actually suggestive of an Irish context. As Matthew Gibson has shown, Styria might well be read as a transposed Ireland; its own politics of Protestant and Catholic conflict and colonial intervention similar to Ireland's political and religious history.[23] Added to that is the more circumstantial evidence of various critical commentators that Le Fanu's English publisher demanded non-Irish settings for his work, and that almost all of Le Fanu's fiction can be seen as Irish-inflected even when superficially unconnected to Ireland. Equally important is that 'Carmilla' includes many tropes regarded as particular to the genre of Irish Gothic: the employment of traditions of folklore and superstition (the vampire and responses to it), the central symbol of the grand house (Laura's father's 'feudal residence'), and the sexualized Irish female figure (Carmilla).[24] It is possible, perhaps even probable, that alert contemporary readers of 'Carmilla' performed a simple geographical substitution of Styria for Ireland and read the story as an examination of the Anglo-Irish, albeit with the added Gothic frisson of Eastern European exoticism.

In particular, though, Le Fanu draws on the historical evidence of the European cholera epidemic of 1832 – which hit Ireland particularly badly and is one of the touchstones for future stereotypes of the Irish as disease-ridden. That epidemic revealed how different theories of the emergence and transmission of disease were routinely bound together, and often misapplied. During the 1832 epidemic, the port city of Glasgow returned all Irish beggars to Derry, fearing that they might spread cholera to Scotland, despite the fact that cholera struck in Ireland while the beggars were in Glasgow. Alternatively, the Irish establishment, reflecting both contagionist and miasmatist rhetoric, blamed, variously, 'manure heaps', 'the poorer quarters of Dublin', 'prostitutes' and working men 'of irregular habits'.[25] In 'Carmilla' Le Fanu similarly overlaps miasmatic and contagionist principles of infection.

After Carmilla's arrival disease spreads rapidly throughout the local area, killing the wife of a pig farmer and the daughter of a forest ranger, both members of the Styrian working class. Laura fears that their deaths signify the beginning of a 'plague or fever' but even when this disease closes in on her home, reaching 'the sister of a young peasant on [her father's] estate, only a mile away' Laura is convinced that disease will not reach her.[26] When Carmilla asks Laura if she is afraid of infection Laura replies, 'I should be very much if I fancied there was any real danger of my being attacked as those poor people were'.[27] This is the language and aetiology of contagion: constructing disease as the spreading plague that attacks individuals randomly. Yet it is also implicitly the discourse of miasmatism; Laura's confident assertion of her safety is largely premised on the difference between a presumed cleanliness of her own home and the polluted environment of the local population. While class difference undoubtedly contributes to Laura's opinion, her perceived separation from the infected locals is also based on ethnicity: the local peasant class are recognizably Styrian, and therefore ethnically different to Laura who chooses to define herself through her Englishness.

With these patterns of ethnic difference and disease in place it is possible to begin to discern the schema of Le Fanu's narrative. Disease, whether contagious or miasmatic, is an attribute of race, of the visibly 'other' black woman or the known (but racially invisible) Styrian. As the progenitor of disease, the vampire is therefore inevitably racially different to Laura. Of course, this is exactly the case: Carmilla, we discover, is Styrian, or Irish, while Laura is Anglo-Styrian (Anglo-Irish). Taking a step further, Le Fanu therefore locates disease in Irishness, with the vampiric Carmilla acting as a synecdoche of the infecting potential of the Irish on the clean, unpolluted Anglo-Irish Laura. The importance of drawing on both miasmatic and contagionist understandings of disease transmission become more apparent in this schema. Miasmatism, after all, focuses on place as the cause of disease, while contagionism privileges the centrality of the body. In Le Fanu's organization of disease both the bodies of the Irish and Ireland itself become sites of infection, and it is the vampire Carmilla who holds these two together; she is both the dangerously contagious Irish individual and the locus for a miasmatic Ireland. In this fictional reworking of historical epidemiology, Le Fanu is additionally commenting on the ease with which microscopic vision led to disease being imagined as part of the identity of both place and person.

Popular fiction like 'Carmilla' undoubtedly had an impact on the public understanding of infection. For example, when in 1873 the *Sanitarian*, one of several periodicals supporting the sanitary science movement in Britain, offered an article on the use of domestic antiseptics, it drew a clear parallel between the spread of disease and the vampire:

> Disease and grim death stalk through our fine dwelling houses ... when we are soundly asleep ... the impalpable subtle enemy, malaria, arises from the outlets of the very utensils introduced for the preservation of our health, for the prolongation of our lives, fanning us into deeper slumbers like the wings of the vampire.[28]

However, the powerful notion of infection as an active agent working invisibly on a specific location could be redeployed by different public groups to counter the view that disease was part of the identity of insanitary spaces and a signifier of the lack of morality of those who lived in them.

Public opposition to the British Institute gives a clear example of how the perceived relationship between disease and identity might be turned back on the scientific community that had constructed it. A Mr Reid noted, for example, the connection between unsanitary conditions and the British Institute's laboratory:

> There were 90 summons to house owners to put their houses in sanitary condition [in Chelsea], and if they were to have this Institute what sort of an atmosphere would they have. They wanted a healthy fine atmosphere to live in and the erection of the Institute would pollute the air and that was what they wanted to get rid of.[29]

Reid claims it was the scientists of the Institute whose ethical identity was questionable. He characterized the work that might lead to the spread of disease as a 'dastardly crime' and said that the root of the opposition to the Institute's new laboratory was 'its moral aspects'.[30] For Reid the invisible germs of disease brought into view by microscopic observations and employed in the public imagination told a different story. They did not signal the visual authority of microscopy and bacteriology in determining the identities of individual citizens or entire communities. Rather, they illuminated the ethical responsibilities of the scientific community (and particularly of the laboratory community) and suggested that the authority to speak on questions of science was not the preserve of laboratory staff but should be claimed also by the lay public.

Regardless of the call for greater public scrutiny of the laboratory the ongoing characterization of the British Institute was as an organization extraordinarily similar to the invisible germs with which it worked and for which it had responsibility. In the public imagination the Institute laboratory became the germ, or even the vampiric Carmilla: secretive, hidden from public view, its practices and motivations invisible to outside observation. In the *Gloucester Chronicle*, for example, the anti-vivisectionist Edith Carrington, wrote that the British Institute's scientists 'are stealthily at work to attain their object [the building of the Institute in Chelsea], and trusting to introduce, under a plausible title, that which would be against the national will and conscience were the public aware of its details'.[31] The Institute was charged with being similarly secretive by the *Star*, who reported on the resistance Chelsea's councillors met with when attempting to gain entrance to the laboratory. Under the headline 'MORE SECRECY!', the *Star* reported that 'the councillors were met at the gate by the legal representative of the Institute, who stated that he had instructions to refuse them admission, as their inspection was considered both vexatious and unnecessary'.[32] These examples, only a few among the many similar articles on the Institute that dominated the popular press during the mid-1890s, confirm Kohler's view that authority is one of the most important issues in laboratory history. They also reveal, however, that it is not only the laboratory's 'cultural spaces that actively shaped what went on inside them' but also the cultural spaces surrounding the laboratory.[33]

Moreover, these examples show how the meaning of microscopic vision can shift as it moves from one sphere of culture to another. What began in microscopy as the elusive microbe or germ, continually difficult to detect, and always invisible to the normal eye, has here become a metaphor for the opacity of laboratory practice. The scopically-resistant germ is made representative of the larger culture of secrecy and invisibility that dominates at the British Institute. Importantly it is not the work of the Institute that creates this representation, but the culture of opposition that emerges in the spaces around it. It exemplifies, as Gooday argues, the laboratory as an 'interstitial' institution that draws 'value and meaning from the people and institutions with which [it is] juxtaposed, opposed, and integrally networked'.[34]

Bram Stoker, whose *Dracula* was already beginning to form itself in his mind during the years of most pronounced public interest in the British Institute, is also interested to examine some of the different meanings of the microscopic vision of infection that emerge when it enters the world of social and cultural value. Stoker uses the character of Lucy Westenra to examine how infection mediated by microscopic vision can have profound effects on individual and institutional identity. Stoker uses Lucy as a marker of Dracula's infective potential: the disease of vampirism is transmitted to her through his bite. A significant portion of the first half of the novel focuses on the aetiology of vampirism as disease as it reduces Lucy from health to illness, and eventually to death. However, at the same time as providing an imaginative view of the progress of infection, Stoker also asks his reader to consider the effects of the disease on Lucy's character.

Critical opinion has consistently argued that Lucy's vampiric infection is a marker of moral laxity leading to sexual transgression.[35] Yet if we consider Dracula's appearance as the onset of disease as well as the more commonly articulated sexual encounter, Lucy's straightforward sexual transgression must be challenged and Stoker's consideration of her character investigated differently. Lucy is first infected by Dracula during an episode of sleepwalking which takes her to the graveyard overlooking Whitby harbour. Her illness begins, then, with an unforeseen expedition to a place of infection, a site of impurity of which she had no knowledge, and which she could not therefore see. This is the vampire as miasma, but a miasma that is situated away from any environment over which Lucy has control. Lucy is not even at fault for her own sleepwalking; Mrs Westenra admits to Mina that it is a hereditary complaint, passed to Lucy by her father who 'had the same habit; that he would get up in the night and dress himself and go out, if he were not stopped'.[36] If Lucy's initial exposure to infection can be explained by her unfortunate inheritance from her father, her continued illness is more readily the fault of her mother. It is Mrs Westenra who mistakenly privileges her knowledge of sanitary science over Van Helsing's more sophisticated understanding of germ theory and further exposes Lucy to Dracula's attacks. In fact the decisive moment in Lucy's battle with disease comes when her mother enters her bedroom and, frightened by Dracula's appearance at the window, 'clutched the wreath of flowers that Dr Van Helsing insisted on [Lucy] wearing round [her] neck, and tore it away from [her]'.[37] Lucy's death is not altogether the necessary conclusion of sexual transgression, but a more general failure of authority: of state and scientific authorities failure to maintain Britain as a place free from disease, and from the symbolic figures of social authority, Lucy's parents.

Nevertheless, the novel does lead us to consider Lucy's sexuality as responsible for her own infection, nowhere more so than when she attempts to seduce Lord Godalming while under Dracula's influence, an episode that reinforces the connection between her diseased state and her sexual assertiveness.[38] Why should this be the case, when Stoker is at pains to suggest an alternative reading for her infection? What the novel is aiming to highlight is the ease with which disease can alter the social position of the infected individual and the ready associations between sexual conduct and propensity for infection. While Lucy can be regarded as 'the embodiment of the positive feminine qualities of "sweetness and light"', as William Hughes has argued, her infection by the vampire has the potential to turn her from a virginal aristocratic girl into a wanton and immoral sexual woman.[39]

Mina Murray recognizes this immediately on discovering that Lucy is predisposed to sleepwalking. Her narrative of Lucy's somnambulism is revealing of the social fear associated with infection, describing the 'fear' of their being a 'witness of Lucy's condition' that might lead to 'exposure' of 'her reputation' as well as her 'bare feet'.[40] Mina's complex response to Lucy's attack offers a clear indication of Stoker's acknowledgement of the effects of infection on public image as well as the easily reversible route from disease back to individual sexual impropriety. Mina's unarticulated fear coalesces both as an antipathy to public discovery and a presentiment of Lucy's infection. Lucy's health and (sexual) reputation are seen to be simultaneously at stake, as though one is dependent upon the other. Yet Mina's greatest concern is not with the 'white face and red, gleaming eyes' of Dracula, but from the 'exposure' of Lucy to the Whitby community and her feet to passers-by.[41] The bared feet are, if anything, most significant, for they register Mina's deeply held fear of being dis-covered, of being made subject to the vision of others. Her own bare feet, resonating symbolically as the feet of the street urchin, are covered in filth so as not to reveal their unclean status. Stoker cleverly plays on the irony of Mina's polluted feet hiding the real pollution of public discovery of disease.

More than that, however, the bare feet suggest nakedness, not only the metaphoric nakedness to public shame but also bodily nakedness associated with infidelity to Victorian propriety and suggestive of the sexual encounter. Mina's surrender of her shoes to Lucy becomes a moment of great sacrifice, risking her own sexual status to save her friend's. Despite Mina's successful efforts to protect Lucy from social stigma it is thereafter that disease begins to determine her identity as sexual transgressor, both in her actions after infection and in suggesting that infection is the result of previous sexual misconduct. While maintaining

Lucy's innocence in contracting the disease of vampirism, the novel also makes explicit her emerging sexuality to trace how infection can culturally shift into other forms of pollution: moving from physical ill health to moral impurity.

That the representation of Lucy appears to offer two contradictory readings of her identity – as sexually anarchic and innocently infected – is an indication of how Stoker's knowledge of the complex interactions between medicine and society accurately reflect the contradictory reception, and continual redeployment, of the microscopic observation of germs. His fiction represents an imaginative construction of the microscopist Jabez Hogg's claim that 'by vision, aided by knowledge, we pierce into the ... interior of bodies ... convinced that the infinitely small ... have for us no bounds'.[42] Indeed the characterization of Lucy is boundless: she is sexually provocative and innocently ill *at the same time*; an oppositional doubling that illuminates the embodiment of disease in late Victorian Britain as indicative of both the failure of authority and individual irresponsibility. Stoker's carefully balanced articulation of Lucy's position as in-between the reactionary sexual politics that characterized the Contagious Diseases Acts and the more liberal view of disease as arising from institutional failure, highlights how capable his fictions are of imaginatively investigating, in Mort's phrase, the 'marked polarization of political positions' from which disease and infection were viewed.[43]

Place and National Identity

Microscopic vision did not just enable Hogg to cross the boundary separating the interior of the human body from the exterior world. It also allowed for a different kind of crossing: the metaphoric transgression of national space and identity. Historians of science usually associate ideas of nation in scientific practice with imperial science. As Roy Macleod set out, in an introductory essay on science and the colonial enterprise, to consider transnationality in science is almost always to consider imperialism.[44] What dominates this type of history are the relationships between colonial power and colonized powerlessness. Even the most recent cultural history of imperial science, Macleod argues, remains concerned with 'the multivalent perspectives coloured by the complexities of contact'.[45] Yet as both Mary Louise Pratt and Bruno Latour, in different contexts, have shown, the nation at the 'centre' of such imperial projects is also always altered by its engagement with other nations and their practices.[46] Laura Otis, in particular, has recognized that imperialism provided a powerful metaphor for certain types of scientific investigation.[47] For Otis it is not the actual contact between imperial powers, or indeed between imperial and non-imperial scientific practices, but the symbolic potential of imperialist discourses that gave certain sciences their particular character. 'Bacteriology', Otis claims, served an 'imperialist ideology' in stressing the threatened boundaries of the cell, leading many to become 'more and more anxious about penetration and about any connection with other people – the same anxieties inspired by imperialism'.[48] Just as microscopic vision gave rise to anxieties about personal and institutional identity (as the examples in the previous section reveal), it also generated further anxieties about national identity, and in particular the potential weaknesses of British national boundaries (political and cultural) in the face of an invisible force not easily seen or identified. For Le Fanu, Stoker and the opponents of the British Institute, potential infection could not be readily disassociated from a potential national identity crisis. In each case, infection is used metaphorically to explore anxieties about the nation's weaknesses.

Le Fanu's symbolic scheme for 'Carmilla' – that imagines disease as the vampire – binds together discourses of science, vision and Irish identity. Just as disease often resides within the body, at the subepidermal level, and is therefore invisible without the aid of visual technologies, so too is the vampire, as a mythological creature, an object of cultural invisibility reliant on belief. While this attenuation of science with myth is a complex enough overdetermination of the symbol of the vampire, the narrative's Gothic coda overlays them with the discourse of Anglo-Irish ascendancy. Carmilla's hidden identity, her association with superstition and myth, and her representation as infectious disease all conform to, and in some respects exceed, what Le Fanu would clearly have recognized as a common pattern (especially in Gothic fictions) denotative of Irishness.

Carmilla's ethnicity is as invisible as her infection, and for the same reasons. While disease was, at this period, understood as existing invisibly inside the body (whether it had begun in the surrounding environment or in the body itself), it was therefore often imagined as hidden from view below the visible layers of the skin. The common medical phrase for such invisibility was 'the latent period of infection' when no 'morbid phaenomena have appeared'.[49] This could, as William Henry claims in his BAAS report on contagion, be a period of weeks or even months.[50] Similarly, Carmilla's Irishness cannot be seen: she does not wear her identity on the skin as did many of the colonial subjects of Victorian Britain. This was always a problem for the Anglo-Irish ascendancy class in their imperialist rule over Ireland: there was no 'epidermal schema', in Frantz Fanon's phrase, from which to see ethnic difference.[51] Like disease, Carmilla's Irishness is subepidermal, invisible to the naked eye. Le Fanu pushes this point at several moments in the text. One of Carmilla's first appearances in the narrative is at a masked ball where the identity of all participants is hidden from view, and at which she stubbornly refuses to reveal her identity. This refusal is continued in Carmilla's discussions with Laura, and made manifest in the dust-covered and therefore opaque portraits of the Karnstein family that are only made visible by thorough cleaning.

While Carmilla's Irishness cannot be detected by visual surveillance, her characterization as the Gothic vampire provides another avenue for recognition. It was one of the central generic tropes of Anglo-Irish Gothic to associate the native Irish population with ancient folklore and mythic traditions. As Jarlath Killeen argues, Anglo-Irish writers were fascinated by 'antiquarianism, folkloric studies and Irish "superstitions"' and often employed these in their Gothic fictions.[52] R. F. Foster goes further, indicating that it was common in Protestant ascendancy Gothic to figure the Irish as 'occult' figures, whose 'Catholic magic' engenders both 'repulsion and envy'.[53] Carmilla is entirely in line with this Gothic tradition: as the vampire she both represents Irish myth and instigates the superstitious fears of the local population. Vampirism is also her own brand of Catholic occult magic, imagined in the narrative as a variety of supernatural characteristics from shape-shifting through dream appearance to vampiric attack. Seeing native Irishness, then, is vitally important for the Anglo-Irish, because to be Irish was to have the hidden potential for violent opposition to the Protestant ascendancy.

To be able successfully to see disease and Irish ethnicity is to become empowered by vision. Clear sight is masterful surveillance, knowledge and power. It is also a disruption of all efforts to hoodwink the observer and is therefore an adumbration of the power of the other. For disease theorists, confidence in the visual veracity of the microscope would lead scientists, as Samuel La'mert believed, 'to detect the latent causes of human suffering with a view to their removal'.[54] For the Anglo-Irish, recognizing the Catholic Irish's propensity for violent intervention in the colonial supremacy of the Protestant ascendancy was a first step towards maintaining that supremacy intact. 'Carmilla' suggests that, for Le Fanu, these two things are inextricable. Read through contemporary disease theory and the politics of Irish Gothic the underlying narrative emerges as a conservative political allegory: where the native Irish are imagined as diseased and violent insurrectionists preying on an unsuspecting and innocent Anglo-Irish ruling class, who, awakened to the danger of their position, mercilessly put down the Irish and free Ireland from its contagion.

While Le Fanu's anxieties about national infection lead him to draw symbolic connections between disease and Irishness, those who opposed the British Institute were more concerned by the potential moral and ethical infection of French and German bacteriology. Focused primarily on vivisection as a key experimental practice in the study of infectious disease, opponents of the British Institute characterized the British scientists working there as harbouring a desire to mirror the dubious actions of French and German science. For many opponents of the Institute this importing of foreign practices was the most problematic aspect of the new Institute's existence, for it suggested a secretive and invisible alteration of the identity of British science.

A particularly fruitful line of attack for those various lobby groups who opposed the building of the British Institute was to suggest that it would not only be a 'great school of cruelty' and a 'market garden of disease' but also a place where the scientific practices of other imperial nations, specifically France and Germany, would be conducted under the guise of British science.[55] This, suggested the Institute's opponents, revealed that those bacteriologists, chemists and pathologists involved with the Institute were colluding in hiding from the public their real identity as practitioners of foreign experimental medicine. Just as Carmilla kept secret her identity as the Irish vampire by being so like the Anglo-Irish, so too were the Institute's scientists attempting to fool the public by representing themselves not as Continental or European but as British. The anti-vivisectionist and feminist, Frances Power Cobbe, argued exactly this case in her polemical article 'Vivisection and Its Two-Faced Advocates' in which she illuminated the difference between the public pronouncements of leading British exponents of experimental medicine and their private advocacy of French and German models of research.[56] For Cobbe, the hypocrisy evident in the different discourses of British scientists led her to conclude that they were consciously maintaining as secret their evident admiration for the work undertaken in France and Germany.

It is important to recognize that these characterizations of the scientists working at the British Institute and other European scientists was, for the most part, the creation of lobby groups and the local population. While the Institute was certainly secretive, and its imposing Chelsea building (see Figure 2.2) only served to reinforce that belief, it did not look like the 'school of cruelty' it was often considered to be. Indeed its laboratory spaces, such as that pictured in Figure 2.3, were ordinary in the extreme; rather dull domestic labs far removed from the images used by Cobbe and others to represent Pasteur's Parisian counterpart.

Nevertheless, the publication of numerous exposés of the condition of Louis Pasteur's Paris laboratory suggested that there was a public appetite for an alternative perspective on experimental medicine, as well as a clear link drawn between what was done in the name of science in France with what would be done in Britain. The humanitarian Joseph Collinson, for example, gave a lengthy account of Pasteur's laboratory to the press in which he describes in Gothic terms the 'howling dog' injected with chloral, who might 'escape, all bleeding and torn' into the local area.[57] Soon, the new British Institute was retitled by its opponents as the 'Pasteur Institute' in Britain, to highlight the close, yet secretive, connection between Pasteur's French medicine and that about to be conducted in Chelsea.

Figure 2.2: The imposing façade of the British Institute of Preventive Medicine's building in Chelsea, *c.* 1899; a site of significant protests in the 1890s. Source: Archive of the Lister Institute, Wellcome Library; reproduced with permission of the Wellcome Trust.

Figure 2.3: The interior laboratory spaces of the British Institute of Preventive Medicine. Source: Archive of the Lister Institute, Wellcome Library; reproduced with permission of the Wellcome Trust.

There was evidence to support their views. While Sir James Whitehead, the Lord Mayor of London, had responded to a lengthy petition against the British Institute by stating that 'there was no suggestion [by Britain's scientific community] as to the establishment of a Pasteur Institute in this country' those closely involved with the Institute, such as Lankester, had initially suggested that any funds gathered for the Institute should be sent to Pasteur in Paris rather than used to support the building in Chelsea.[58] Indeed Lankester also believed that the creation of the British Institute 'would go far towards the object which many of [his fellow scientists] had in view', to 'wipe out the insult' given to Pasteur by anti-vivisectionists in Britain.[59] Clearly, there was an important connection between the British Institute and Pasteur's laboratory in the minds of those scientists involved with its organization, even if the financial support for it never materialized.

Not content with representing the British Institute as covertly introducing a Pasteur laboratory to Britain, the Institute's opponents also charged the scientists working there with following Continental laboratory practices. As Nicolaas Rupke has shown, to associate the Institute with Continental laboratories was to suggest that immoral or even 'horrific' experiments were likely to be conducted there.[60] The *Daily Chronicle*, giving the views of one group of protestors, reported that German experimental methods were very likely to be conducted out of sight in the British Institute's laboratory, leading inevitably to vivisection being performed on humans as well as animals: 'whatever may be the truth about human vivisection in England', the *Daily Chronicle* concluded, 'there seems no doubt that experiments upon human beings – deliberate, repeated, painful, injurious experiments – are accepted by medical men in Germany as falling entirely within their right'.[61] Moreover, as other opponents of the Institute claimed, the spread of disease would not only be a product of continental practice, but a conscious aim. *The Field* reported that the dog protection advocate Lloyd Price 'suggests as probable the practice of turning out dogs inoculated with the rabies virus, in order to increase the number of cases of rabies, and so encourage what is called the "Pasteur craze"'.[62] Invisibility is the keynote here; the hidden methods of foreign science would lead directly to secretive human vivisection and to the emergence of invisible bacteria onto the streets of Chelsea.

What occurs in these tactics of opposition is the construction of a series of representations of the British Institute, all of which serve to define it as a secretive organization whose true identity remains invisible to the general public. It is from this inability to see the Institute's true identity that its danger emerges, for such observational blindness denies the possibility of seeing that its national (British) identity has been corrupted by foreign influence. The discourses of imperialism that can become attached to infection are clearly apparent here. If the British Institute is a danger to the local population in Chelsea, it is not only because it may heighten the risks of disease but because disease itself will be the product of foreign experimental methods colonizing the (presumed to be) moral codes of British science. Lloyd Price's fear of the rabid dog is actually a fear of immoral French medical practice: the disease transmitted is not hydrophobia but the Pasteur craze. Recent reconsiderations of the history of health and its relation to place have suggested that it is not uncommon to find a 'particular imperialist imaginary' emerging in the construction of that relationship.[63] In particular, as Gregg Mitman, Michelle Murphy and Christopher Sellers argue, it is important to recognize that local histories of health are often constituted by their engagement with a wider imperial politics, by the 'movement, tensions and connections' between the local and the international.[64] It is important to recognize how the extended cultural influence of microscopic vision plays its part here. While there is no immediate connection between microscopical ways of seeing and Continental research practices, it is the relationship between the seen and the unseen that microscopy had created which allows the British Institute's opponents to characterize it as sympathetic to foreign experimental practice. Microscopic vision had revealed the potential for identity to be hidden or invisible, carried out by agents of disease usually unseen. Transferring this paradigm from the microbe or germ to those who investigated them, the Institute's opponents were able to construct the identity of British biologists as a similarly hidden agency of disease causation. This tactic employed the metaphoric power of microscopical vision to see the British Institute's employees as diseased experimenters. And the most powerful existing vision of such an identity was the Continental experimenter who was already characterized as immoral and dangerous. It is, then, on the extended continuum of microscopic vision that opponents of the Institute built their programme of active propaganda.

Stoker's *Dracula* also employs disease as a metaphor to consider the issues at stake for Britain's health in engaging with other nations. The novel investigates disease and national politics in their relationship to imperialist economics, and with particular attention to the relationship between the Count and his legal advisor, Jonathan Harker. It is relatively straightforward to make connections between Dracula's status as a foreigner and his role as carrier of disease, but his arrival in Britain, representing the transmission of disease to Britain from abroad, is only achieved with the help of Jonathan Harker.[65] It is Harker who discovers and purchases British property in Count Dracula's name, travels to Transylvania to aid the Count in his British business dealings, helps the Count master the intonations of the English language, and master his performance as the English gentleman. Harker's help is varied and profound, and yet he actually conducts his business with Dracula with the intention of achieving his own personal advancement and economic success.

Harker plays the role of the superior colonialist from the moment he opens his narrative account of his business dealings with Dracula. As Stephen D. Arata has previously noted, Harker's sly castigation of the inaccuracy of the train timetable and his derogatory comments on the map-making skills of Transylvanian cartographers construct a dynamic in which Britain is perceived as intellectually and technologically advanced while the East is a land of anarchy and superstition.[66] Harker's imperialist attitude continues while he is Dracula's guest, and leads him to contract a disease that he may well have been able to avoid. Dracula, conscious of the disease properties of his castle which are personified in the three female vampires, warns Harker against trespassing on certain areas, concluding that 'we are in Transylvania; and Transylvania is not England'.[67] Yet Harker displays an extraordinary disregard for Transylvanian custom and its domestic environments, believing himself imprisoned by the locked doors rather than protected by them. His panic at being so incarcerated leads him to abuse Dracula's trust and forcefully enter a suite of rooms in an older section of the castle. Here, Stoker develops his theme of imperial disturbance of ancient culture in association with the revival of ancestral infection:

> This was evidently the portion of the castle occupied in bygone days ... the windows were curtainless, and the yellow moonlight, flooded in through the diamond panes, enabled one to see even colours, whilst it softened the wealth of dust which lay over all and disguised in some measure the ravages of time.[68]

The infusion of light supporting an improved view of colourful dust is a clear signifier that this is a moment of microscopic vision that Harker fails to interpret correctly.[69] It is surprising that Harker is not more visually adept. His professional training should mean that his observational power is substantial; his lawyerly instincts for patient sifting and interpretation of evidence, his instinctual preference for fact over fancy, all suggest that he might be an ideal observer.[70] Yet Harker's vision is ideological rather than ideal. He interprets what he sees with the eye of the English imperialist, forcing the alien other to conform to his own mythic narrative of how Transylvania might be defined. For example, Harker turns his intrusion (visual and physical) into a romance narrative: 'I determined not to return tonight to the gloom-haunted rooms, but to sleep here, where of old ladies had sat and sung and lived sweet lives whilst their gentle breasts were sad for their menfolk away in the midst of remorseless wars'.[71] Harker recreates Transylvania imaginatively, according to his own cultural traditions, in this instance of the stories of chivalric romance that make up England's cultural heritage.[72]

Vampiric disease's long journey to the heart of the British empire begins, then, with Harker's problematic microscopic vision, his cultural desecration of Transylvania and his economic exploitation of Dracula. These imperialist transgressions abroad have consequences for both Harker and Britain's nation state. Harker falls victim to the vampire having unearthed its ancestral roots in Dracula's castle and aided its cause by preparing the ground for Dracula's journey to Britain. Britain, too, suffers from Harker's decision to opt for the increased status in business and society he feels he will gain from Dracula's wealth and patronage. His professional offices enable Dracula to plan an efficient assault on Britain, spreading the disease of vampirism from the property that Harker has helped him purchase. Harker – having learned from Van Helsing how to observe appropriately – eventually recognizes this in a momentary epiphany while stood over Dracula's coffin: 'This was the being I was helping to transfer to London, where, perhaps, for centuries to come he might, amongst a teeming millions, satiate his lust for blood, and create a new and ever-widening circle of semi-demons to batten on the helpless'.[73] It is here that Stoker implicitly poses an important question: who actually represents impurity borne of moral corruption expressed in the novel as infectious disease? In the purchase and exchange of Carfax Abbey we witness Stoker's meditations on these questions, which ask us to revisit our understanding of the location of disease and immorality.

Jonathan Harker deals both with the purchase of Carfax Abbey and the mapping of the house and its grounds, although he does so only partially and with the aid of a specific visual technology, his Kodak camera.[74] This is significant for it highlights the limitations of Harker's observations of the English house, which, had he undertaken his duties properly, might have given him an indication of its diseased environment. Stoker implicitly compares the lens of the camera with the lens of the microscope in this instance: the scopic power of the camera lends Harker nothing more than the limiting eye of the landscape photographer, able to capture a 'view' but not to give the observer, as Edwin Lankester puts it, 'a greater knowledge of the nature of all objects' as the microscope may have done.[75] Indeed, this moment confirms what we know of Harker's ideological vision. His Kodak views are, like his reading of the chivalric spaces of Dracula's castle, romantic and picturesque, representative of a particular narrative of England's past that was always fictional, and far away from the microscopic penetration required to read Carfax as a site of disease.

The novel makes clear, too, that Carfax Abbey is a place of infection. Its 'faint, malodorous air' reminds Harker of finding Dracula in his earth box in the cellar of the castle, although here at Carfax 'the long disuse had made the air stagnant and foul' and its enclosed space 'intensified its loathsomeness'.[76] It would be easy to view Carfax simply as infected by Dracula's presence but Stoker is methodical in not allowing such a straightforward connection to be made. Instead, he draws comparisons between Dracula's castle and Carfax in terms of the miasmatic atmosphere, the dust, and the age of the building. If Dracula's castle is a site of disease precisely because it is filled with ancient dust and foul air then so too is Carfax. Seeing Carfax as a place of ancient infection is highly significant within the structure of disease causation that Stoker creates. It reveals that disease is to be found in Britain as well as abroad, and that it may well be in the character of the British imperialist, Harker, that moral infection lies.

Like both Le Fanu's 'Carmilla' and the discourses of opposition to the British Institute of Preventive Medicine Stoker links infection to specific forms of national and moral identity. And also like those other narratives he, too, imagines the danger that arises from the scopic difficulties that such identity provides. As invisible as the microbes of infection that define them, the national identities of Carmilla, Dracula and the experimenters of the British Institute remain hidden from view, and are entirely invisible to the eye of the observer. Yet as Stoker also argues, just as Le Fanu and the protest groups had done, observational perseverance can prove fruitful. Although microscopical research had revealed that the materials of infection were difficult to see, it had also provided a new ideology of visual surveillance where invisibility became a marker of the presence of something rather than its absence. In arguing that microscopic vision could, and indeed should, be transferred into scopic encounters with the naked eye, microscopy revealed a new form of ocular power that determined meaning and invited action from imagined rather than actual perception. It is this kind of vision we see at play in 'Carmilla', where the vampire is discovered by the eventual acceptance of a myth. It is also the form of vision employed in the narratives of protest against the British Institute, as its opponents construct a believable fiction of what might take place within its laboratories. Stoker, likewise, reveals this vision in Harker's imagined version of Transylvania, and ironically contrasts it with a different kind of visualization, the images seen through the lens of the Kodak camera.

Yet in their evocations of infection and identity *Dracula* differs from 'Carmilla', which in turn offers a different perspective to the opponents of the British Institute. 'Carmilla' highlighted the threat to Britain and British identity from the immoral, diseased Irish. Although the British Institute's opponents also stressed such a threat, in their characterization of French and German scientists as immoral and likely to bring their infections to Britain, their narratives also suggested that the Institute's scientists may themselves have the potential to be similarly immoral. Stoker goes further, by probing the question of the infective potential of Britain and the British in his portrayal of the actions and motivations of Harker. In these three different studies of disease, then, there is a clear movement of the metaphors of infection from a focus on disease and its dangers to British identity to a representation of British identity as itself diseased.

The case of the opposition to the British Institute sits at the centre of these shifting cultural representations. Indeed, placing the narratives of the British Institute and its Chelsea laboratory within such a set of discourses illuminates its connections to wider cultural perspectives. It shows that it is not just the 'practices' of the bacteriological laboratory that are important but also, as Michael Worboys notes, the 'representing of germs' more broadly; or, as Nancy

Tomes and John Harley precisely state, it shows that germs are 'a construction, the product of many communities engaged in vigorous debate'.[77] It confirms, too, what Bruno Latour argued was important in his work on Pasteur's laboratory, that by not 'reducing the sciences to a few authorities that stand in place of them, what reappears is not only the words of human beings, as in Tolstoy, but also the "nonhuman," eternally banished from the critique'.[78] Moreover, situating the laboratory firmly in its geographical place reveals how the cultural metaphors in play at a local level also come to bear upon its definition. To that extent, the literary vampire, too, played its role in how the British Institute and its scientists were cast. J. Andrew Mendelsohn, in his excellent investigation of infection in Paris in the 1880s, has shown how that city, its fictions, its intellectual culture, and the scientific workers, all defined the processes and procedures of bacteriological research. As Mendelsohn cleverly reveals, in Parisian novels 'the laboratory, the clinic, the city' were 'blurred ... into a continuous space of knowledge'.[79] Additionally, bringing into dialogue fictions of infection and observation, microscopic and bacteriological discourses, and laboratory science is one way in which 'interactions between professional and popular epistemologies' can be explored. It offers one answer to the question that Gregg Mitman, Michelle Murphy and Christopher Sellers asked in their recent revisiting of the state of health and environment studies: 'how do we ... navigate alternative and conflicting accounts of a phenomenon when laypeople and experts, or different groups of experts, disagree?'[80] For these writers, such divergent accounts 'confront a politics not just of epistemology but of ontology as well'.[81] That is, it asks questions not about types of knowledge but about ways of looking at that knowledge. However, as this chapter has shown, asking questions about ways of looking should not necessarily lead to philosophical distress; tracing microscopic vision has not revealed an occluded view of bacteriology's interaction with culture, but has rather brought it further into focus.

PART II: LARGE

3 OPTICAL SHATTERING: PERCIVAL LOWELL, MARS AND AUTHORITIES OF VISION

In the summer of 1894 the astronomer Percival Lowell began to construct his new observatory on a peak above the Arizona town of Flagstaff. Lowell's explicit aim was to study the planet Mars in search of the canals described by the Italian astronomer Giovanni Schiaparelli. From the moment of Mars's near approach to Earth later in 1894 until his death twenty-two years later, Lowell looked at the planet and sought out its canals. His work was widely publicized, much debated and continually criticized, yet its impact on Martian astronomy, as well as on a broader public understanding of Mars, was significant and enduring. One of the most consistent keynotes of Lowell's investigations was the role played by vision in the practice of astronomy. In an essay on common errors in Martian observation, for example, Lowell wrote of the problems associated with telescopy:

> The matter of having observed the planet with the full aperture of a large telescope is in itself not only not conducive but actually deterrent to seeing the canals of Mars ... 1. There are atmospheric waves which interfere always with the performance of a large glass more than with that of a small one, necessitating diaphragming down for definition. Only in the very best air can the full powers of a large glass be used. Such is not the case at Meudon [Observatory], in the immediate vicinity of a large city, the kind of location where definition is notoriously poor. 2. With a large glass there is too much light for Mars causing large spurious disks of every part of the planet which cover over the detail. This light must be reduced either by limiting the aperture or interposing a darkening glass in order to permit the detail to show. 3. The eye also suffers for the like reason and demands a darkening glass to take off the glare and so allow of definition.[1]

From that first Martian opposition in 1894 until writing this unpublished essay in 1909 Lowell was continually striving to solve problems of seeing wherever they arose; in the technical complexity of telescopic equipment, with atmospheric interference, or in the limitations of the human eye.

Similarly, the narrator of H. G. Wells's speculative fiction on Mars, *The War of the Worlds* (1898), writes of the surprising difficulty of viewing Mars through a powerful telescope:

> Looking through the telescope, one saw a circle of deep blue and the little round planet swimming in the field. It seemed such a little thing, so bright and small and still, faintly marked with transverse stripes, and slightly flattened from the perfect round. But so little it was, so slivery warm – a pin's head of light! It was as if it quivered, but really this was the telescope vibrating with the activity of the clockwork that kept the planet in view. As I watched, the planet seemed to grow larger and smaller and to advance and recede, but that was simply that my eye was tired. Forty millions of miles it was from us – more than forty millions of miles of void. Few people realize the immensity of vacancy in which the dust of the material universe swims.[2]

From the dust that interferes with the narrator's view of the planet, to the vibratory influence of the telescopic machinery, and the fallibility of the imperfect human eye, Wells captures precisely the optical misprisions of astronomical seeing that vexed Lowell. The parallels are not coincidental: Wells's imaginative impetus came directly from the astronomical debates on telescopic vision – including Lowell's – that were commonplace in a number of scientific texts in the period from 1896 to late in 1897 when *The War of the Worlds* was completed.[3] It is, however, too reductive to define literary responses to these debates as merely reflecting scientific knowledge. Indeed, as these debates were ongoing at least until the 1920s, it is more accurate to argue that fictional interventions were contributions to an active set of discussions about visual authority and the role of the observer of, or witness to, Mars's planetary structures.

This chapter will concentrate on the observer as witness in the context of the Martian canal controversy of which Lowell was a central figure. Lowell perceived of Mars as a planet geo-structurally dominated by an enormous canal network that must indicate intelligent life. In arguing such a case Lowell was only the latest contributor to the centuries old astronomical debates on the plurality of worlds. The specific moment of Lowell's involvement, however, saw the emergence of new sets of questions about vision as grounds for conflict. Whether in the visual technology of the telescope, the placement of observatories, the clarity of the atmosphere or the frailty of the eye, vision became the focal point both for challenging the authority of witnesses to Mars's canals and for defending their accuracy. Fictional accounts of Mars have an important role to play in this supposedly scientific controversy. While the work of Wells and others are clearly not the resultant publications of scientific investigations, they do share some generic features with those books, articles, proceedings, annals and other texts that are clearly and consensually defined as science. In particular they treat Mars as a scientific object to which observers must first bear witness and then place within the context of relevant scientific and cultural knowledge. Their similar interrogation of what it means to see at a distance, as the parallel examples from Lowell and Wells reveal, places fictions of Mars on a continuum with Martian astronomy when it too asks important questions of perception and truth.

Astronomy was always, to some extent, concerned by optical knowledge. Just as microscopical researchers both valued and were suspicious of the microscope's power of magnification, so astronomers vacillated between ontological and epistemological positions, sometimes readily accepting the evidence obtained through the use of the telescope, and at others beset by optic nervousness. The most influential Victorian astronomer, John Herschel, in arguably his most influential astronomical work, *Outlines of Astronomy* (1849), devoted a chapter to the errors of instruments, claiming that 'with regard to errors of adjustment and workmanship, not only the *possibility*, but the *certainty* of their existence, in every imaginable form, in all instruments, must be contemplated'.[4] Herschel goes on to argue, however, that by recognizing the potential for instrument imperfection the astronomer can 'derive available and practical means of destroying and eliminating the influence of such imperfections' and so come to trust the truths the telescope reveals.[5] Historians of astronomy focused on earlier periods have also noted that the history of the existence of intelligent life on other planets raised questions about observational accuracy. Michael J. Crowe, for example, shows how that history of pluralism 'contains numerous instances of astronomers being misled not only in theories but also in observations'.[6] Similarly Stephen J. Dick argues that the 'metaphysical framework' of pluralist theory had a significant influence on 'which observations were made, and which may have affected the interpretation of the evidence'.[7]

Lowell's astronomy has received only sparse critical attention in the context of observation, and less that considers the implications for vision more generally. Crowe usefully contextualizes Lowell's work on Martian canals within a longer history of pluralist astronomy, focusing his attention on its scientific personalities, rather than their contribution to the observational culture of late nineteenth- and early twentieth-century astronomy. Both William Sheehan and Robert Markley pay significant attention to Lowell's part in the history of Martian observation, considering his role from the perspectives of psychology (in Sheehan's work) and ecology (Markley).[8] Both offer interesting interpretations of Lowell's observations and those astronomers who challenged them. Sheehan argues that it is Lowell's psychological perception that is most interesting in this context, drawing on William James's contemporaneous ideas about suggestion to contrive an argument that saw Lowell as subject to a broader suggestibility, prevalent in Victorian culture, of life beyond the confines of Earth. Taking a different perspective on ideas of suggestion, Markley claims that Lowell's observations were influenced by his imaginative response to both the nebular hypothesis and Darwinian evolution, from which he constructed (inductively) a Mars on the brink of ecological collapse. By contrast, Lorraine Daston and Peter Galison draw on Lowell's work directly to consider his relationship to emerging definitions of scientific objectivity in nineteenth-century experimental practice.[9] Daston and Galison view Lowell's extensive writings on Mars as textual (and later photographic) examples of what they call mechanical objectivity: a reliance on visual technologies such as the telescope to provide 'images uncontaminated by interpretation' which are closer to the truths of the natural world than can be achieved by any individual human observer.[10] Jennifer Tucker's discussion of Lowell occupies similar territory in examining the photographs of Mars taken by Lowell's assistant Carl Lampland.[11] For Tucker this episode in scientific visual culture epitomizes the 'persisting significance of scientific skill, judgement, and aesthetics in debates over the evidentiary value of mechanical reproductions'.[12]

Of all the critical interpretations of Lowell's observations only Markley introduces those fictional responses to Martian astronomy that are an important part of the cultural reception of the canal theory. Markley understands the fiction as receiving scientific ideas which it then puts into play in an entirely fictional Martian adventure. As he argues, 'the emergence of a Mars as a world of "seas" and canals created a narrative frame of reference for science-fiction writers'.[13] This diffusionist hierarchy – in which science provides material for fiction – is a common element of the critical work on fictions that respond to astronomy. Anna Henchman, for example, in an interesting analysis of Thomas Hardy, identifies the 'ways in which Hardy brings astronomy into his literary works'.[14] However, Hardy is not simply a passive recipient of astronomical knowledge: by 'teasing out the intellectual implications of the contradictions found in astronomy' Hardy's poetry and fiction 'goes well beyond scientific writers in thinking about how the mind moves between sensory and theoretical versions of the universe'.[15] Pamela Gossin makes a similar claim, but on the broader landscape of culture, of which she regards literary fiction as one element. For Gossin, Victorian astronomy's new findings presented fresh 'existential questions for human individuals' that writers like Hardy and Tennyson attempted to answer.[16] This was, however, just one example of the 'deep and lasting impressions upon the visual imagination of the Victorian public' that astronomy engendered.[17]

Centring the discussion on Lowell's astronomical work and the fiction directly related to it brings greater specificity to the debates on the cultural significance of vision in astronomy. It is clear that the existing scholarship recognizes the importance of the visual and the textual in astronomy's nineteenth-century history but has yet to identify the concrete modes of interaction that provide evidence for an unbroken continuum between astronomical practice and the literary text (acting as a key marker of cultural authority). This is not to argue that literature straightforwardly reflects or responds to new astronomical discoveries. Rather it reveals how literature takes part in astronomical discovery contemporaneously with astronomical communities; just as the Gothic fictions of Stoker and Le Fanu contributed to ongoing debates about vision and disease, as Part I argued. Similarly, the specific contexts for Lowell's claims about Martian canals will highlight the significant influence of elements of visual culture not only contained within science but also at play in the social, cultural and political world. Lowell's arguments may well have been countered by an address to their astronomical weaknesses but they were challenged too on the grounds of his status, his eyesight, his artistic skills, his mechanical ineptitude and his choice of location. By considering in detail Lowell's visuality – how he observed, perceived and interpreted – the often too opaque relationships between science, culture and fiction can themselves be brought into the light.

Air and Apertures

While the telescope and the microscope may have marked similarities in conception and structure – both rely upon optical knowledge of reflection and refraction and both use lenses organized within a tube to magnify a specific object – they differ from one another in two particular ways. First, the telescope is used to look at objects a great deal larger than the mechanism which brings them within the human visual field. That is, the objects of telescopic observation can be seen by the ordinary human eye, as long as they are close enough to allow this. Second, the distance between the telescopic lens and the object under observation is far greater than the same distance when using a microscope. These differences lead to considerable variation in the epistemological questions such instruments asked of their users. Microscopists were concerned to know whether the objects they saw actually existed, as they were invisible to ordinary human vision, and also in what ways the object might have been transformed by their preparation of it upon a microscopic slide. Astronomical observers, by contrast, knew that many of the objects they looked at were real and unadulterated, but worried whether any elements of the spatial distance intervening between the object and the telescope had interfered with the accuracy of their seeing. Telescopic vision did not produce the 'phantasmagoria' that Jonathan Crary finds in some nineteenth-century visual technologies but rather engaged the observer in a strobic contest to collect data in the brief moments when both the telescope's apertures and the atmosphere through which it had to penetrate allowed optical clarity.[18]

Lowell's own observations at his Flagstaff observatory were beset by problems of telescopic vision: from the difficulties of using the complex mechanisms of his new telescopes, to their size and adaptability, as well as the atmospheric disruptions in Northern Arizona. In his first book detailing the Martian observations he and his staff undertook in 1894, Lowell began his preface with an acerbic comment on the necessity of good atmospheric conditions and the appropriate knowledge, and size, of telescope:

> This book is a result of a special study of the planet [Mars] made during the last opposition, at an observatory put up for the purpose of getting as good air as practicable, at Flagstaff, Arizona. A steady atmosphere is essential to the study of planetary detail: size of instrument being a very secondary matter. A large instrument in poor air will not begin to show what a smaller one in good air will. When this is recognized, as it eventually will be, it will become the fashion to put up observatories where they may see rather than be seen.[19]

Lowell's sharp tone is a result of the opposition he encountered to his theories of the Martian canals. From his first set of observations and publications in 1894, Lowell discovered he had several combative foes in the astronomical community whose denial of the existence of the canals led to attacks on his methods and conclusions. One of the earliest points of contention was the placement of Lowell's observatory at Flagstaff, Arizona. Working from generally agreed astronomical knowledge that in 'the mountains or table-lands in the western and south-western regions of North America, the atmosphere is clear and steady in an extraordinary degree' Lowell experimented with several sites in Arizona before deciding upon Flagstaff. [20] As early as September 1894 he was proclaiming in print its excellence as an observatory site: 'The results already obtained are very encouraging, amply confirming the importance of choosing as good air as possible for an Observatory site'.[21]

The clarity and steadiness of atmospheric waves ('good air' or 'good seeing' in the astronomical parlance of the period) were vital components in telescopic observation. What Lowell crystallizes in his early pronouncements on Flagstaff's excellent conditions is the importance of the relationship between observation and location in uncovering astronomical truths. In the concluding section of the first volume of the *Annals of the Lowell Observatory* (1895), Lowell and his staff argue that observational accuracy is determined by place:

> Certain as it is to be looked at askance by astronomers trained in the impossibility, from the positions of their observatories, of seeing the phenomena upon which it [the canal theory] is based, a time is no less sure to come when other observatories shall be put up in favourable places and the phenomena generally recognized.[22]

The explicit link between seeing the Martian landscape as it actually is and the location of the observer derogates appropriate telescopic vision to *where* phenomena are seen rather than *by whom* they are seen or *with what* instrument they are seen. Visual authority no longer rests in the organic lens of the human eye or the glass lens of the telescope but in the passage of vision through space. The observer's geographic situation, and the line of sight this allows, is instantiated as the primary mode of astronomical seeing.

Although Lowell's claims about geographical location did not go unchallenged his continued insistence on Flagstaff's authoritative visual position did find imaginative correlatives in several of the fictions of Mars published between 1896 and 1911.[23] In Edgar Rice Burroughs's *A Princess of Mars* (1911), the narrator John Carter begins his journey to Mars from Arizona.[24] Carter describes the Arizonan air as 'clear' and 'pure', and its landscape as so unusual that looking at it felt 'as though one were catching for the first time a glimpse of some dead and forgotten world'.[25] This comparison between Arizona and Mars is fully explored in the moment of Carter's transference from one to the other, when his looking leads to profound reflection:

> As I stood thus meditating, I turned my gaze from the landscape to the heavens where the myriad stars formed a gorgeous and fitting canopy ... My attention was quickly riveted by a large red star close to the distant horizon. As I gazed upon it I felt a spell of overpowering fascination – it was Mars ... My longing was beyond the power of opposition ... [I] felt myself drawn with the suddenness of thought through the trackless immensity of space ... I opened my eyes upon a strange and weird landscape. I knew that I was on Mars ... It was midday, the sun was shining full upon me and the heat of it was rather intense ... yet no greater than would have been true under similar conditions on an Arizona desert.[26]

Burroughs's juxtaposition of Arizona with Mars is an imaginative rendering of Lowell's contention that only from his observatory could the Martian landscape be properly seen. Moreover, Burroughs argues that clear observation of Mars is inextricably linked to superior investigation. His fantasy of travelling to the planet, to see it at close quarters, is a metaphor for the observational detail required to provide grounds for further planetary study.

H. G. Wells's *The War of the Worlds* also offers an analysis of location in relation to seeing, found most commonly in the narrator's determination to find the most appropriate site from which to view the Martians recently landed on Earth. He first tells the reader how delighted he was to become 'one of the privileged spectators' to the Martian arrival, and subsequently spends considerable time either 'seeking some point of vantage' or, having found it, describing both the clear 'view' and those moments of obscuration when 'a haze of smoke ... hid the Martian shapes'.[27] Like Burroughs, Wells regards good seeing as a necessary precursor to thoughtful reflection and analysis. This is embodied in his narrator's response to a moment of particularly clear observation:

> With a queer feeling of impersonal interest I turned my desk-chair to the window, sat down, and stared at the blackened country, and particularly at the three gigantic black things that were going to and fro in the glare about the sand-pits ... I began to ask myself what they could be. Were they intelligent mechanisms?[28]

The narrator's arrival at a moment of interpretation follows from careful and sustained observation; clarity of vision is here regarded as the foundation for the pursuit of astronomical knowledge. While the evidence presented to the narrator's eyes (of invading Martians) is almost beyond comprehension in being so supernatural, the clearness of his vision underscores an epistemological certainty about what is seen.

Lowell put forward the same argument for the Martian canals. In the *Annals* for 1894, he noted that 'the better the means of observation, the stranger the whole system appears', adding that 'good air leads to supernatural regularity in the canals'.[29] At the same time he argued that poor atmospheric conditions (found at other observatories) could very easily, and credibly, fool an observer. Atmospheric disruptions of small wavelengths could readily obscure the detail of the canals while leaving the larger view of the planet distinct. This, Lowell, claims, is 'semblance masquerading as reality': it is understandable that 'to one not versed in air-waves the conclusion should seem simple that no such detail exists'.[30] There is an interesting reversal in this construction of observable reality. Lowell argues that the strange and supernatural are more real than the mundane, and that the very fact of the canals' strangeness is evidence of their existence. Good air performs an astonishing role here; it makes the uncanny familiar and the supernatural ordinary. In fact, such reorganization of observational imperatives parallels one of the key generic tropes of Martian fiction, to make the fantastic appear commonplace, or as Wells puts it, 'to domesticate the impossible hypothesis'.[31] As fictions of Mars draw very clearly on Lowell's pronouncements about Flagstaff's superior seeing to construct observational principles in their narratives, so Lowell mimics their fictional strategies in his analysis of the visual properties of good and bad air.

What is also certain is that Mars at this historical moment is emerging as an exemplary scientific object. For Bruno Latour the scientific object is a thing that exists because it is embedded in a range of systems that supports its continued presence.[32] While for Latour these systems are generally scientific, Daston argues that scientific objects are also defined by their further embeddedness in other practices: 'they grow more richly real', Daston claims, 'as they become entangled in webs of cultural significance'.[33] Lowell's Mars had certainly, at the end of the nineteenth century, been accepted as a 'new' object of scientific scrutiny both within but also beyond the boundaries of astronomical communities. Mars becomes enmeshed, certainly, in the culture of fiction, where Lowell's conception of it as an inhabited world dominated by a giant canal system was the impetus for an imaginative rendering of his vision. More significant, however, is that as a scientific object, Mars was also connected with broader issues related to late nineteenth-century visual uncertainty. The quality of the air through which the telescope had to penetrate was not only a scientific problem; it was an issue, too, of geo-political status.

This is apparent in Lowell's assaults on the Italian astronomer E. M. Antoniadi's observations at the Meudon Observatory in Paris, which he classified as 'notoriously poor' because of its location in a major city. Lowell was also, and for the same reason, severely critical of the Lick Observatory in Los Angeles, California, where, he argued, the air was considerably poorer than at the rural location of Flagstaff. The Lick's director, the astronomer W. W. Campbell, strenuously opposed Lowell's efforts in the first decade of the twentieth century to undermine his telescope's geographical position. In a series of increasingly bitter essays, Lowell and his assistant V. M. Slipher argued that the Lick's urban situation gave rise to such poor atmospheric conditions that any observations undertaken there were necessarily suspect. Campbell, faced with uncomfortable data that appeared to confirm the Lick's inadequacy, took the extraordinary step of deprecating his own observational skills in order to avoid admitting the inferiority of the Lick's air. When Lowell and Slipher argued in 1908 that 'so much purer was the air at Flagstaff than at Mount Hamilton ... that more stars were visible with a 24-inch glass there than with a 36-inch one at Lick' Campbell replied that this was due entirely to Lowell's superior observational skill, having spent 'more than ten years in observing difficult objects'.[34]

Lowell targeted not only the Lick Observatory but also the air at Harvard, and at the Naval Observatory in Washington DC. Although ostensibly an issue of astronomical significance, the disagreements about air quality and location were actually engendered by Flagstaff's political and geographical status relative to these other observatories. Marginalized by its rural situation and its place outside the traditional institutions of astronomical research, Lowell sought advantage in visual superiority where he could not attain it in either geographical location or political influence.

In his fictional sequel to Wells's *The War of the Worlds*, entitled *Edison's Conquest of Mars* (1898), Garrett P. Serviss parodies the influence of geo-political power on astronomical research. Readying an international mission to Mars, the collaborating nations come together to disburse the necessary funds and find themselves unable not to outbid one another:

> Germany, coming in alphabetical order just before Great Britain, had named, through its chancellor, the sum of $500,000,000, but when the First Lord of the British Treasury, not wishing to be behind the United States, named double that sum ... the Emperor William looked displeased. He spoke a word in the ear of the chancellor who immediately raised his hand. 'We will give a thousand million dollars,' said the chancellor.[35]

Burroughs displaces specific conflicts between astronomers onto the landscape of the nation state, yet still captures the political consequences of astronomical competition.

That Lowell's claims about the excellence of Flagstaff's air were as much political as they were astronomical arguments is made more apparent in the

Lowell Observatory *Annals*. There, in the detailed examination of the extensive observations made by Lowell and his assistants, it is clear that the Flagstaff air caused major optical problems and that Lowell's publications that championed its purity and steadiness were to some degree propagandist. In 1895 Lowell writes that 'even the best air is far from perfection and varies greatly from second to second. Only at intervals is a detail seen in anything like its best.'[36] This had significant repercussions for astronomical knowledge, as Lowell accepts when noting the number of misidentifications that resulted from poor atmospheric conditions.[37] Despite making great claims for Flagstaff's air, at the expense of the air at other observatories, it is clear that the Arizona atmosphere might as easily deceive the observer as at any other location. Several Martian fictions dramatize the ability of the air to deceive, and they do so, intriguingly, through almost identical descriptions of clouds that obscure Mars and the Martians. Serviss, for example, describes a 'great smoke barrier' which 'rose high in the air, and carried by invisible currents in every direction ... blotted out of sight everything'.[38] Likewise Mark Wicks, in his fiction *To Mars Via The Moon* (1911), a work dedicated to Lowell, places his characters in the midst of a 'sand-cloud' through which they could 'see nothing else'.[39]

It is likely that both these fictions drew inspiration from Wells's articulation of the 'heavy, inky vapour' that the Martians fire upon Earth to quell any insurrection, which the narrator describes as a 'huge and ebony cumulus cloud' of 'black smoke'.[40] Wells's is the most sophisticated interrogation of the effects of poor atmospheric conditions upon observational accuracy. In an episode that directly reimagines the brief moments of clarity followed by periods of obscuration that Lowell describes, Wells's narrator finds himself close enough to a Martian to observe it but obstructed by the tumultuous air of a storm:

> And this Thing I saw! How can I describe it? ... A flash, and it came out vividly, heeling over one way with two feet in the air, to vanish and reappear almost instantly as it seemed, with the next flash, a hundred yards nearer. Can you imagine a milking-stool tilted and bowled violently along the ground? That was the impression those instant flashes gave. But instead of a milking-stool imagine it a great body of machinery on a tripod stand.[41]

This strobic vision clearly parallels the experiences of Lowell at his telescope. As he describes in his notebooks, seeing occurred in very brief moments, lasting from fractions of a second to 'a second or more' when clarity was at its zenith. For Wells such fragmentary observation raises a question of interpretation: how does the observer piece together what has been seen into a single image? Wells's narrator chooses to offer a comparison with what Daston calls 'quotidian objects', in this case the milking-stool.[42] Yet it is a milking-stool that is transformed almost immediately into a 'scientific object', the great tripod machine. As Wells's narrator later acknowledges there is a considerable threat here from an unannounced subjectivity (the narrator's imaginative rendering of a Martian as akin to a milking stool) which reduces and makes static our understanding of a scientific object that by contrast should allow knowledge to 'broaden and deepen'.[43]

For many astronomers the most obvious way to avoid errors of subjectivity was to rely upon the objectivity of the telescopic equipment they employed. The aperture of the telescope offered no mediating consciousness nor did it have an imagination that might reduce a new scientific object to the level of a quotidian object. The telescope was a technology that offered observation 'uncontaminated by interpretation'.[44] Daston and Galison claim that Lowell exemplified the turn to mechanical objectivity which emerged in the mid-nineteenth century as technologies of vision improved. In their reading of his Martian observations over two decades, Lowell proved himself an indefatigable worker who demanded objectivity to such a degree that he wilfully refused to impose himself on the images the telescope provided for him; he typified 'the annihilation of the self by the self' that characterized the objective scientific observer.[45] Yet the evidence of Lowell's many writings on telescopy, several of which see him drawn into conflict with the same adversaries he encountered when aggrandizing Flagstaff's air, suggests that the technology of the telescope gave rise to serious concerns about instrumental deception.

It was commonplace to assume that the larger the aperture of a telescope the greater detail one was likely to see in the object observed. The British astronomer Simon Newcomb, in a chapter on the Martian canals in a work of popular astronomy from 1902, made this assumption in claiming that the Lick Observatory's findings were to be most trusted as 'its telescope is the largest and finest in the world that has ever been especially directed to Mars'.[46] Antoniadi, working at the Meudon Observatory in Paris where he had access to another large aperture telescope, agreed that his most important observations of Mars were made with that large aperture when the seeing conditions were 'excellent'. When these large apertures were trained on the Martian surface, however, the canals were noticeably absent. At the Lick Observatory, Barnard did not observe any canals, but instead 'a vast number of minute and very faint markings'.[47] These observations were confirmed independently by Antoniadi, who saw only 'a tessellated series of dots' rather than a series of linear canals.[48]

The authority inherent in the large aperture telescope (and of the astronomers who used them) was a direct challenge to Lowell's claims of having witnessed the canals of Mars on numerous occasions with his smaller 24-inch reflecting telescope. The accumulated testimony of Barnard and Antoniadi implicitly suggested that Lowell had been deceived by his own visual technology. Lowell would not have been surprised to discover that the accuracy of his telescopic observations were under scrutiny. From his first observations in 1894 he had critically reflected upon the images of Mars his telescope had shown him. In part, his recognition that deception played a role in astronomical observation was founded on the fact that telescopes were extremely difficult visual technologies to manage. His correspondence with Alvan Clarke, for example, testifies to the anxiety Lowell felt in building and using his new equipment on arrival at Flagstaff. These fears of optic inexperience did not diminish as his career advanced. When he took delivery of a new spectroscope from the master telescopist John Braesher in July 1901, Braesher noted in a letter that 'should you have any difficulty in putting the instrument together we will be glad to write you full instruction'.[49] Lowell replied by telegram in less than a month: 'Please wire address best book on adjusting spectroscope answer paid'.[50] In a more thoughtful piece of writing, a sonnet written in his first years at Flagstaff, Lowell considered in metaphor the risk of observing through an aperture:

> As brief as when the maid Greek poets knew
> In flinging wide the casements of the day
> A moment barred the light free passage through
> The lattice of her fingers till its hue
> Rose-pink appeared; yet nought but common grey
> Shone forth revealed when she her hands withdrew.[51]

In this second part of the sonnet, Lowell examines the changeability of the light drawn through the aperture of the window which appears red when seen through the smaller aperture made by the maid's hands, and then grey when the aperture widens to the size of the window once again. The writer of the sonnet seems uncertain which light is more true to nature, or whether both might be, yet in its final clause its air of resignation suggests that the rose-pink colour is the product of romance rather than reality. The sonnet is a carefully crafted moment of detailed poetic observation which should be read not just in the context of Lowell's astronomical work, but as an example of it. Although clearly in a different genre of writing, the sonnet deals with precisely the same issues of observation that animated his reply to Braesher; an optic nervousness borne out of the difficulties that telescopic seeing engendered.

In his private writing and correspondence, then, Lowell questioned the ideal of mechanical objectivity. In his public astronomical writing, however, he remained a robust defender of his own instruments and the access they gave to truth. Lowell's response to those who claimed superior seeing for larger apertures was to draw once again on atmospheric conditions. Air waves, he argued, disrupted the image received by larger lenses far more than they did with smaller ones. In an unpublished manuscript that later became part of his summative article on 'Mars in 1909' Lowell claimed that it was a serious error to assume that 'a large aperture necessarily defines better than a small one'. Although more light entered a larger aperture 'the air waves entirely vitiate their performance' and 'this fundamental fact ... far outweighs any consideration of size of instrument'.[52] He then rounded on his critics, especially Antoniadi: 'Now it will prove to students of the subject that this optical shattering due to a large glass is precisely what M. Antoniadi observed at Meudon ... His observed mosaic effect is the exact theoretic effect that a large aperture should produce on continuous lines such as the canals'.[53] In his important essay 'Small Apertures Versus Large Ones' Lowell attacked the users of larger apertures more vociferously for their failure to recognize that air waves effected observation: 'even the performance of the Lick glass has been ... disappointing, while objectives and even mirrors of from 6 to 10 inches go on cheerfully portraying' the Martian canals. In conclusion, Lowell claimed that 'the larger the aperture the more it hides, by virtue of the air waves, the very detail it is supposed to show'.[54] By contrast, the smaller aperture (which Lowell used) was more flexible in its ability to reduce the amount of light entering the telescope by the use of screens and diaphragms that could reduce the object glass from 24 inches to '18, 12, 9 and 6-inch diameters respectively'. Such a device was important, Lowell argues, to 'counteract nature', but the practice of using it was a method of 'assisting nature'.[55]

Lowell's confused claims to be simultaneously counteracting and assisting nature appear self-contradictory, but the implications of his argument are clear. While his use of the smaller aperture contradicts the theoretically superior vision enjoyed by larger telescopes, the practical reality of telescopic seeing necessitates the kind of assistance to vision that the flexible smaller glass allows. The greater mechanical objectivity that might be assumed to exist in the keener vision of the larger telescope gives way to a more pragmatic objectivity when the instrument is required to be used in the real world, in a specific place and under particular conditions. This is, in itself, a form of 'objectival consciousness' that has its roots in late eighteenth-century scientific practice when, as Daston and Galison show, savants wilfully intervened in the natural world in order to synthesize it with certain ideals of truth.[56] Lowell's description of the processes of observing the Martian canals highlights not his subservience to the mechanics of the telescope but rather the employment of his own will and ability to reason. At the same time, however, he recognized that his own seeing might be deceived by the conditions in which he was observing and that he should ideally set aside his own metaphysics in favour of a more objective telescopic vision. Lowell's observational practice therefore combines truth-to-nature with mechanical objectivity; his representations of the canals were therefore not entirely 'after life',[57] as truth-to-nature would have it, but instead within life, cognisant of the individualized and differentiated images one might encounter with an objective eye under specific conditions.

This combination of Lowell's own will and the instrumental vision of the telescope is often illustrated by the images Lowell chose to accompany his popular articles on Mars and its canals. On several occasions he chooses images that show him at work at the telescope rather than the telescope itself. Such iconography settles the viewer's attention on the observer, not on the observing equipment: it privileges the personal over the technological. In a related context the work of Simon Schaffer, and a corrective to it by Christoph Hoffmann, are useful here. In an article on the personal equation in astronomical observation Schaffer argues that the errors of individual observers were corrected by the disciplinary regime of the astronomical work.[58] The unruly individual observer is controlled by the organizing principles of the observatory. This, as Hoffmann has argued, is used as evidence by Daston and Galison to support their theory of objectivity. Yet Hoffmann claims that this was not the case: it was more often likely that 'the observer ended up aligned with his instruments' rather than reorganized by any disciplinary hierarchy.[59] Certainly this is true of Lowell's astronomical observations, as he would have wished his audiences to recognize. For Lowell it was the interconnection between his expert seeing and the technology of the telescope that was the observational ideal.

Writers of Martian fiction also investigate epistemologies of objectivity, often in recognition of the entirely subjective position that such imagined versions of the planet hold. Burroughs's *A Princess of Mars*, for example, offers a fantasy of an ideal telescope which provides images of such clarity that any questions of objectivity become moot. These telescopes 'throw upon a screen a perfect image of what is transpiring upon any planet and upon many of the stars. These pictures are so perfect in detail that, when photographed and enlarged, objects no greater than a blade of grass may be distinctly recognized'.[60]

Wicks, by contrast, allows his narrator to detail the technical difficulties of observing Mars by reference to Lowell and Barnard's differing opinions on the evidence of the spectroscope:

> The plates of the spectrum of Mars show a much more definite darkening of the 'a' band, and Professor Lowell contends that this can only be due to water vapour in the atmosphere of Mars. Professor Campbell has, however, made similar experiments, and is of opinion that Professor Lowell has been deceived by the water vapour in our own atmosphere.[61]

Wicks's aim is to illustrate the problems of deception and interpretation that persist in objective observational evidence. Nevertheless, this short lecture is a precursor to landing on Mars, where new observations will confirm objective reality and sweep away such conflicting interpretation. For Wicks's narrator, like Lowell, objective analysis within life is the ideal observational practice.

It is, however, Wells who offers the most nuanced reading of objectivity. In one of the most important scenes of *The War of the Worlds*, the narrator finds himself privileged to be able to look on the Martians through 'a triangular aperture' created by a hole in the rubble of a house.[62] Wells's use of the technical vocabulary of the telescope to describe this hole is designed to remind the reader that this is a fictional version of astronomical observation. The narrator's comparison of his own observations of the Martian machines through this aperture to images produced by others is implicitly a discussion of objectivity:

> People who have never seen these structures, and have only the ill-imagined efforts of artists or the imperfect descriptions of such eye-witnesses as myself to go upon, scarcely realize that living quality. I recall particularly the illustration of one of the first pamphlets ... The artist had evidently made a hasty study of one of the fighting-machines, and there his knowledge ended. He presented them with an altogether ... misleading monotony of effect ... They were no more like the Martians I saw in action than a Dutch doll is like a human being.[63]

Although the narrator recognizes that his own observations are likely to be imperfect, the difference between his vision of the machines and that of the artist is clearly a contrast between the objective seeing achieved through the use of the small aperture and the subjective sketch. Wells argues, as did Lowell, that any observation, however objective it may appear, can be subject to error, but that image creation drawn from observations done within life (Wells says 'in action'), as the narrator's is, has the potential to reach closer to an ideal of objectivity than other representational practices. Importantly, Wells also recognizes the role played by the observer in the observational economy of image-making. The narrator does not simply see better than the artist, he has experience of looking at them and his observations are done when he is 'motionless ... and under no urgency of action'.[64] By contrast the artist is 'hasty' and has only a single observational moment from which to work. For Wells, the observer's own visual accoutrements – his perception, experience and the time allowed for looking – are key components of authoritative visual knowledge.

Eyes and Authority

Lowell, too, understood the importance of the observer's role as witness. If his canal testimony was to achieve the status of evidence then his own position as a witness to their existence had to be without suspicion. In the early years of his career at Flagstaff Lowell wrote extensively of the relationship between the eye and the brain in an effort to construct for himself a robust observing personae rooted in thoughtful, experienced perception. Taking a lead from the mid-to-late nineteenth-century investigations of optical researchers, Lowell knew that the eye was not solely responsible for vision but worked in concert with the brain to achieve perception. Yet he also recognized that it was possible for the mind of an observer to alter an image by imposing on it a set of preconceived ideas:

> Minds have their familiar ideas, which an incoming idea is sure to rouse ... If we expect to meet a certain person, an approaching figure will most deceitfully take on his garb. The mere idea of a man walking finds the expectation ready instinctively to endow it with the attributes of our friend. But this may happen truly as well as falsely. The expert sees what the tyro misses, not from better eyesight but from better mechanism in the higher centres.[65]

How one perceives the gait of a walking figure was a favourite example of explanatory works on optics from the mid-nineteenth century, and it is likely that Lowell was directly paraphrasing that experimental field's most prominent figure, Hermann von Helmholtz, especially as Helmholtz employed the image of the telescope to aid his description:

> The spectacle of a person in the act of walking is a familiar sight. We think of this motion as a connected whole, possibly taking note of some of its most conspicuous singularities. But it requires minute attention and a special choice of the point of view to distinguish the upward and lateral movements of the body in a person's gait ... But look through an astronomical telescope at a crowd of people in motion far away. Their images are upside down ... then there is no trouble whatever in noticing the peculiar motions of the body and many other singularities of gait ... [but] it is not so easy to tell whether the gait is light or awkward, dignified or graceful, as it was when the image was erect. Consequently, it may often be rather hard to say how much of our apperceptions as derived by the sense of sight is due directly to sensation, and how much of them, on the other hand, is due to experience and training.[66]

Helmholtz's view that perception is a combination of the image received into the eye and the accumulated knowledge of similar images carried in the brain by the expert is exactly the argument Lowell was aiming for to construct an authoritative observing self. This is further confirmed in his early publications when he notes that his reading of the Martian landscape emerged from seeing 'first with the brain and then with the eye'.[67] Certain fictions of Mars also accepted Lowell's view on the perception required for expert seeing. Serviss, for example, describes in *Edison's Conquest of Mars* how a Martian chosen for a scientific career does not improve his eyes but rather 'had his brain developed into ... an instrument of observation'.[68]

Lowell's opponents in the astronomical community, however, regarded him as the tyro rather than the expert. In critiquing his Martian observations, the British astronomer E. W. Maunder undertook an experiment in 1903 to prove that Lowell's canals were the product of a naive perception which turned a series of random markings into more organized lines. Maunder's experiment consisted of inviting a group of schoolboys to draw what they thought they saw on a sphere some distance away. Many of the schoolboys drew a series of lines when the spheres were actually covered with random dots. While clearly this challenged Lowell to prove that his canals were not simple optical illusions, it also rather archly suggested that his astronomical seeing was no better than a schoolboy's.[69] Lowell, in an essay ostensibly about Schiaparelli but clearly responding to the optical illusion critique, argued that 'the geometrical character of the canals was forced upon him by the things themselves instead of being, as his critics took for granted, foisted on them by him'.[70] Nevertheless, the potential for the canals to be optical illusions that had tricked Lowell's perception gained credibility. Simon Newcomb, for one, wrote of it as though it were proven fact in his popular work on astronomy in 1902. Comparing Lowell's observations of Mars with an amateur looking at a stipple-engraved portrait, Newcomb argues that 'as the eye makes an assemblage of dots into a face, so may it make the minute markings on the planet Mars into the form of long, unbroken canals'.[71] In Lowell's own copy of Newcomb's book he has made a marginal note against this comment that reads 'might be so, but is not true on Mars'.[72] Again, in private correspondence, Lowell's tone is less certain than was ever the case in his public writing.

Lowell's tactic in answering these criticisms of his perceptive faculties was to offer an alternative reason for the authority of his vision. Moving from perception to the eye itself, Lowell argued that his own eye was better suited to planetary observation than the eyes of his astronomical opponents. In 'On the Kind of Eye Needed for the Detection of Planetary Detail' Lowell claimed that there were two types of eye, which he called sensitive and acute. The essay reveals Lowell's shift in argument from perception occurring in the mental faculties to an embodied vision rooted in the organ of the eye itself:

> So far as concerns man the eye is the portal to perception; through it is determined what shall enter the brain. But there are two ways in which eyes may admit information; by being sensitive to light or by being acute to form. The two qualities are quite distinct and in the study of planetary detail, the one is as vital as the other is fatal.[73]

Lowell proceeds to define the sensitive eye as 'one which is responsive to faint light or to delicate shades of contrast'. This type of eye, he says, 'perceives readily but particularizes ill'. The acute eye, by contrast, 'defines minutely' and 'perceives objects with great precision'. It is this acute eye 'that detects planetary detail' and it is, of course, the type of eye which Lowell and other observers of the canals possess: 'That the two traits are seldom, if indeed, ever, found together has been the writer's experience not only in his own case but in that of numerous assistants and he was interested to learn from Schiaparelli that the same was true with him'.[74]

There are two striking differences in making an argument for a specific type of eye rather than the experience of perception. First, expertise is no longer required when the power of vision resides in the actual material structure of the eye. As Lowell argues, it is on 'the size of the cones' on the retina that 'keen-sightedness' depends.[75] Second, authoritative vision also comes to depend upon surveillance of the eye of the observer, who must prove the quality of his eyesight to be accepted as a suitable astronomical witness. It would be absurd, Lowell argues, 'that where everything depends upon accuracy of eyesight the testimony of any man ... shall be received according to the prejudice of an unpracticed jury of the moment, without that eyesight having previously been subjected to the slightest scientific test'.[76] What emerges here is a distinct 'calibration of the eye' not yet noted in history of science scholarship.[77] While Daston and Galison, for instance, trace the evolution of the disciplinary, period and tutored eye of scientific observers, Lowell's epistemology of the eye is of a different order.[78] His invention of the acute eye that better sees planetary detail is a kind of idealized eye that gives the observer an inherent objectivity disconnected from training, expertise or environment. As the idealized eye is embodied in the physical attributes of the observer's ocular structure it must be discovered and carefully monitored, but it goes to the heart of the observer's ability to bear witness to astronomical phenomena. By associating the ideal eye rather than perception with visual authority Lowell argues that it is only specific observers (himself among them) who can inhabit the role of witness, because the necessary tools (the retinal cones) for doing so inhabit them.

Wicks's *To Mars Via The Moon* spends considerable time exploring some of the same complexities both of bearing witness and inhabiting the role of witness. Wicks was in correspondence with Lowell in the years prior to the publication of his novel and was a fervent supporter of Lowell's canal theory. In September 1907, for example, he wrote to Lowell of his efforts in 'combatting [*sic*] the sceptical views as to the reality of the canal lines on Mars'.[79] Wicks conducted this campaign through 'the press and by means of lectures' but writing in support of Lowell is also one of the central aims of his fiction.[80] For the most part his conviction in the truth of Lowell's Martian observations are obvious in *To Mars Via The Moon*. On an exploratory journey around Mars his narrator notes, for instance, the existence of towns which Earth astronomy has named 'oases' and 'which, like many other Martian details, have been described as illusions. I only wish we had a plentiful supply of such illusions in our own old country!'[81] Yet, when Wicks's characters are called to bear witness to their Martian observations in front of an audience of professional astronomers they encounter a far greater degree of scepticism: 'I gave my address, which lasted about half-an-hour; but it was received even more chillingly than I had anticipated, and the few comments made by the members were nearly all indicative of scepticism of my statements and unbelief in my *bona fides*'.[82] Wicks astutely portrays astronomical doubt as the product of distrust in the narrator's authority; it is his credentials as a witness as much as his statements as witness with which they find fault.

Similarly, Wells's narrator reflects on his own authority, or lack of it, when he too inhabits the role of witness to Martian phenomena. In the epilogue to his story of the Martian invasion, which in being a retrospective narrative is written as though it were a long witness statement completed after the events have taken place, the narrator becomes concerned that the images he has described may be regarded as nothing more than 'unproven speculations' that 'shall certainly provoke criticism'.[83] For Wells's narrator this sudden lack of confidence in his position as witness comes from a perceived paucity in the testimony itself. 'I cannot but regret', he says, 'how little I am able to contribute to the discussion of the many debatable questions which are still unsettled'.[84] The final word is important in this context, for it hints at the narrator's discomfort in being regarded as an authoritative observer. By capturing the fluctuating concerns over the slipperiness of visual evidence and personal authority Wells cleverly encapsulates the shifting ground of the Martian debates on eyesight and perception. Yet Wells perceptively recognizes that it is not only in eyesight that authority was vested but also in professional identity. His narrator is certainly concerned that his own profession marks him as unsuitable for one inhabiting the role of witness to scientific phenomena. The narrator explains that 'his particular province is speculative philosophy', which suggests neither the necessary expertise nor talent for accurate description of the Martian body.[85] Rather resignedly, he notes that his 'knowledge of comparative physiology is confined to a book or two'.[86] Wells stresses here the importance of trained judgement in the description of scientific objects, arguing that the authority of a scientific witness is accumulated through professional experience and the continued exercising of expert observational techniques.

Unable to contradict Lowell's claim to have the ideal eye for observing Mars, his critics instead questioned his professional training. There was certainly ample fuel for such a fire to be lit under Lowell. His university training, while certainly distinguished, was in mathematics, and until setting up the Flagstaff Observatory he had no expert knowledge of astronomy.[87] Lowell was deliberately vague about his expertise in print, noting only, for instance, in an 1894 article in *Nature*, that he was 'a man of scientific training'.[88] Yet even in such an ambiguous statement it is clear that Lowell recognized the importance of professional practice. Members of the astronomical community were certainly suspicious of Lowell's status and sought to exploit his lack of astronomical expertise in order to undermine his canal theory. Their efforts were generally surreptitious and often proceeded by drawing unfavourable comparisons that implicitly suggested Lowell's lack of training. Simon Newcomb, for example, drew a contrast between Lowell and Barnard, whom he called a 'cautious observer', while a further 1894 article in *Nature* characterized the Martian observations of French astronomer Javelle, clearly *inter alia* of Lowell's, as 'careful work'.[89]

Despite having made significant efforts to create an ideal observing personae based on the material structure of his eyes, Lowell clearly recognized, as Wells had done in *The War of the Worlds*, that observational authority also depended upon the extent to which the observer was recognized as an expert. This is not only a response to the specific context of astronomical observation but also to wider ideals in visual culture which had been in place for over a century. Peter de Bolla identifies two methods of access to visual culture in his study of late eighteenth-century art. The first is being able to respond appropriately to an image (through feeling in the case of art objects) which enables the observer to take up his position both within the 'exhibition room' and as 'a subject within the society of visuality'.[90] The second is the 'superiority of the knowing viewer' who has been 'trained in the correct ways of looking and legitimated by the institutions of cultural evaluation'.[91] While de Bolla recognizes only the second of these methods as appertaining to science, both are relevant to Lowell. As a man of private financial means with experience of, and access to, the knowledge culture of scientific institutions, it was relatively easy for Lowell to enter the astronomical field and to seem to conform to appropriate observational standards. His choice of former Harvard astronomer W. H. Pickering as his chief assistant, for example, automatically gave him access to the kind of cultural authority that enabled him to enter the particular 'society of visuality' found in astronomy. However, a defence of his canal theory required not just entry into the visual culture of astronomy, but a superior position within it. That is, Lowell needed to prove that he had, in de Bolla's phrase, been 'trained in the correct ways of looking'.

What exactly was the correct way of looking in the astronomical culture of the period? For Daston and Galison appropriate seeing required 'trained judgement'.[92] This was built on the experience of observing, a familiarity with observational technologies and with the objects under scrutiny, and the cultivation of a patient, attentive and keen observational practice. Over time such training gave rise to an intuitive recognition of images, which Daston describes as the 'all-at-once-ness' of properly expert visual detection.[93] Lowell's cultivation of himself as an appropriately trained observer during the 1894 Martian opposition reflects these definitions. In the *Nature* article written by his supporter the astronomer and journal editor W. J. Lockyer in September 1894, after approximately four months of constant observing, Lowell is described as 'watching attentively ... continually and for a long period' the changes in Mars's topographic character.[94] Such sustained and systematic observation led Lockyer to describe Lowell as a 'keen and patient observer', the very epitome of trained judgement.[95] By the time Lowell writes 'Means, Methods and Mistakes in the Study of Planetary Evolution' in 1905, he has the confidence of a decade's experience to write that 'practice in any kind of observation is at least for half in the result. Everyone is aware how acquaintance with the subject sharpens to an al[m]ost incredible degree one's eyesight in it'.[96] This necessary familiarity, Lowell argues, requires 'apprenticeship ... in learning how to see'.[97] Without any apparent self-consciousness about his own lack of training when he began his Martian observations, Lowell draws a comparison between the amateur writer and the untrained astronomer: 'just as almost every man has an impression he can write good English because he knows how to spell so because every observer can look through a telescope he considers himself qualified to pass judgement upon visual observations'.[98]

How far Lowell had assumed the identity of the expert observer, and how this was still resisted, can be seen most clearly in his fractious correspondence with Antoniadi in 1909. The context for their letters was Antoniadi's observations of Mars first at the observatory at Juvisy and then at Meudon in the final months of 1909, and his growing conviction that the canals were illusions brought on by observational errors. In these letters Antoniadi is responding to the drawings of Mars (and its canals) made by Lowell from 1895 onwards. One example, in Figure 3.1, shows Lowell's observation of Mars's northern axis, with the straight lines of the canals clearly visible as they leave the northern ice-caps and proceed towards Mars's interior. Antoniadi begins the sequence of letters by noting Lowell's expertise in Martian observation, particularly commending him for his 'exhaustive scrutiny' of the planet over a period of fifteen years.[99] He also addresses Lowell as 'Professor', an assumed title, but one that recognizes his status as a fellow member of the professional, institutionalized astronomical community. Having apparently confirmed Lowell's trained judgement, Antoniadi proceeds to question it by inviting Lowell to comment on the apparent discrepancies between a series of photographs of Mars and Lowell's drawings of the same area. Although Antoniadi is careful not to implicate Lowell's observational accuracy in highlighting the differences between drawing and photographic image, it is clear that he regards the photograph as a source of greater objectivity than the man-made map. Lowell responds by pointing out that the grain of the photographic plate led to some inconsistencies in the gradation of canal lines, from which his own draughtsmanship was free. For Lowell, then, objective accuracy resided in his own expertise rather than in the mechanically-produced image of the camera.[100]

Between these letters and the next Antoniadi had the opportunity to observe and draw Mars for himself. The tone and content of his subsequent letter is markedly different from those that came before. Attaching four drawings of Mars for Lowell to consider, none of which show canals, Antoniadi wrote:

> These drawings are by far the best I have ever made ... With the exception of the linear canals ... all seen by flashes of ⅓ of a second, all the other markings I show were held steadily, and the tremendous difficulty was not to see the detail, but accurately to represent it. Here my experience in drawing proved of immense assistance, as ... I sat down and drew correctly, both with regard to form and intensity, all the markings visible ... But what an incredible change in Syrtis Major [a canal] from what you have discovered 15 years ago!'[101]

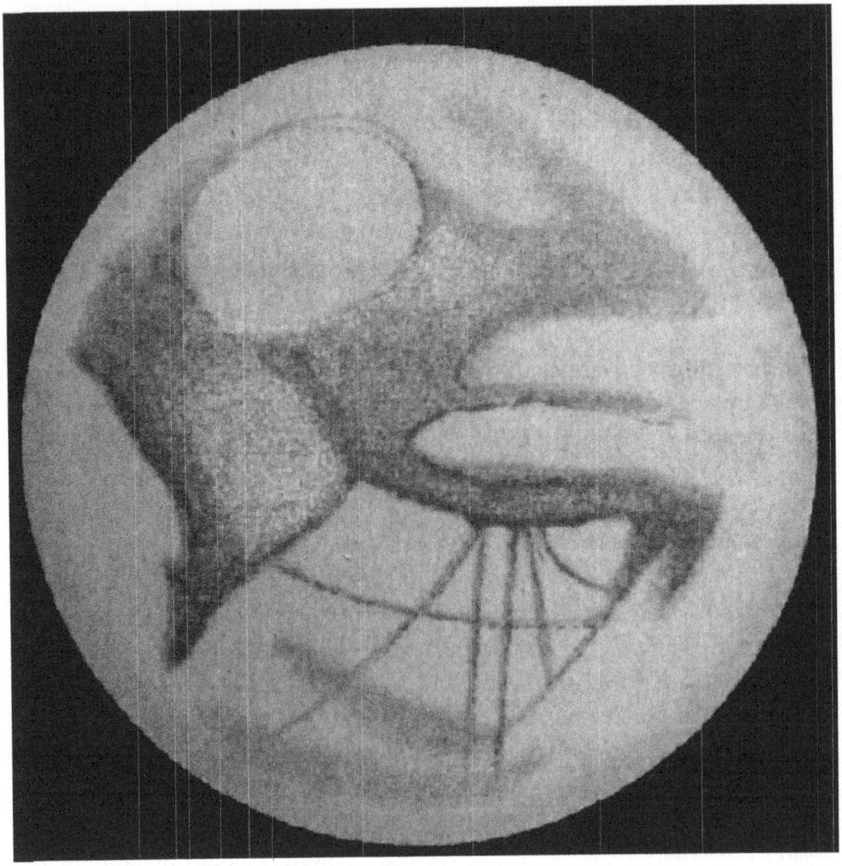

Figure 3.1: One of Lowell's celebrated drawings of the 'canals' he saw on Mars, *c.* 1895, which received severe criticism from his astronomical rivals. Source: Percival Lowell Archive, Lowell Observatory; reproduced with permission of the Lowell Observatory.

Implicit in Antoniadi's description of his observational experiences is a forceful criticism of Lowell's judgement. Antoniadi privileges his own 'experience', as well as his greater ability to see all-at-once, in those brief moments of clear seeing that characterize the strobic vision of astronomical observation. The 'incredible' difference between his and Lowell's representations of Mars is to be read as evidence of Lowell's poor observation and, consequently, of Antoniadi's own superior expertise. Other aspects of the letter reinforce Antoniadi's diminishing respect for Lowell's 'judgement-inflected vision'[102]; most especially his change of nomenclature from 'Prof. Lowell' at the beginning of the letter to 'Mr. Lowell' near its conclusion.[103]

Figures 3.2 and 3.3 show examples of Antoniadi's drawings of Mars, used by him to discredit Lowell's earlier images (such as in Figure 3.1).

(4)

your lakes, strung on Eumenides-Orcus). When, an hour or two later, Mare Cimmerium was well in view, the pale yellow material N. of it showed no markings, Trivium Charontis, Cerberus and Elysium being invisible in the yellow mass. On September 6, $\omega = 80°$, the S. polar cap was

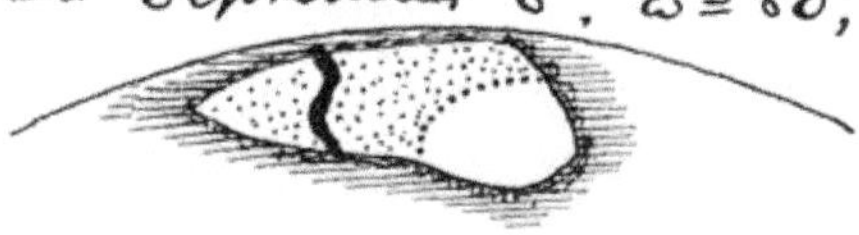

seen as on annexed figure. It was most brilliant white to N.-W., duller all over the rest of the snow and split by a dark rift.

M. Deslandres has kindly invited me to observe Mars from Meudon with the most powerful refractor in Europe, and I hope shortly to have the pleasure of sending you an account of my work there.

With every best wish,
Believe me, Dear Prof. Lowell,
to remain, with the most genuine admiration and the greatest respect,

Your most obedient
Servant

EM Antoniadi

Prof. Percival Lowell,
Director of the Observatory
at Flagstaff
Arizona (U. S. A.).

Figure 3.2: A page taken from a letter from Antoniadi to Lowell showing details of Mars directly contradicting Lowell's findings. Source: Percival Lowell Archive, Lowell Observatory; reproduced with permission of the Lowell Observatory.

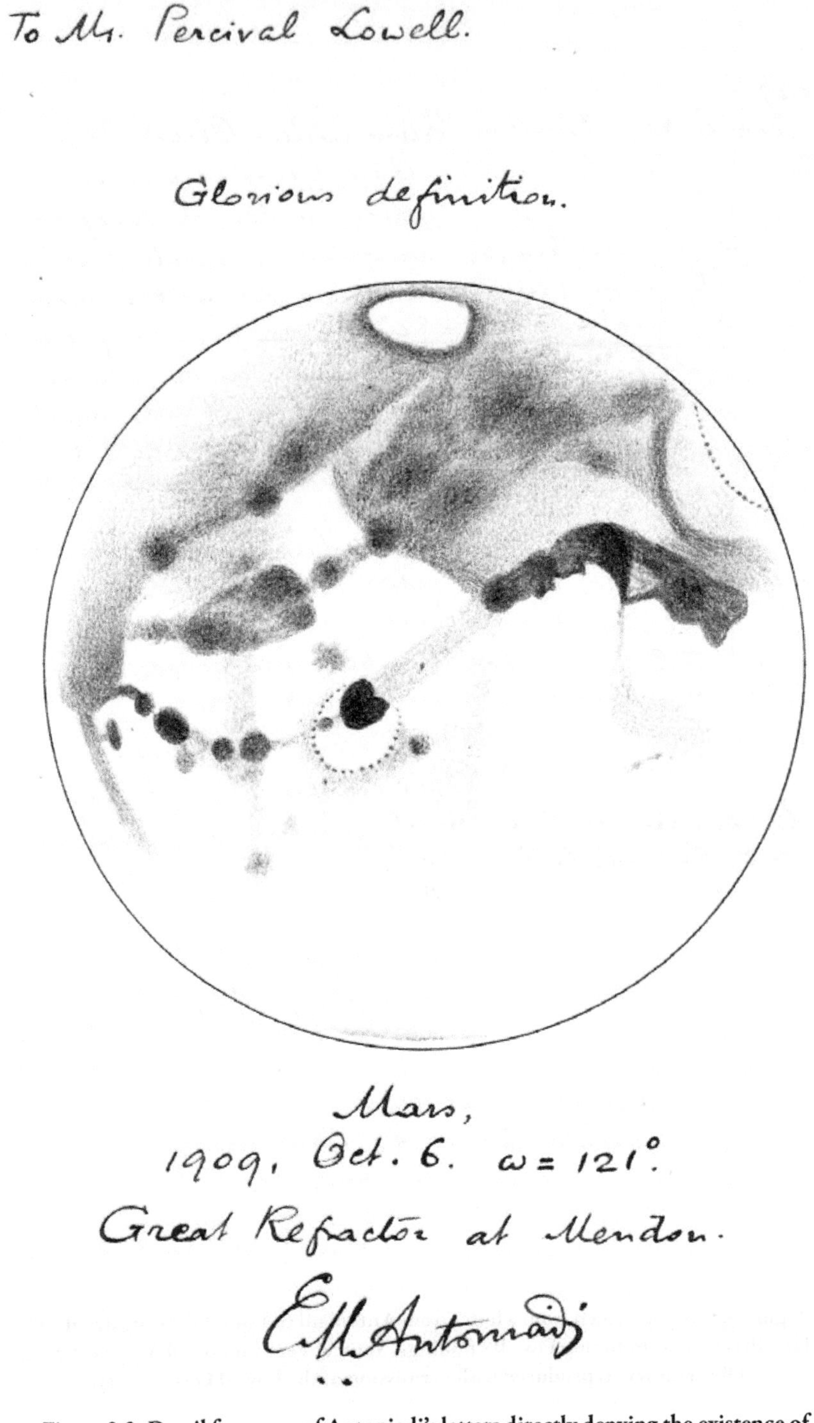

Figure 3.3: Detail from one of Antoniadi's letters directly denying the existence of Lowell's canals. Source: Percival Lowell Archive, Lowell Observatory; reproduced with permission of the Lowell Observatory.

In Figure 3.2 Antoniadi draws one small section of Mars's south polar ice-cap, without the regular lines that would suggest a canal structure, and instead with the stippled or dotted lines that Maunder had claimed was the true picture of Mars's topography. In Figure 3.3, by contrast, Antoniadi offers a more artistic representation of Mars's landscape, and of the same section of Mars as Lowell's 1895 drawing. In Antoniadi's drawing (done with 'glorious definition', he claims) the canal lines had been replaced by stippling and shadows suggestive of natural rather than man-made geographical structures.

Lowell wrote what was, for him, a restrained reply, noting that he felt the drawing Antoniadi had completed under 'tremulous conditions' rather than in 'glorious definition' was best, probably because it could be interpreted as showing canal detail, while the others were 'not so well defined' and were likely to be the product of using a large aperture which gave rise to 'a fine imperceptible blurring which transforms detail really continuous into apparent patches'.[104] Less restrained was Lowell's attack on Antoniadi in a publication written soon after this epistolary exchange. Throughout 'On Some Fallacies in Observing Mars' Lowell attempts systematically to dismantle Antoniadi's reputation as an expert on Martian observation. He calls Antoniadi an 'occasional observer' who is 'ignorant of the optics of observation in the use of a large instrument', inexperienced in 'the different kinds of seeing' and without the 'endless patience' necessary for accurate observations.[105] These presumed weaknesses in Antoniadi's observational practice reflect the criteria for trained judgement: familiarity with equipment and techniques, consistent practice and the patience required to reach an intuitive understanding of the observed image. However similar the grounds for the conflict between Lowell and Antoniadi, what is interesting is that they disagree on trained judgement at all. In one of the most recent and comprehensive analyses of the philosophy of scientific observation Daston argues that scientific ways of seeing are by necessity 'grounded in trained, *collective*, cultivated habit' [my emphasis].[106] The case of Lowell and Antoniadi shows that in moments of real conflict this collective can fracture. In fact, observational practices previously held as community epistemologies become instead weapons of that conflict, wielded against those who seem to share them. Lowell and Antoniadi draw upon the criteria for trained judgement in order to contest the truths that each has reached by the employment of that judgement. While they may agree that each has the 'tutored eyes' of the expert this allows them only to see rather than to see well.[107] Yet seeing well can only be confirmed when the community as a whole agrees that an observed object, such as Mars, can be defined in one way and not in another. There exists, that is, an aporia at the heart of any conflict of observation in science which will be closed only when a truth has been agreed. In the meantime, as the conflict continues, observational practice becomes unstable as trained judgement is itself interrogated and shown to be, potentially at least, prone to failure.

Fictional accounts of Mars are themselves products of this instability in astronomy. While the truths of Mars and of Martian observation were still under examination, fiction writers had an opportunity to offer their own version of Martian investigation. At the same time, however, the genre of speculative fiction that most writers used as their vehicle for doing this demanded that instability be constrained for fear of the fiction over-reaching into fantasy. In the Martian fictions written during the period of Lowell's observations there is a fascination with scientific authority and its challenges and with the acceptance or denial of trained judgement. Wells, Serviss and Wicks make a great deal of their respective narrators' relationship to the scientific community. Wells's narrator, for example, spends considerable time early in his story relating his observations alongside 'Ogilvy, the well-known astronomer' and is then invited to join a group led by the 'Astronomer Royal' investigating the Martian cylinder that has arrived on Earth.[108] Serviss takes this entry into a community of scientists even more seriously, placing his narrator in the flagship of the spacecraft headed to Mars alongside Thomas Edison, 'several leading men of science' and a group including 'Lord Kelvin, Lord Rayleigh, [and] Professor Roentgen'.[109] In fact, Serviss's novel makes its case for scientific authority before the narrative has even begun. Its title page notes that the novel has been written 'in collaboration with Edison' and that the author, Serviss, is a 'well-known astronomical author'.[110] Statements such as these – which find a parallel in the early pages of Wells's work where his narrator displays his knowledge of contemporary astronomy by mentioning observations undertaken at 'the Lick Observatory', others by 'Perrotin of Nice', and finally an 'issue of *Nature*' – are designed to alert the reader to the work's close relationship to current scientific investigation but also to persuade the same reader that the speculations to be found in the novel have a solid foundation in actual astronomical knowledge.[111]

While these strategies are used to stabilize the fictions within their genre as serious works of speculation that happen to employ the imagination as their creative impulse, within the works there is a greater interrogation of the kinds of conflict over judgement in which Lowell was both victim and aggressor. Wells, for instance, implicates astronomical disagreement about Martian life in the disaster that befalls Britain when the Martians invade. The astronomer Ogilvy, used by Wells to exemplify the views of one branch of the professional astronomical community, 'scoffed at the vulgar idea of [Mars] having inhabitants' and is therefore entirely unprepared when the Martians invade.[112] The punishment for his failure to interpret his observations of the planet appropriately is to be the first person killed by the Martian's technological weaponry. By contrast, a 'speculative writer of quasi-scientific repute' who had written a short work on Mars and the Martians is proved right in his assessment of Martian life despite being dismissed as 'foolish and facetious' and having his claims satirized in *Punch*.[113]

The narrowing of distinctions between the professional and amateur is also central to Wicks's *To Mars Via The Moon*. His narrator is undoubtedly an amateur astronomer, yet he takes the title 'Professor' (just as Lowell did). Recognizing the issue of taking on professional accreditation without training the narrator explains that he does so in order to remind himself to 'to live up to his reputation, and be most scrupulously careful to make sure of the accuracy of the information which he desires to impart'.[114] It is clear why both Wells and Wicks would wish to privilege the amateur's speculations over the professional's trained judgement. Both, after all, were writing as amateurs who felt they had a contribution to make to contemporary Martian astronomy. Nevertheless, their articulation of the instability of trained judgement reflects their understanding of the uncertainties that existed in astronomy over the true character of Mars. Indeed what their fictions reveal is how ontological instability in astronomy of the kind revealed by Lowell and Antoniadi's conflict opened a gap for new kinds of testimony from outside astronomical communities altogether.

Fictional Witnessing

The fictions of Mars by Wells, Serviss, Wicks and Burroughs can readily be dismissed as the kind of fantastical science-inspired narratives that always emerge at the outer reaches of scientific debate. Aspects of their fictional content are certainly best placed there, especially those themes and episodes that arise from genres at odds with scientific writing such as the adventure story or Western. Yet there are elements of these fictions that are clearly more closely connected to science. They all draw on current astronomical knowledge to provide material for their imaginative rendering of Mars. Each of the fictions pictures Mars as traversed by a system of canals, for example. The fictions also address themselves to specific astronomical sites and to individual astronomers. The Lick and Flagstaff observatories, Lowell, Barnard and Antoniadi are all given due credit for their contribution to the present state of Martian astronomy. More interesting, of course, is how these fictions respond to the specifics of astronomical debate; on the pros and cons of observational sites, the problems of telescopic seeing and the conflict over trained judgement.

It is unsatisfactory, however, to regard these fictions as nothing more than passive recipients of scientific knowledge whose limited ambition is to reflect the current state of Martian astronomy. This does nothing more than set up a less than useful, and entirely artificial, hierarchy of knowledge in which the elite work of astronomers is diffused to grateful writers of fiction. There is instead a much greater interaction between the fictional text and astronomy. Astronomical writing and Martian fiction employ some of the same generic tactics to persuade their readers of the plausibility of their claims. Both also recognize the importance of locating their observations of Mars within life; taking account of how their organization of vision is part of a much broader, active and participatory visual culture. Finally, and most significantly, both are concerned to interrogate the role of the witness to scientific phenomena, and especially how such witnessing might be a route to the dissemination of astronomical knowledge.

In generic terms Lowell's efforts to normalize the supernatural appearance of the canals within the ordinary routines of his familiar observational habits parallels fiction writers' strategies of making their speculative claims appear quotidian or commonplace in the context of the world they have created. From a critical standpoint the similarity is more striking when we acknowledge that the canals are themselves a fiction, although Lowell was convinced of their scientific reality. Further, both astronomical writing and Martian fiction utilize similar narrative techniques to communicate authority. Both reinforce the expertise of the observer through whom the portrayal of evidence is focalized. Lowell stresses his own powers of seeing, his experience and his training, often introducing corroboration from previously published works or former colleagues. Likewise writers of Martian fiction give their narrators keen powers of observation or opportunities for expert viewing, often placing them within expert groups who can give credence to their evidence. These techniques are again traits of shared generic principles that, on a textual level at least, bring the work of science and fiction into closer contact.

Like the astronomy of Mars that gave them their first impetus, Martian fictions are keen to interrogate Martian observations within contemporary culture rather than exclusively in the context of scientific experimentation. For Lowell and his various astronomical interlocutors observation was not merely a set of theoretical or experimental principles but took place within the practical conditions of specific moments in time and space. Lowell's various writings on his Martian observations are always characterized by his insistence that good seeing is about the recognition that scientific observation takes place within life: the life of specific instruments, the living atmospheric conditions of a particular night at a single site and the active life of the observer's eyesight. Placing observations of Mars within a vivified and individualized context is precisely the strength of fictional representations which employ the artistry of the imagination to place astronomy in the real world of human social, cultural and political behaviours.

In a rarely referenced essay on scientific illustration and its relationship to artistic perception, Roland Mortier discusses the use of illustrations to exemplify scientific objects in eighteenth-century encyclopaedias of arts and crafts. Mortier notes their innovative use of *tableaux vivants*: illustrations of scientific objects in use in the real world, often set alongside illustrations of the same scientific objects in objective detail.[115] These *tableaux vivants*, Mortier argues, 'stand midway between the two meanings of the French word art in the eighteenth century: fine arts and mechanical arts'.[116] Although Mortier does not develop the significance of the *tableaux vivants* any further they are emblematic of a continuing relationship between science and artistic representation in the nineteenth century, characterized here as representation within life, which can clearly be seen in Lowell's astronomy and Martian fiction. A novel such as Wells's *The War of the Worlds* is itself a *tableau vivant* of Lowell's astronomical observations and as such bears the same relationship to that astronomy as the *tableau vivant* does to the objective illustration placed next to it. In fact, the relationship is closer than that, for as Lowell's work highlights, astronomical writing also tried to illuminate the real-world context in which scientific objects (canals, for example) were being seen. Just as with generic tropes, there is overlap here between Lowell's astronomy and Martian fiction. Despite the fact that astronomy and fiction approach the *tableau vivant* from very different ontological perspectives – the former objectively and the latter subjectively – that they collide in the presentation of scientific observation within life is significant. In particular it suggests that if the *tableau vivant* is a form of scientific representation then, as an example of that, so too is Wells's *The War of the Worlds* or Wicks's *To Mars Via The Moon*.

It would be an abuse of this logic to conclude that fiction is a form of science. That is not the point of illuminating their moments of connection. What is at stake is a recognition that literary texts might well offer access to knowledge of the natural world that science alone cannot deliver. This does not turn fiction into science, far from it, but it does acknowledge that fiction has particular strengths in interrogating scientific knowledge, especially when that knowledge is still in flux. Investigations of the visual culture of science is one of the central ways in which fiction has a role to play in understanding the emergence of scientific knowledge. The case of Martian astronomy discussed here has at its core the authority of the observer as witness. This is also a common thread in fiction, especially nineteenth-century fiction; writers are often interested in creating narrators whose authority as witness to the events of the narrative is in doubt or unstable. For Wells, Wicks, Burroughs and Serviss, whose fictions are responding very particularly to questions of observational authority the testimony of observers is even more critical.

How, though, might the relationship between Lowell's astronomical work and Martian fiction be structured in this context? First, it is useful to consider the role that scientific writing plays in questions of observational witnessing. In their landmark study, *Leviathan and the Air-Pump* (1985), Steven Shapin and Simon Schaffer characterize scientific texts as providing 'virtual witnessing' effects for their readers.[117] Through a study of Robert Boyle's texts they conclude that 'the technology of virtual witnessing involves the production in a *reader's* mind of such an image of an experimental scene as obviates the necessity for either direct witness or replication'.[118] Shapin and Schaffer argue that this is a 'literary technology', supported by 'pictorial representations' which supplement 'the imaginative witness provided by the words in the text'.[119] A further keynote of this new genre of scientific writing was its prolixity, the provision of 'circumstantial detail' that gave the impression 'that the reader had been present at the proceedings'.[120] While Shapin and Schaffer make the point that such 'verisimilitude' was a method of mimicking the accompanying illustrations, it is clear that it is also a literary technique both for the provision of authority by reaching towards realism and for the replication of a *tableau vivant* of experimental practice that happens within life.[121] Shapin and Schaffer's important study helps us to recognize that the authority of having observed well, so as to become a credible witness, is dependent upon a number of strategies employed to reveal (or to create) a particular witness's credentials.

Lowell's astronomical work on Mars shows us that such strategies included technological mastery, recognition of the individuality of scientific sites, an idealizing of seeing itself and the construction of the self as expert. While these are strategies that Lowell's texts *use*, they are also the strategies that fictions of Mars *interrogate*. Lowell constructs his arguments for the existence of Mars's canals around the idea of himself as an ideal witness and inculcates his readers to become virtual witnesses to his observational prowess. Martian fictions, on the other hand, invite their virtual witnesses (and this might be their narrators as well as their readers) to ask questions of how appropriate witnesses might be constructed. That is, they begin to probe at the very nature of the creation of scientific knowledge. Wells ponders the relationship between perception that begins in actual seeing (at the telescope, for example) as opposed to that which springs from speculation. He wonders, too, about the difference between eyesight and the imagination. Burroughs, for example, questions the role of familiarity (the problem of the preconceived idea) in Martian observation, suggesting that the Lowell's Martians owe as much to the Arizona apache as it does to any observed 'facts' of Martian life. The literary text has an advantage over the scientific text here, for it is freer to undermine any objectivity it might first claim. While it uses the technique of virtual witnessing it can readily destabilize it without endangering its production of knowledge.[122] Yet it is in that destabilizing that it can offer ways of understanding observers and observation – witnessing and testimony – that the scientific text, and the experiments which precede it, would not wish to identify.

Although this chapter has shown how Lowell's defence of the canal theory led him into numerous conflicts, and that these conflicts reveal uncertainty and different opinion on observational accuracy and the authority of witnesses, no single letter, article or journal entry expresses the ambiguities that emerge in the case as a whole. It is only by reading across several texts that its elements can be brought into dialogue. Yet in the fiction these elements can be found together. Martian fiction provided the opportunity to reflect upon the complexity of astronomical observation by seeing it dramatized in real-world situations and considered by a variety of witnesses claiming different kinds of authority. This is, as Richard Cunningham argues of virtual witnessing, a shifting from the singular of science into the multiple of 'reader, text, author, world'.[123] Such a consideration of the role of observers and their observations is one way in which fiction comes to occupy the same territory as science and why a consideration of the history of controversies in scientific observation cannot afford to ignore its important role as cultural artefact.

4 LOWELL'S MINIMUM VISIBLE: WONDER, IMAGINATION AND POPULAR SCIENCE

While fictions were clearly interventions in scientific debate, as argued in the previous chapter, they were also, in certain specific texts and on particular occasions, conscious creative efforts to contribute to scientific knowledge more directly. As this chapter will show, popular fictions did make attempts to become popular science and to employ certain epistemological categories of knowledge-making in the same way as professional scientific practitioners. At the same time, scientific knowledge-makers employed creative fictional tactics to enhance their understanding of scientific objects under scrutiny.

Sometime during the opposition of Mars in 1894 Percival Lowell began to write a long verse poem entitled 'Mars'. The poem draws on his experiences of nightly observations of the planet and although it is not obviously a work of science it should be read as one part of his many reflections on the dissemination of astronomical knowledge. In the early stages of the poem, where Lowell's untidy versification suggests a process of speedy writing under the influence of an immediate inspiration, Lowell stresses his desire to know the answers to his many questions about Mars:

> We know just enough to long to know more
> Of that first habitable shore
> Across the ocean of the sky,
> Ocean whose aether-waves of light,
> Buoyant to nought more gross than sight,
> To thought alone give passage o'er[1]

Using 'we' and 'our' Lowell regards his quest for understanding as one shared by everyone. The 'limitless yearning' to know more of astronomical objects was an emotion felt not only by professional and amateur astronomers but also by a broader general public.[2] As Lowell notes, this desire for astronomical knowledge arose in part from the very simple fact that many objects in the sky could be seen. Vision, for Lowell, leads to a type of thought that can be defined as curiosity: a curiosity that popular astronomical demonstrations sought to alleviate.

In *Ways of the Planets* (1912), a work of popular astronomy for a general audience written during Lowell's tenure at his Flagstaff Observatory, Martha Evans Martin describes her encounter with an itinerant showman offering the opportunity to 'look at Mars and Saturn' through a six-inch telescope.[3] While Martin puts her eye to the telescope an audience gathers around her 'slightly curious' of what she is seeing.[4] When they, in turn, also look, the showman recites a series of astronomical facts about the planets' size and distance from Earth. For Martin this did nothing to 'make them alluring'.[5] What was required, she argues, was for 'someone fairly to cry out: "Come here and look at this planet. It is different from anything else you have ever seen, or ever will see."'[6] Martin's argument focuses on two complimentary ideas: that popular astronomy should engage an audience's desire, and that it should illuminate the uniqueness of the visual experience astronomy affords.

A century earlier, William Wordsworth had dealt with a similar experience of meeting an astronomical showman in his poem 'Star-gazers' (1806) which opens with 'What crowd is this? what have we here! We must not pass it by, / A Telescope upon its frame, and pointed to the sky'.[7] Despite this initial excitement, Wordsworth notes that those who have looked through the telescope 'seem to meet with little gain'.[8] The poem ponders the discontinuity between an initial 'impatient' desire to look and the 'dissatisfied' response when having done so.[9] Wordsworth asks whether it might be the telescope, the eye, the spectators or the sky itself that is to blame for this fracturing. Alternatively, Wordsworth hopes, it may be that such looking leads to 'deep and earnest thought' that 'admits no outward sign' of the 'grave and steady joy' felt by the telescopic observer.[10] This belief in the power of astronomical observation compliments Martin's view that popular astronomy relies upon the unique qualities of vision to inspire. It also suggests, in common with Lowell, that observing should lead to thinking and reflecting. Finally, Wordsworth's concern that the experience of looking may have been disappointing hints at the importance of appropriate explanation that should accompany star-gazing. What is missing from the poem's description of the astronomical demonstration is any intervention from the showman. Observers are left to look alone and become lost (unless in reverie) in their inability to order what they have seen, or to recognize its significance. How much more might have been gained by observers had Martin or Lowell been alongside to guide them?

In late nineteenth- and early-twentieth century popular astronomy vision was a foundation for thoughtful reflection on the meanings of astronomical objects. But the passage from seeing to thinking required careful navigation. The role of the writer of popular astronomy was to activate readers' desire for observation and then to guide them towards perception. As this chapter will show, works of popular astronomy employed several techniques to achieve this. First, they evoked for their readers a sense of wonder; that curiosity for knowledge that begins in marvelling at the skies and is a vital component of scientific observation. Second, they worked to construct foundations in fact from which the imagination might rise to engage with new perspectives and speculations on astronomical knowledge. Third, and running counter to the others, popular science commodified vision within a system of value closely related to consumer capitalism, where visual culture was also dominant. These techniques are significant in their own right, but illuminating them in popular astronomy also sheds light on their existence in works that would not normally be defined as 'popular': that is, in astronomical works addressing a professional audience as well as in works of fiction. The same sources considered in the previous chapter – Lowell's observatory annals, the astronomer Antoniadi's correspondence with Lowell and fictions of Mars by Wells and Wicks – all address the wonder that emerges from observations and the imagination needed to understand them. Each, too, seeks to place a value on vision and to show how observational knowledge might be bound up in systems of capitalist economics. The importance of finding such specific connections points to the need for a reorganization of the relationship between popular science writing, professional science discourse and fiction. For if the techniques and themes of these genres are overlapping it raises questions about the viability of continuing to think of them only through their differences.

Fiction as Popular Science

While for some time history of science scholarship considered popular science to be the reductive, anti-intellectual and somewhat superfluous product of serious and influential elite science, there has, over the past two decades and more, been a significant reconsideration of its place across all the sciences.[11] Recently James Secord, in an essay on the circulation of scientific knowledge, argued that 'we need to shift our focus and think about knowledge-making itself as a form of communicative action' and that this 'means eradicating the distinction between the making and the communicating of knowledge'.[12] While this is not an argument solely focused on popular science, it clearly has considerable repercussions for the relationships between popular, professional and fictional representations of scientific knowledge. In particular Secord is keen to stress the necessity of being conscious of the textuality of some of this knowledge, so that texts are recognized as being written with 'a narrative voice aimed at a particular horizon of expectations'.[13] This is a particularly literary-critical approach to new histories of science – Secord mentions reader-response theory – that has been championed also by Ralph O'Connor. In *The Earth on Show* (2007), O'Connor details the importance of genres of writing, developed in literature, that had an influence on the writing of popular geology in the first half of the nineteenth century.[14] In a subsequent essay on popular science, O'Connor goes further in calling for greater engagement between history of science and literary studies, which would be especially helpful in better understanding 'the network of cultural exchanges' that defines popular science writing in the nineteenth century.[15] For Andreas Daum, writing in dialogue with O'Connor, scholarship on popular science needs to become 'a broader and more interdisciplinary endeavour'.[16] Histories of science dealing with popular science have perpetuated the idea of the exclusive scientific community, argues Daum, and in doing so have failed to recognize that most of its members were 'always members of multiple communities'.[17] Therefore disciplines other than science must be studied 'if we are interested in finding out how societies generated and publicized knowledge to make meaningful statements about their world'.[18]

There is no need to rely only on new historiographies of popular science to uncover the close ties between different disciplines. Exponents of popular astronomy in the second half of the nineteenth century saw their mission as bringing to a popular audience the synergies between astronomical and literary knowledge. Amedée Guillemin, for instance, opens his popular 1867 work on contemporary astronomy by noting that 'the heavens should be read with something of the charm of the romance'.[19] Astronomical study, he continues, invites many new questions and is not always able to discover their answers:

> But the coldest mind, the mind least accessible to the suggestions of fancy, cannot entirely banish them. In spite of itself, there comes a moment, an hour of reverie, when it too propounds the same problems; and truly we cannot wish it otherwise. Does it not afford one proof the more, of the truth of what day by day becomes more evident, that science borders on poetry?[20]

Guillemin's work was not by any means on the outer margins of popular astronomy, as was sometimes true of the more metaphysical studies of his colleague Camille Flammarion, for example. It was translated and edited by J. Norman Lockyer, soon to be the first editor of the journal *Nature*, published in London by the well-established publisher Richard Bentley and warmly received by astronomical societies across Europe. The president of the Royal Astronomical Society of London, Warren de la Rue, wrote to Guillemin to congratulate him on its success, as did Sir John Herschel.[21] Similarly, the Cambridge astronomer, Sir Robert Ball, opened his 1886 work of popular astronomy by stating that 'The Story of the Heavens is the title of our book [and] we have indeed a wonderful story to narrate'.[22] Likewise Mary Proctor, daughter of the highly-regarded popular astronomy writer Richard Proctor, evoked the connection between astronomy and literary genre in the title of her popular work in the 1920s, *Romance of the Planets*.[23] Even Lowell, in *Mars as the Abode of Life* (1908), wrote of astronomical study in literary generic terms as the 'epic of our solar system'.[24]

Most often these works of popular astronomy come from the pens of writers who would identify themselves first as professional astronomers. In creating popular narratives of astronomy they are therefore approaching the literary from the disciplinary ground of science. Popular science writers of this ilk are those commonly in the mind of historians when discussing new methodologies for the study of textual popularization. O'Connor, for example, in calling for increased attention to literary genre notes that this 'lends coherence to the variety of literary forms *inhabited by* science since 1800' [my empahsis].[25] What remains absent from these important reconfigurations of the place of the literary in scientific culture is a recognition that certain literary texts themselves attempted *to inhabit* science. Both Wells's *The War of the Worlds* and Wicks's *To Mars Via the Moon* exemplify the efforts of fiction writers to claim their imaginative texts as works of popular science. Whether these works are accepted as such in present critical readings is perhaps moot: their authors conceived of them as popular science and their audiences (at least in part) received them as examples of it.

Wells began to consider recent astronomical debates on Mars as the source for a new novel in the summer of 1895, after the success of his earlier work *The Time Machine.*[26] By 6 November 1895, when he wrote to his friend Grant Richards, he considered it likely to be 'a big scientific story'.[27] In January of 1896 a complete synopsis of the novel was completed, and by 25 March Wells told his agent Morris Colles that 'the final instalments of the War of the Worlds are now in the hands of my typewriter'.[28] Although Robert Crossley argues that Wells had never read any of Lowell's work before completing the novel, it is very likely that he encountered Lowell's theories of Mars in the numerous articles both by and about Lowell that were written between 1894 and the end of 1895.[29] In *Nature*, for example, a journal which Wells read regularly, there were four significant articles on Lowell's work between 14 June 1894 and 8 November 1894. During that same period Wells also published two of his own articles in *Nature*. One of these was an important essay on popular science that provides an insight into Wells's own definition of popular science writing and the relationship of his work to it.

'Popularising Science' appeared in *Nature* on 26 July 1894.[30] In it, Wells defends popular science writing as a necessary act of scientific communication. He recognizes that it is often viewed as an inferior form of science that 'conveys a flavour of contempt to many a scientific worker' but argues that its importance in giving science a public face should not be underestimated.[31] However, the keynote of Wells's essay is his defining of the criteria which produces good popular science writing: beginning with the importance of a well-structured narrative (an 'orderly progression and development').[32] In particular Wells argues that science writing should follow the literary genre of romance or detection. 'The fundamental principles of construction that underlie such stories as Poe's "Murders in the Rue Morgue" or Conan Doyle's "Sherlock Holmes" series', Wells writes, 'are precisely those that should guide a scientific writer'.[33] Additionally, the popular science book should aim for 'philosophical' rather than 'technical interest'.[34] The reader, Wells argues, is interested in 'the balance and interplay of life' or the 'relationship of [scientific] phenomena to the thinking man', not the minute technicalities of a scientific discipline.[35] Finally, popular science writing should be 'absolutely serious' and avoid humour at all costs.[36] For Wells, then, popular science writing would be best exemplified by a serious, well-organized narrative that mixes romance with philosophy and develops its evidence so as to reflect upon science's role in the human world. This essay, despite its obvious importance to the genre of popular science writing in the late nineteenth century, is rarely discussed. Jennifer Tucker, in her work on Lowell and photography, does give it a brief mention, but rather oddly stresses that what Wells calls for is 'no flights of imagination', when in fact his introduction of literary techniques and genres, as well as his desire for greater philosophical writing (for which we can also read speculative, a word Wells often substituted for philosophical in his writing) tends to increase the role of the imagination rather than reduce it.[37]

What is most striking about this essay, though, is how closely its definition of popular science writing resembles Wells's definition of his own fiction, especially *The War of the Worlds*. In a private letter to his fellow fiction-writer Grant Allen in the summer of 1895 (when *The War of the Worlds* was conceived) Wells reflected upon his ongoing fictional projects, attempting to give them definition. 'I am trying to cultivate', said Wells, 'this field of scientific romance with a philosophical element'. Certainly the structure and content of *The War of the Worlds* reflects this aim: it is a serious work, organized around a strong narrator whose narrative is positioned retrospectively, just as Conan Doyle's Sherlock Holmes stories are narrated after-the-fact by John Watson. The novel is, in generic terms, a romance, but it withholds elements of the plot so as to provide surprising denouements, creating the same kinds of suspense through lack of knowledge that Poe's and Conan Doyle's work thrived on.

While Wells's fictional project appears to bear significant parallels with his definition of popular science writing such similarities remain within a rather closed hermeneutic circle. That is, it is difficult to draw any significant conclusions when the definitions for each form are drawn up by one individual. How Wells's readers defined his use of astronomy in *The War of the Worlds*, and whether they saw any parallels between his speculative novel and the work of astronomers, might well offer a clearer picture of its relationship to popular science. John St Loe Strachey, editor of the *Spectator*, wrote one of the earliest reviews of *The War of the Worlds* in January 1898. He saw Wells's imaginative version of Mars as grounded in scientific fact: 'This is the age of scientific speculation, and scientific speculation, rightly or wrongly, has declared that if there are living and sentient creatures on Mars they will be very different from men. Mr.

Wells, whose knowledge of such speculations is obviously great, has followed the prevailing scientific opinion'.[38] The reviewer of the journal *Academy*, writing on the same day, went further:

> Mr H. G. Wells has probably a greater proportion of admirers among people actively engaged in scientific work than among any other section of the reading public. It is not difficult to understand the reason for this. Nothing irritates a man of science more than incorrect assertions with reference to natural facts and phenomena; and the writer who essays to use such material must obtain information from Nature herself, or he will provoke the derision of better informed readers. Mr. Wells has a practical familiarity with the facts of science, and this knowledge, combined with his imaginative mind, enables him to command the attention of readers who are not usually interested in romance.[39]

Both Strachey and the *Academy* reviewer commend Wells's scientific knowledge and regard his imagination as springing from a 'concrete and specific' foundation in contemporary astronomy.[40] Moreover, the *Academy* reviewer makes an interesting comment on Wells's novel's ability to claim a new readership for a new kind of genre: the romance which combines the facts of science with the imagination. One such reader may well have been Sir Richard Gregory, later editor of *Nature* and one of the foremost Victorian science journalists.[41] In a review of Wells's novel he wrote that 'upon a groundwork of scientific fact, [Wells's] vivid imagination and exceptional powers of description enable him to erect a structure which intellectual readers can find pleasure in contemplating'.[42] For Gregory, then, *The War of the Worlds* was the kind of philosophical speculation that Wells was aiming for, and which also defined, at least in Wells's own opinion, the best of popular science writing. In concluding his review, Gregory made explicit his understanding of the relationship between the novel and science: 'it is worth remark that scientific romances are not without value in furthering scientific interests; they attract attention to work that is being done in the realm of natural knowledge, and so create sympathy with the aims and observations of the men of science'.[43] For Wells's readers, therefore, *The War of the Worlds* was very much of a new generic formulation that combined a keen understanding of scientific fact with the creativity of the literary professional. It was aimed, in their view, at an educated reader who wished to know more of contemporary astronomy (and not to fear that they were being sold 'incorrect assertions') but who also wanted to be invited to think for themselves about its philosophical significance. Wells's novel was defined by his audiences in the same terms that he had defined popular science. That a work of *fiction* can occupy this territory so decidedly reveals how popular science writing was much more open to innovation from disciplines traditionally regarded as outside of scientific communities.

Wicks's *To Mars Via The Moon* shows even more clearly the potent combination of literary creativity and scientific knowledge. Wicks's 'astronomical story', as the subtitle contends, directly responds to Lowell's observations of Mars.[44] Its dedicatory page, shown in Figure 4.1, is addressed to 'Professor Percival Lowell ... to whose careful and painstaking researches, extending over many years, the world owes so much of its knowledge of the planet Mars, this little book is respectfully inscribed by one who has derived infinite pleasure from the perusal of his works on the subject'.[45] In several respects Wicks's book is generically situated as a work similar to the larger body of popular science textbooks. It includes, for example, a series of plates and tables, properly referenced and attributed, and re-drawn (often by the author) from the existing work of Martian astronomers. One such example, shown in Figure 4.2, provides a chart of the Moon indexed with topographical features extraneous to the narrative, but in keeping with reader expectations of works of popular science. It confirms what Bernard Lightman has noted as common to popular astronomical writers in the later nineteenth century, that 'they presented themselves as scientific experts who had the ability to interpret images produced by the ... telescope'.[46]

TO

PROFESSOR PERCIVAL LOWELL

A.B., LL.D.

Director of the Observatory at Flagstaff, Arizona

TO WHOSE CAREFUL AND PAINSTAKING RESEARCHES, EXTENDING OVER MANY YEARS, THE WORLD OWES SO MUCH OF ITS KNOWLEDGE OF

THE PLANET MARS,

THIS LITTLE BOOK IS RESPECTFULLY INSCRIBED BY ONE WHO HAS DERIVED INFINITE PLEASURE FROM THE PERUSAL OF HIS WORKS ON THE SUBJECT

Figure 4.1: Mark Wicks's dedication to Percival Lowell in *To Mars Via The Moon*. Source: M. Wicks, *To Mars Via The Moon: An Astronomical Story* (1911; n.p.: Dodo Press, 2009); reproduced with permission of the Lowell Observatory.

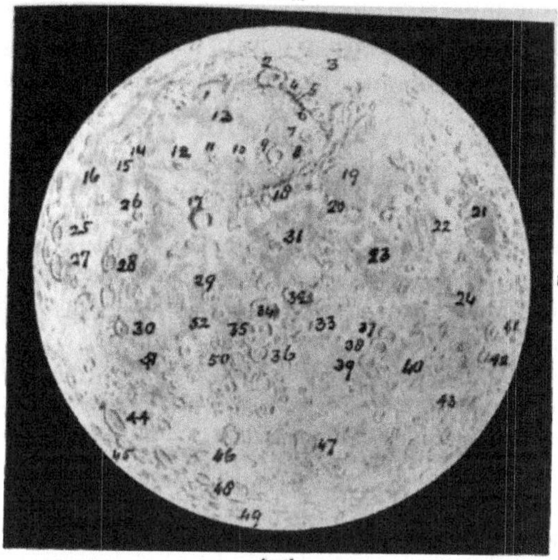

Plate IV

INDEX MAP TO CHART OF THE MOON

1. Bay of Rainbows	18. Apennine Mountains	35. Alphonsus
2. Plato	19. Sea of Serenity	36. Arzachel
3. Sea of Cold	20. Haemus Mountains	37. Theophilus
4. Alps Mountains	21. Sea of Conflicts	38. Cyrillus
5. Great Alpine Valley	22. Proclus	39. Catherina
6. Cassini	23. Sea of Tranquillity	40. Sea of Nectar
7. Autolycus	24. Sea of	41. Langrenus

Figure 4.2: A chart of the Moon in Mark Wicks's *To Mars Via The Moon* combining fiction with popular science. Source: M. Wicks, *To Mars Via The Moon: An Astronomical Story* (1911; n.p.: Dodo Press, 2009), p. 43; reproduced with permission of the Lowell Observatory.

Yet Wicks's narrative is largely a fictional romance, the design of which is clear from its early chapters. Interweaving the fictional adventure with Wilfrid Poynders's lengthy answers to questions about Mars from his co-travellers, Caxton and M'Allister, Wicks constructs a form of Socratic dialogue that informs the reader about contemporary Martian astronomy. One example, as the trio approach Mars, will give a clear picture of the text's strategy:

> Mars was rapidly growing in size and brightness ... We had noticed on several occasions a mistiness on some parts of the planet, which I attributed to the vapours rising from the canals by the heated atmosphere.
>
> On the 21st of September, when we were all enjoying a smoke in the 'evening' and conversation had dragged somewhat, John started us off on a fresh track and gave us something to talk about for a very long time.
>
> He winked at M'Allister and, looking at me with a knowing smile, said "Professor, as we are nearing our destination it might perhaps be week if you now gave us some detailed information, respecting the planet ...
>
> 'Yes, Professor,' chimed in M'Allister, 'I'm quite ready to learn something definite about Mars, for I can't say I really know much about it at present.' ...
>
> So I began – 'Mars, as no doubt you are aware, is a much smaller planet than the earth, its diameter being only 4220 miles, which is a little less than twice the diameter of our moon.'[47]

This begins one of Poynders's lengthy lectures on Mars's relation to the other planets, the controversy over the existence of water on the surface of Mars, and Lowell's contention that the surface markings represented canals. Wicks's story, therefore, uses the vehicle of a fictional trip to Mars to explain contemporary astronomy, explore the various debates about Mars's geography, and to provide a portrait of the astronomers involved in the canal controversy. While this strategy is rather clumsy at times, it does illuminate the textual mechanics of combining scientific knowledge with fictional narrative, and reveals how these can be used to complement one another. The fictional narrative gives Wicks the context for introducing scientific fact, while the long discourses on astronomy ground the fiction in scientific reality.

Wicks stresses in his preface to the book that he chose to combine fiction with fact in order to fulfil a perceived desire for a new work of popular astronomy that dealt with its subject in 'plain untechnical language'.[48] 'It occurred to me', Wicks says, 'that it would be much more useful and appeal to a more numerous class if, instead of writing a book on the usual lines, I wrote a narrative of events which might be supposed to occur in the course of an actual voyage to Mars'.[49] To this end, Wicks explains, 'I have endeavoured to convey by means of natural incidents and conversations between the characters portrayed, the most recent and reliable scientific information respecting ... Mars'.[50] Wicks concludes with the expectation that his work will 'nurture a love of the sublime science of astronomy, and at the same time provide some food for thought'.[51] There are considerable similarities here with Wells's vision of a more dramatic popular science. Like Wells, Wicks calls for a shift from technical language to a more literary style, for the discussion of scientific knowledge within a more creative narrative genre, and for the evocation of a more philosophically reflective response to science.

Wells's and Wicks's work illuminate the extraordinary diversity of popular science writing about astronomy, and its close ties to fictional literature. Indeed *The War of the Worlds* and *To Mars Via The Moon* confirm the truth of Guillemin's rhetorical question that science might border on poetry. While it is important to recognize the influence of other disciplinary networks in defining popular science writing it is also vital to consider in detail those connections between creative works of popular astronomy and astronomy's more professional discourses. These should also be identified if the 'border crossings' that both Topham and Secord identify as part of the reorganization of our understanding of science 'as communication' that erases the (false) differences between popular and elite forms are to be properly secured across the multiple communities that inform science in culture.[52]

Vision Evoking Wonder

An intense fascination with vision is one of the key identifiers of the correspondence between professional astronomical discourses and the various genres of popular astronomy. Popular science writers, as Lightman has argued, 'made spectacular visual images central to their work': writers of popular astronomy such as Agnes Clerke and Richard Proctor were in the vanguard of a newly-professional group who took advantage of the 'developing mass visual culture' to establish themselves as 'reliable guides to nature through their use of new visual technologies'.[53] Lightman draws this conclusion by focusing on the pictorial images that these writers employed in their work, but it was also the case that writers of popular astronomy were interested in dealing with vision in textual form. Guillemin, for example, hopes that his work on the night sky will leave the reader believing him to be 'a painter before a beautiful landscape' who has been 'able to portray beauties of the grandest of all scenes, that of the infinite variety of the stars'.[54] Guillemin's writing was intended to stimulate in the reader 'a strong desire to see for oneself' the 'marvels' of astronomical objects.[55] In this he was part of a tradition of popular science writing which, Lightman shows, aimed to 'bring into view hidden worlds of wonder in nature'.[56] For Lightman this was in contrast to the scientific professionals who 'ignored' them.[57] Although Lightman stresses the ontological gap between professional and popular discussions of scientific illustrations, there was no such difference in textual evocations of wonder initiated through observation.

In his first book on Martian observations Lowell discussed the historical emergence of the canals and the recognition of them by the astronomer Schiaparelli. In his dramatic narrative of Schiaparelli's observations, Lowell writes of how wonder is stimulated by sight:

> At the next opposition [Schiaparelli] looked to see if by chance he should mark a repetition of the strange event, and went, as he tells us, from surprise to surprise; for one after another of his canals proceeded startlingly to become two, until some twenty of them had thus doubled. This capped the climax to his own wonderment, and, it is needless to add, to other people's incredulity; for nobody else had yet succeeded in seeing the canals at all.[58]

Lowell's drawing of a key difference between wonder and incredulity is linked to vision. Wonder arises from having observed, incredulity from having failed to observe. For Lowell, wondering is the first path on the journey towards understanding; it is, as Daston and Park argue, a 'cognitive passion' that allows for analysis and reflection.[59] In opposition to this is incredulity, a sense of disbelief that comes from having not seen and therefore disengaging the process of thought that might lead to the emergence of scientific knowledge. Daston and Park claim that wonder disappeared from the processes of scientific observation at the end of the eighteenth century, giving way specifically under the pressure exerted in science to accept only objective facts.[60] Nevertheless, Lightman sees wonder continue into the nineteenth century in the work of scientific popularizers who championed natural theology. To show the natural world as wonderful was to illuminate the work of God.[61] For Lowell, however, wonder was a form of ratiocination, a mode of understanding what was seen. It was not, therefore, merely the evocation of astronomy as romance that led him to introduce the idea of wonder in his popular astronomical work.

Lowell also considered the role of wonder and its relationship to observation in his professional writing. In the first edition of the *Annals of the Lowell Observatory*, he also discusses Schiaparelli's observations and again expresses the sense of wonder that the sight of the canals produce. Schiaparelli did not, Lowell argues, mistake the 'surprising geometrical character of the lines' visible on Mars: 'The truth is that the facts exceed the account. The better the means of observation, the stranger the whole system appears.'[62] Later in the same description, Lowell calls these lines 'supernaturally regular' and notes that 'such unnatural regularity is certainly not what one would expect to see' but again dismisses the idea that the wonder felt by Schiaparelli had impinged on his objective interpretation: 'The first observer might be given unconsciously to recording marvels, but the second would as unconsciously be inclined to deny them'.[63] As he had in his popular work, Lowell argues in his professional writing that Schiaparelli's sense of wonder was an appropriate analytic response with its roots in an originating observation: 'It is patent from the foregoing observations that in the "canals" we are confronted by a set of phenomena as intrinsically strange as at first sight they appear inexplicable'.[64]

Wonder also finds a place in private correspondence between professional astronomers. Lowell's and Antoniadi's letters, especially those of Antoniadi, are filled with the kinds of data exchanged by astronomers that make most sense within professional communities. Yet there is space, too, for the expression of wonder at observations made. Antoniadi, for example, writes to Lowell of his experiences at the telescope on a night of particularly clear seeing:

> After my excitement, and the bewildering amount of detail visible, was over, I sat down and drew correctly … all the markings visible … I shall not expatiate on the description of my drawings, as they speak by themselves to an aerographer like yourself who knows the planet better than anybody else. But what an incredible change in Syrtis Major from what you have discovered 15 years ago … I thought I was losing my senses.[65]

In this brief insight into the actual practice of astronomical observation Antoniadi reveals to Lowell the wonderment that can emerge from vision. His narrative of observation oscillates between excitement and bewilderment (as corollaries of wonder) and the exactitude of the drawings he was thereafter able to make. Antoniadi may have felt he was losing his senses – and it is clearly the visual sense to which he refers here – but he is convinced that his work is correct and valid. Antoniadi's observations have provoked wonder and allowed him to begin the process of interpretation.

Wonderment emerged from observation because what was being seen appeared unusual, even extraordinary, to the eye of the spectator. For Antoniadi the extraordinary features of the Martian landscape were unusual because he was looking at them for the first time. Astronomers recognized that part of the reason for their wonder was their unique position as explorers of new worlds whose privileged eyes often looked upon things never before subject to human vision. In a lecture Lowell gave late in his life, in Boston in 1916, he began by acknowledging that 'to make the acquaintance of a new world is a wonderful experience'.[66] This was a sentiment Lowell had held throughout his career. In his 1894 'Mars' poem he responded to some of his own first sightings of the planet's markings by writing of an imagined exploration:

> Till as I peer myself I see
> Sailing some landlocked Martian sea
> Scaling some Martial mountain crown
> To gaze from its vantage summit down
> On a landscape whose strange mein
> Ruddy red where we look for green
> Pricks perception to grow more keen[67]

In these lines Lowell reveals the connection between initial sight, exploratory vision and the feeling of wonder that stimulates better perception. An obvious inter-text for this poem was Keats's earlier poetic parallel between the exploration of new worlds and astronomical vision. In his 1816 verse, 'On First Looking into Chapman's Homer' Keats had compared his feelings of wonder at, and admiration for, Chapman's translation of Homer's *The Iliad* to the same feelings in the astronomer or adventurer:

> Then I felt like some watcher of the skies
> When a new planet swims into his ken;
> Or like stout Cortez when with eagle eyes
> He stared at the Pacific.[68]

Lowell returned to the image of the astronomer as an explorer discovering new worlds in his popular work, *Mars*. In the introduction to a chapter on Mars's geography, Lowell tells his reader that 'we may thus make a far journey without leaving home, and from the depths of our arm-chairs travel in spirit to lands we have no hope of ever reaching in body. We may add to this the natural delight of the explorer, for we shall be gazing upon details of Martian geography never till last summer seen by man'.[69] Here, as in Lowell's poetry and lecture, wonder emerges from unique observation and is romanticized as exploration. In turn, the astronomer is characterized as the explorer whose curiosity (aroused by seeing and focused analytically by wonder) leads to discovery.

Those fictions that placed themselves in the genre of popular science also linked wonder to sight and to exploration. Wells's *The War of the Worlds* makes a clear distinction between the narrator's 'blank wonder' when the Martians are only 'invisible terrors' and the 'sort of wonder' he experienced as he struggled 'bitterly for that horrible privilege of sight' which marks his exploration of the Martian's preparations for their attack on the earth.[70] For Wells, cognition is amputated from wonder when there is no observational data from which to proceed. Adversely, any useful analysis of the Martians requires them to be looked at carefully. The artilleryman whom the narrator meets in London bases his plan for the overthrow of the Martians on painstaking observation: 'We must keep up our science – learn more. We must watch these Martians.'[71] Wicks's *To Mars Via The Moon* makes more of the wonder that looking at new worlds inspires. As the explorers near the moon, the narrator describes his view:

> There it was below us, but still slightly ahead of the *Areonal*; and its magnificence was so overpowering, that it almost seemed to take my breath away, although I was fairly well prepared for the sight. Many times when viewing it through the telescope I have almost lost myself in admiration of the sublime spectacle it presents; but what I had seen on those occasions could not be compared with the splendour of the view now before us ... John and M'Allister told me they had both been gazing upon the

> splendid scene for a very long time with astonishment and delight equal to my own; and the latter went on to say ... 'Mon, its worth many a journey like this to see such a bonnie thing!'[72]

Like Lowell and Wells, Wicks sees wonder as a product of observation, but he also considers it to be intensified when combined with exploration. Indeed for Wicks's astronomical adventurers their Martian exploration is explicitly designed to uncover 'new subjects for wonder'.[73]

For professional and popular astronomy alike wonder was a key generic theme. That it could be used in either context highlights its suppleness; wonder was both a foundation for perception leading to knowledge and a romanticized response to new sights uncovered by exploration. This is not to suggest that wonder works only in such binary mode, with its route to knowledge acknowledged in professional discourse and its stimulation of romance a trope of popular writing. Rather, wonder's meanings crossed and re-crossed the boundaries between professional, popular and creative; where professional discourse drew on the romance of exploration and creative popular astronomy evoked wonder to reach towards keener perception. This undermines Daston's and Park's analysis of wonder as coming to an end-point at the beginning of the nineteenth century, while at the same time upholding their analysis of how wonder operates. In addition, though, wonder's role in astronomical exploration stimulated a further trope common to all astronomical writing – the important role of the scientific imagination in deciphering the complex visual experience of astronomical observations.

Vision and Imagination

Guillemin's popular history of astronomical exploration, which he hoped would be read with the same 'powerful interest which belongs to traveller's tales of unknown lands', invites the reader to draw on their imagination in order to understand the complex objects found in the solar system:[74]

> What a vast field, moreover, what a magnificent horizon is presented by the Heavens to the most active of human faculties, to the imagination! When our sight, aided by the most powerful instruments, dives into the depths of space ... a thousand questions rise to our lips. We find ourselves involuntarily making in thought a hundred travels, more interesting, more strange, more marvellous than those the scene of which lies on our own planet.[75]

Guillemin taps into a broader set of discourses which recognize astronomy's dependence upon the imagination as a vital tool of perception, similar to wonder in its ability to help the observer towards interpreting and therefore understanding the images of celestial objects. Walt Whitman had hinted at this role for the imagination in his derogation of popular astronomical demonstrations by professional astronomers in his 1865 poem 'When I Heard The Learn'd Astronomer'. Whitman contrasts the experience of the lecture – where 'proofs' and 'figures' were 'ranged in columns' that left him 'tired and sick' – with his star-gazing: 'I wander'd off by myself, / In the mystical, moist night-air, and from time to time, / Looked up in perfect silence at the stars'.[76] Whitman prefers to reach for understanding through the imagination rather than in the evidence of geometry, and his sense of the rupture between what can be noted as fact and what might be imagined exemplifies Daston's understanding of the breach between objective knowledge and the imagination that occurred in the nineteenth century.[77]

However, it would have been ironic, to Whitman at least, that professional scientists would not necessarily have disagreed with the trajectory his poem takes from the truths of the lecture room to the imagination. John Tyndall, in his important address on the scientific use of the imagination given at the annual meeting of the British Association for the Advancement of Science in 1870, argued that the scientific imagination was a powerful instrument when properly used:

> There are those in science who regard Imagination as a faculty to be avoided rather than employed. They observe its action in weak vessels and are unduly impressed by its disasters. But they might with equal justice point to exploded boilers as an argument against the use of steam. Nourished by knowledge patiently won; bounded and conditioned by co-operant Reason, Imagination becomes the mightiest instrument of the physical discoverer.[78]

Tyndall limits the imagination in two ways. It is first prefigured by the steady acquisition of knowledge and, second, is always constrained by an active, yet supportive rationality. Whitman's depiction of gathering astronomical data before employing his imagination in a silent yet sustained study of the stars might easily be a poetic representation of Tyndall's lecture. Certainly the lecture had a wide influence. Lowell, in the second lecture of a series on Mars in 1916, characterized the imagination in exactly the way Tyndall had done forty-six years earlier: 'For all great work imagination is vital', wrote Lowell, 'just as necessary in science and business as it is in novels and art'.[79] Yet imagination could not work alone, it needed to be 'harnessed to Reason' which supplied the 'guiding rein' to imagination's 'motive power'.[80] Similarly, Proctor's *Romance of the Planets* exemplifies Tyndall's view of knowledge as a foundation for imaginative thinking in considering how we might imagine the Martian landscape as based upon our knowledge of the earth: 'We see in imagination, as actually takes place in the parched plains along the banks of the Nile, the growth of various forms of luxuriant vegetation'.[81]

To 'see' in imagination, as Proctor puts it, indicates the closeness of the relationship between vision and imagination. Tyndall also recognized this in his address, claiming that when observers have 'forced upon [their] attention phenomena which no eye has previously seen' it is then that the 'creative power in which reason and imagination are united has, we believe, led us into a world not less real than that of the senses, and of which the world of sense itself is the suggestion and justification'.[82] Imagination begins, then, with seeing, although for astronomical writers it is often the difficulty of that seeing that allows the imagination space to flourish.

Lowell wrote on several occasions of how his observations of Mars were restricted both by distance and various forms of distorting effect. Lowell called the point at which these restrictions took hold the 'minimum visible', a term he used in several of his papers on Mars.[83] Imagination works at this threshold, taking over when vision reaches its minimum. Several scholars have noted that visual information was, as Anna Henchman puts it in her perceptive study of Thomas Hardy's relationship to astronomy, 'a jumping-off point for the imagination'.[84] Pamela Gossin, too, in a discussion of the Herschels's cultural influence, argues that their astronomical work 'made deep and lasting impressions upon the visual imagination'.[85] For Thomas Hankins and Robert Silverman, in their analysis of scientific instruments, the imagination was an organizing faculty which mediated between vision and reason.[86] Even George Henry Lewes, writing in the mid-century, noted that while looking through the microscope was observation, for the telescope it was imagination.[87] There is, therefore, an already existing understanding of the imagination's pivotal role in observation. For Lowell and other writers on astronomy at the end of the nineteenth century, however, the imagination had a very specific place in the observational economy. It offered a way of extending the visual capacity when observation reached its end point, that is when seeing dropped below the minimum visible. Imagination was not simply an aid to sight, but a form of it; a way of seeing that went beyond the limitations of the eye and the telescope.

In an article on astronomers sceptical of his canal theory, and therefore of his telescopic observations, Lowell made clear his understanding of the imagination as extending sight beyond the 'limit of vision'.[88] Arguing against those who had claimed that the canals were created by Lowell's imagination rather than existing in reality, Lowell denied that they were anywhere near that 'limbo of doubt' where the imagination might begin to take hold.[89] His observations did not, therefore, 'fulfil the prerequisite to the supposed illusion'. Although this article reveals Lowell's scepticism about the imagination's usefulness, it does highlight his belief that its role is to offer an alternative way of seeing when the more usual observational formulae have been exhausted. In his poem on Martian observation, Lowell is more explicit about the astronomer's imaginative faculty. Continued Martian observation has already been 'luring imagination on', Lowell tells his reader, before he comments more fully on the loss of clear sight and the emergence of alternative vision:[90]

Then suddenly, without a stir
The while you gaze, a greyish blur
Films the spot where the markings were;
A great amorphous colorless doubt
Blotting all your landmarks out
Till coasts and bays but now so clear
Grow dim, then dimmer, then disappear.
Martial cloudmasses have erased
The ground where in her happy haste
Fancy already her footprints traced.[91]

Beginning with knowledge founded on clear visual information Lowell traces the shifts in his observation as atmospheric conditions draw him first towards the minimum visible, then beyond it into visual blindness and finally into 'fancy', a common poetic word for the imagination. It was not only Lowell who believed the imagination to begin where vision ended. Martha Evans Martin, in her book of popular astronomy, also believed that it was the 'limit of vision' that stimulated an imaginative response that she calls 'opinion' or 'speculative theory'.[92]

Astronomical discourses, both popular and professional, acknowledged the utility of the scientific imagination in efforts to organize visual phenomena into meaningful truths about celestial objects. Like wonder, the imagination was a way of exploring the heavens, but while wonder stimulated astronomers to begin the work of interpretation, the imagination worked after the fact to extend its reach, beyond the limits of observation, into reasoned speculation. Guillemin's *The Heavens* (1867) exemplifies how popular science employed this scientific imagination to appeal to an audience's own imaginative powers:

> Basing our work on the facts already acquired, we set ourselves to build up our neighbouring worlds; the configuration of their continents and seas, the rivers which water them, their mountains, which are the very skeletons of worlds, the living inhabitants, animal and vegetable, which people them, all present themselves before us in the most varied forms.[93]

What may seem here like a rampant subjectivity is, in fact, a popular representation of Tyndall's controlled imagination. Guillemin may indeed be inviting his readers to give themselves over to speculation but it is speculation 'based on ... the facts already acquired'. In this context such a phrase does not appear a simple caveat to allow the writer to dispense with objectivity, but rather an important acknowledgement that it is the scientific imagination which is to be employed in presenting a series of hypotheses about extraterrestrial life.

In this light, Wells's *The War of the Worlds* appears to conform rather faithfully to the employment of the scientific imagination as one of the communicative acts of professional and popular science writing. Certainly Wells characterizes his narrator as subject to the same imaginative impetus as either

Lowell or Guillemin. One of the clearest examples of this similarity arrives at a point of relative calm after the Martian invasion of the Earth, when the narrator has an opportunity to observe the Martian machines at work:

> They seemed amazingly busy. I began to ask myself what they could be. Were they intelligent mechanisms? Such a thing I felt was impossible. Or did a Martian sit within each, ruling, directing, using, much as a man's brain sits and rules in his body? I began to compare the things to human machines, to ask myself for the first time in my life how an ironclad or a steam-engine would seem to an intelligent lower animal.[94]

The narrator's questions begin with observation, but soon take on an imaginative perspective as he considers several hypotheses for the apparent intelligence of the machines. Just as Mary Proctor imagined the Martian landscape by comparing it to the Nile flood-plain, so the narrator compares the Martian machines to terrestrial technologies. Like Guillemin's *The Heavens*, this episode invites the reader to imagine extraterrestrial life in its difference from human life.

Wells's original readers saw that this use of the scientific imagination was a tactic of his work as a whole, as well as in individual instances. The *Academy* reviewer believed that 'upon the scientific imagination depends the structure, the plot, of the whole thing'.[95] The reviewer proceeded to draw a direct comparison between Wells's novel and Lowell's views on the canals, based on observations that can make out 'just enough of the planet's surface' to allow his reasoned imagination to offer some interpretation.[96] 'The view that the Martians', the reviewer concluded, 'would look towards our Earth with longing eyes is thus quite within the bounds of legitimate speculation'.[97] Richard Gregory, writing in praise of Wells's controlled imagination, appreciated his 'ingenuity ... in manipulating scientific material' to create a narrative that 'was not only familiar with scientific facts, but ... knew them intimately enough to present a view of the future'.[98] For members of Wells's original audience, then, *The War of the Worlds* was a fine example of the possibilities of the scientific imagination.

For writers who wished their work to be popular there was a more pragmatic reason for using the scientific imagination to expand the vision of their readers. They hoped, as Wicks puts it in the preface of *To Mars Via The Moon*, to satisfy 'a widespread desire for fuller and clearer information' by offering something that was not 'on the usual lines' but rather in 'an interesting form' that readers could 'really grasp and understand'.[99] By giving readers a more acute vision of Mars writers such as Wicks hoped to sell more books. Using the scientific imagination was not only, then, a method for increasing scientific rigour, it was also a way of turning vision into a commodity for the sake of a market.

Commodifying Vision

Aileen Fyfe and Bernard Lightman have shown how popular science was inextricable from the new markets fostered by nineteenth-century consumer capitalism. They argue that 'the nineteenth century saw an immense growth in consumer culture' and that by the second half of the century many more people had 'access to the wide range of commodities and experiences available'.[100] These included a range of options connected to science, from 'museums and exhibitions to lectures and books'.[101] In particular, they note the importance of 'the language of the marketplace' as a key influence on the 'novelty and innovation' to be found in popular science in this period.[102] It is this language that most marks astronomical writing on Mars, but it is not always employed self-consciously as a way of attracting consumer attention. In fact the ways in which the language of the commodity comes to permeate Martian astronomy suggests that astronomy was becoming subjugated to commodity culture as well as availing itself of that culture to reach new audiences.

The latter was, of course, especially true in the most creative popular astronomical works, where better vision (evoked by imaginative narratives) was being given value as the commodification of access to scientific knowledge. Wells, for example, provides his narrator with greater authority to write on Martian life because he was a witness to it. The whole narrative is constructed as an authoritative retrospective account which the narrator writes for public consumption. This is clear in his sustained addresses to the reader which begin on the very first page of the novel, and from the descriptive details the reader is given of the narrator's position as a professional writer.[103] Although the narrator reminds the reader that he is neither an astronomer nor a physiologist (two scientific careers that would have given him most authority to speak on Mars) he was privileged enough to have seen the Martians at first-hand. Such visual authority gives him access to the consumer who is now ready to receive an account from someone able to offer a 'broadening of men's views'.[104] Wicks, in addition to the clear aim he set down in his preface of giving his reader a popular science narrative that would stimulate their own imaginative vision, concludes his story by stressing his characters' development of things they have studied on Mars into a successful and profitable business from which they will be 'millionaires before very long'.[105] Guillemin also stresses the economic advantages that can come from successful commodification of observation, although in his case it is his own success in capturing the imagination of his reader. In the preface to the second edition of *The Heavens*, Guillemin notes that his aim of 'obtaining the support of those for whom it was specially written … the general reader' had 'succeeded beyond his expectations'.[106] This was due, he felt, to his efforts in creating 'a faithful picture of the phenomena offered by the Heavens to man's intelligent admiration'.[107] As Wells, Wicks and Guillemin exemplify, the commodification of astronomical vision, often through the imagination, was a successful tactic for the branding of astronomy as an innovative new entrant to the science marketplace.

While these writers sought actively to commodify astronomy for consumption, consumer culture was in its own way already taking hold of the visual perception of Mars. The sociologist Georg Simmel, looking at the late nineteenth and early twentieth century in 1907, had recognized that there was a need to provide a sociology of the senses to articulate the new ways in which the human sensorium interacted with the world. In that important essay Simmel argued that of all the senses the eye was most powerful, but also most vulnerable. Part of the weakness of sight was the potential for vision to be commodified: 'in general', wrote Simmel, 'one can possess only the "visible" whereas that which is audible is already in the past in the moment of its present and provides no "property"'.[108] Unlike other senses, vision could become property and enter the system of capital as a commodity. In Martian astronomy vision was already undergoing evaluation in language that was as clearly economic as it was scientific.

Lowell's creation of a set of values for vision is one clear example of the commodification of sight within astronomy. When called upon to defend his canal theory, Lowell turned to vision as one way of exerting control over opinion. His Flagstaff Observatory, he claimed, offered significantly better seeing conditions than any other astronomical site. Across numerous articles Lowell made the case for Flagstaff's clearer air and less disruptive atmosphere.[109] To quantify this superiority he created a scale of seeing to be used universally. In a notice entitled 'A Standard Scale for Telescopic Observations', published by the Royal Astronomical Society in 1902, Lowell argued that 'at present there exists no criterion among astronomers for the weight to be attached to any given observation' and 'in consequence no absolute value is assignable to any man's work'.[110] At Flagstaff, however, Lowell had conducted a series of experiments to allow such criterion to be set, and 'have thus enabled an absolute scale to be constructed'.[111] Lowell is attempting nothing less here than to give value to vision, to make it subject to a numerical account (from 0 to 10 on his original 'scale of seeing'). His own notebooks, which also evaluate observation, make this clear. In his diary of Martian observations for 1896–7, Lowell notes the value of seeing as though he were entering figures in a business ledger. One set of typical entries reads:

Jan 9–26	Not seen	Seen
0° 0°–15°	!!	!!!
30° 15°–45°	!!!!	!!!!!!!
60° 45°–60°	!!!	–

In these procedures of visual accountancy, seeing is made into a matter of numerical fact, stripped of its nuance so as to be made subject to an artificial scale of value that has no apparent relevance to the actual quality of the observations. Additionally Lowell was also claiming that the value of vision could only be determined from his own observatory site at Flagstaff, where the scale had been created. Vision was, therefore, Flagstaff's – and Lowell's – property. In essence Lowell was claiming that astronomical vision could be given abstract value and become property; that is, vision could be turned into a commodity.

Simmel's fellow sociologist Max Weber would have recognized this standardization of visual quality as entirely the product of the bureaucracy of capitalism which had reduced science's enchanted knowledge-making to organized, disenchanted processes beholden to the consumer market. Indeed this was his argument in a key essay, 'Science as a Vocation'.[112] This disenchantment is also achieved by Lowell's scale of seeing, which makes Flagstaff's air the key marketplace against which the 'value' of any observation must be judged. In this way both vision and the observatory become commodified.

Simmel offers a useful further context for the relationship between abstracted value and vision in his analysis of modern metropolitan life. For Simmel, an ever-changing landscape means in visual terms 'the rapid crowding of changing images, the sharp discontinuity in the grasp of a single glance, and the unexpectedness of onrushing impressions'.[113] Simmel is attempting to capture the contrasts found in the new modern city, but his description closely resembles the observational experiences of Lowell and other astronomers who studied the landscape of Mars. They, too, had to deal with 'abnormal' and 'swarming' details, as Antoniadi referred to them, or discontinuous 'glimpses' that were 'really startling' to the observer, as Lowell noted.[114] Simmel goes on to argue that the observing subject protects himself from this perceptual chaos by privileging his intellect and developing a 'matter-of-fact attitude' that relies entirely on the objectively measurable. Lowell's scale of seeing was targeted at the same conclusion; to make capacities of vision measurable and, on his own terms, objective. Simmel goes further than this, however, in arguing that the 'money economy and the dominance of the intellect are intrinsically connected'.[115] In fact, claims Simmel, 'the matter-of-fact attitude is obviously so intimately interrelated with the money economy ... that nobody can say whether the intellectualist mentality first promoted the money economy or whether the latter determined the former'.[116] In Simmel's schema, then, Lowell's creation of the scale of seeing is also entirely bound up with commodity culture; not simply as a parallel for it, but as part of the same mentality that 'reduces all quality and individuality' to questions of abstract value.[117]

The interrelationship between Martian astronomy and commodity culture extended beyond the observational practices of individual astronomers into other cultural arenas. As Mars became 'public knowledge', to use Andreas Daum's phrase, it also became subject to the influence of the 'consumer society at large' that Daum identifies as one of the most powerful narratives of the modern world.[118] It is where the language of the market pervades public discourse on Mars that this becomes most apparent. The geologist Eliot Blackwelder, for instance, in his critical review of Lowell's final work of popular astronomy, *Mars as the Abode of Life*, compared Lowell's misrepresentation of Martian geology to the outlandish claims made by unscrupulous purveyors of consumer goods. 'The misbranding of intellectual products is just as immoral as the misbranding of the products of manufacture', claimed Blackwelder, in his effort to compare Lowell to the archetypal villain of consumer capitalism.[119] Here, scientific evidence has been transformed into a commodity and the interpretative function of the scientist reduced to the creation of brand identity. More often, commodity culture provided an opportunity for Martian astronomy to be treated with humour. The local Arizona newspaper, the *Coconino Sun*, offered a report on an evening of public viewing at Lowell's Flagstaff Observatory: 'A number of ladies saw the old man in the moon and criticised the fall styles in Mars at the observatory last Friday night'.[120] Such light-hearted satire was not always received well by the astronomers. One of Lowell's assistants wrote (rather humourlessly) in the margin against this article that 'no-one was there that night'.[121] A further newspaper report linked Martian observation with problems in Earth's capitalist economy:

> Astronomers of the imaginative school begin to have the impression that the inhabitants of Mars are trying to open communications with this planet. The possibility is a very pleasing one. They may be able to give us information upon important subjects about which we are all at sea. We may be able to find out from them how to deal with a gigantic Sugar Trust, or how to prevent the Standard Oil Company from putting a blanket mortgage on the planet![122]

Such reports were not ignorant of the astronomical community's discussions about Mars. Both of those above implicitly recognized that the present state of astronomical knowledge believed Mars to be a more mature planet than Earth, with the possibility that its inhabitants would be far in advance of Earth in all aspects of culture.

The poet May Kendall also drew on this knowledge in her 1894 comic poem about astronomers on Mars observing Earth. She creates a Martian telescope technologically superior to those presently available to Earth's astronomers and imagines what Mars's astronomers' observations might discover about terrestrial culture:

> And the two chief astronomers
> Controlled the huge machine,
> And first of all the universe
> They sought our orb terrene ...
>
> The elder first our orb discerned,
> And gazed; and by his mien,
> When mute with ecstasy he turned,
> They knew that he had seen.

'It speaks, it speaks!' he cried at last,
'That highly favoured sphere!
But is,' he said, 'the problem vast,
Or the solution, here?'

Then in exultant, anxious awe,
While all the throng were dumb,
He traced the characters he saw,
The sign 'Linoleum'.[123]

Kendall's poem, entitled 'The Fatal Advertisements' satirizes the extent to which the commodity has taken over Earth's visual spaces. The astronomers on Mars discover first an advert for Linoleum and next for 'Sapolio', a brand of soap.[124] The discovery of these two products creates a schism in Martian astronomy, with one group proclaiming Linoleum as a real characterization of Earth, and another Sapolio as the 'new solution'.[125] Kendall's poem, written in 1894, evokes commodity culture to parody the canal controversy; Linoleum, a brand of floor covering which was often patterned with linear or geometric shapes, stands in for and commodifies the canals while Sapolio offers an alternative interpretation that denies their existence. Kendall's poem concludes with these advertisements replicated on Mars:

But ere the whole discussion passed,
Each zealous faction wrote
Its watchword, graved in letters vast,
For other orbs to note.

And future spheres will sure behold,
When pondering on the stars,
Two signs imprinted on the cold,
Dejected sphere of Mars.[126]

By the poem's end Mars itself has fallen under the influence of Earth's commodity culture: the language of the marketplace becomes the language of astronomy. Clearly the poem is commenting on consumer capitalism's domination of visual culture but it is also making clear that any form of observation is susceptible to its influence. In fact, Kendall's poem provides an excellent example of how the commodification of astronomy and the commodification of vision are inextricably connected.

By excavating the language of the commodity in professional science, various forms of popular science, and in imaginative fictions, it becomes clear that scientific discourse, of whatever genre, is never isolated from the cultures within which it operates, nor from their acts of communication. Nor are these different forms of science separate from one another. As an observational science astronomy is defined by its visual culture and this culture permeates each of its modes of representation from the discourse of professionals to the creative writing of novelists and poets. In their employment of the trope of wonder, the prominence of the imagination and the influence of the language of the marketplace, the different forms of writing that contribute to the science of astronomy reveal their similarity. More than this, they illuminate the permeability of their boundaries, revealing their potential for entanglement with one another and with other cultural formations. It is through the visual sense that such border crossings are made possible. In discussing popular science scholarship, Andreas Daum advises the historian of science to remember that scientists are always members of multiple communities.[127] It is also worth remembering that members of multiple communities can also be scientists, and that their eyes, like scientists' eyes, can be used to look past the edges of scientific images, at the world.

PART III: PAST

5 LOOKING AS TOURISTS AND SCIENTISTS: AMELIA EDWARDS, FLINDERS PETRIE AND THE ARCHAEOLOGY OF THE EGYPT EXPLORATION FUND

Archaeology is often considered one of the most visual of the sciences to emerge in the nineteenth century. Its dependence on description and illustration, the use of photography as a tool to support its practice, its centrality in museums and exhibitions, and the employment of its discoveries in popular entertainments, as well as in pictorial art, attest to the many cultures of visuality with which it interpenetrated and inscribed mutual influence. Yet unlike other scientific disciplines its accumulation of knowledge did not rely upon technologies that extended visual capacity. In contrast to medicine and astronomy which needed microscopes to look at the very small and telescopes to look at the very large, archaeology could look deep into the past with the human eye alone. The most privileged of archaeologists was therefore the fieldworker, who could look with their own eyes at archaeological objects *in situ* as well as study them in the museum, university or other 'centre of calculation' as Bruno Latour has named these sites of knowledge presentation.[1] Yet the embodied vision of the archaeological fieldworker – embodied not only within the organs of perception but also in the cultural milieu of the archaeological site – was far from simple. The fieldworker was caught up in a series of overlapping ways of seeing that drew influence (often unwonted and unnoticed) from the social, political and personal relations with which the archaeological site was necessarily involved.

One of the most influential modes of observation was that of the tourist. It came to dominate late nineteenth-century British views of the East, and in Egypt (its most popular location) it coincided with the beginning of the key British archaeological investigations of the period, the work undertaken by the newly-formed Egypt Exploration Fund (EEF). Tourism and touristic observation was woven into the EEF from its foundation. The driver behind its existence was the travel writer and novelist Amelia Edwards, whose experiences as a tourist in Egypt, and her subsequent book *A Thousand Miles Up the Nile* (1876), fuelled her desire to do something to preserve Egyptian artefacts.[2] While the work of the EEF was wholly archaeological, and conducted in the field by professional, remunerated archaeologists, the ways of seeing both Egypt and Egyptian artefacts was often strikingly similar to touristic observations detailed in travel narratives.

Focusing on the travel writing of Edwards, her predecessors and successors, and the archaeology undertaken on behalf of the EEF by William Matthew Flinders Petrie, this chapter will explore the relationship between archaeological fieldwork and travel writing. What happens to archaeological knowledge in the 'contact zone' of Egypt as touristic location?[3] How do touristic ways of seeing come to impact upon scientific observation? What similar epistemologies of vision are found in the written narratives of archaeology and travel? Each of these questions reaches towards greater understanding of the role played by specific cultural activities in scientific investigation. In particular they suggest that the ways scientists observe and perceive are never wholly objective but rather bound, at least partly, to alternative ocular horizons that emerge from the dominant cultural practices that surround them.

The Egypt Exploration Fund and Tourism

Amelia Edwards's involvement in creating the EEF was only the beginning of its relationship with tourism. While Edwards's experiences as an (independent) traveller in Egypt were her inspiration, commercial tourism was to be of invaluable support to the EEF's activities from its first excavation in the Nile Delta in the first months of 1883.[4] The EEF relied heavily on the Egyptian operations of travel company Thomas Cook for communications, the management of money, and, inevitably, travel to and from archaeological sites. The transactions with Thomas Cook were extensive: in the first three months of 1889, for example, the EEF spent the significant sum of £200 on Cook's services, approximately the same amount as the cost of Cook's forty-day round trip from Cairo to Luxor on a large *dahabiyya* yacht, with a crew of ten, a cook and a guide.[5] Petrie records in his excavatory journals the extent of his involvement (as an EEF archaeologist) with Thomas Cook. He relied upon Cook's representative to find him suitable transport for his river trips, to negotiate with Arab suppliers for the necessary provisions of this voyage, to act as a sorting agent for Petrie's post to and from Britain, and to take on the role of a currency exchange for the EEF's supply of Egyptian money. From this fund Petrie would pay his workers, obtain licences, employ a specialist guide, purchase the domestic necessities of excavatory life and at times also buy Egyptian artefacts for distribution to British museums. Without Thomas Cook's services, the EEF could not have undertaken its research in the ways that it did. Indeed, Cook's operation in Egypt was a key factor in allowing professional archaeology to take place.

The relationship between Thomas Cook and the EEF was sometimes strained. It often took the EEF several months to clear its debts with Cook, especially if these debts had accumulated at the end of the excavatory season. In 1889 it took the EEF six months to repay a debt to Cook, despite a series of invoices of increasing impatience from Cook's accountant to the EEF's secretary, Herbert Gosselin.[6] However, Cook's services were sometimes as erratic as the EEF's reimbursements. Petrie complained bitterly in his diary and journal for January 1884 of Cook's failure to send him his correspondence.[7] However it was the role played by other tourists and travellers that most exercised Petrie's relationship with Cook. Excavatory sites were of particular attraction to the tourist in late nineteenth-century Egypt. Cook's relationship with Petrie (and other archaeologists) meant that Cook's agents were well placed to advise their tourists of the present sites of excavation or to organize trips for particular travellers to visit them. For the most part Petrie found the attention of tourists a distraction from his work. On occasion, though, they very directly interfered with it. In his journal for the excavation at Gizeh from November 1881, Petrie recalls the visit of three consecutive groups of guided tourists to the interior of the pyramid: 'Three parties of travellers came in, and for one some fires were burnt, which obliged me to beat a retreat from the sulphurous gas'.[8] If tourists were sometimes problematic for Petrie's archaeology, his own practices could endanger the earning potential of indigenous Arab tour guides. During the period of his excavation at Gizeh, a site often visited by Western tourists, Petrie was asked by his own specialist guide, Ali, not to enter the pyramids without the company of his Arab guides, as his explorations were 'setting too independent an example for travellers to behold'.[9]

The relationship between the EEF and travel extended beyond the active engagement with professional tourist companies or with specific tourist cultures in Egypt. One impetus behind the EEF's activities (and also behind its successful drive for subscriptions) was to investigate possible biblical locations in Egypt. In particular the EEF hoped that its archaeologists might shed further light on the locations specified in Exodus. The route of the Exodus therefore became a focus for archaeological investigations from the beginning of the EEF's activities in Egypt, placing the journey taken by the Israelites at the very heart of its excavations. Edouard Naville, the EEF's first archaeologist, recorded his involvement in discovering the route of the Exodus in a lecture delivered to the Victoria Institute in 1891. In this lecture he noted that the EEF's initial archaeological sites were chosen to 'throw much light on several points of the Exodus, especially on the first days of the journey'.[10] In as far as the EEF undertook to dig in Egypt, then, its interest was as much in the earliest example of a narrative of travel as it was in Egyptological research.

A further, if less vital, example of the connections between the EEF's work and larger cultural narratives of travel is in the associations that its archaeologists drew between their work in Egypt and the eighteenth-century grand tour. The grand tour was used as a point of comparison to highlight the relevance and interest to Western audiences of the art and culture of the ancient Egyptians that was emerging from the new knowledge that excavations were providing. Petrie, for example, used the grand tour as a conceit in a popular article on Egyptological research in 1888. He opened the article by asking his readers whether they had 'been up the Nile and "done" the ancient Egyptians', perhaps on a Cook's tour, before arguing that 'to walk around Thebes with open eyes is to make the grand tour of the ancient world'.[11] For the EEF's archaeologists, then, travel and tourism were valuable in placing their excavations within larger historical and cultural contexts that might give their work increased resonance with national audiences and potential supporters.

The relationship between organized tourism, travel and the EEF's activities was extensive and complex. Although it was predominantly economic, in the work undertaken by Thomas Cook on the EEF's behalf, it extended beyond that. The EEF's excavatory sites were continually under scrutiny from tourists and travellers; and looking sometimes became active intervention in archaeological activity. Similarly, the activities of the EEF were influential on the ways in which tourists responded to ancient Egyptian sites and to the indigenous Arabs involved in the tourist industry. Finally, the Arab workers employed by the EEF had opportunities to shape that work in ways that best suited their various activities as digger or as guide to Egypt's monuments. What this extensive range of connections between tourism and archaeological work should illuminate is that the tourist or traveller, the travel writer and the archaeologist are subject positions with a certain fluidity. When the traveller discovers and opens a tomb, as Edwards did at Abu Simbel, they take on part of the identity of the archaeologist.[12] Likewise, Petrie's stroll around Alexandria on his arrival there in December 1880 sees him temporarily inhabit the role of the Western tourist, as does his employment of an Arab guide to show him potential sites of interest at the pyramids of Gizeh.[13] It is the employment of different ways of seeing that gives potency to these altered subject positions: Edwards looks into a tomb from the observational position usually taken by the archaeologist while Petrie looks at the Alexandrian columns from the common perspective of the tourist. It is the eye that disciplines the subject. To what extent, then, do subject positions remain porous when Edwards is enacting her role as travel writer, or when Petrie is being archaeological? Are there ways of seeing in which the tourist and archaeologist combine or have influence over one another?

Touristic and Archaeological Ways of Seeing

Travel narratives construct a very specific set of conditions for seeing. While they purport to be a personal biography of travel actually undertaken, they are always constructed stories close to fiction. Indeed, as Nigel Leask has argued, travel writing often includes fictional elements.[14] The observations of the travel writer are, therefore, a blend of the seen and the imagined, and the ways of seeing they construct are often a combination of preferred viewing drawn from experience and a perceived ideal view drawn from the imagination. Across the nineteenth century there were numerous travel narratives focused on journeys through Egypt. Their efforts to understand Egypt and to communicate that knowledge to their readers involved the use of vision as a route to comprehension. How Egypt allowed them to look, and at what they could direct their looking, was a cornerstone of their narratives' realization of Egypt for a British readership. Edward Lane's celebrated travelogue of Egypt, written during a visit in 1825, is a good example of this.[15] For Lane, the most satisfactory view of Egyptian monuments is from a distance, so as to take in a scene in its completeness. At Abu Simbel he notes that 'from a distance, this temple has an imposing appearance; but it loses somewhat of its grandeur on a near approach'.[16] Indeed the closer the view, the less Lane feels comfortable. At the temple entrance Lane finds the perspective 'rather violent' from 'my position being too near the object', and he laments the fact that 'this I could not avoid'.[17] His most enthusiastic observations take place, however, in the temple's interior which initially obstruct his vision because their 'darkness prevented my seeing anything'.[18] To combat this Lane lights a series of fires. These give him 'a most magnificent sight ... altogether like a grand and solemn scene of enchantment'.[19] For Lane, then, the most natural view of Egypt is the complete vista which offers the observer both distance and the opportunity for the eye to capture something in its entirety. Equally satisfactory, however, is the enchanted view which constructs Egypt from the perspective of the Western observer and fulfils the function of making Egypt look as the British traveller (and reader) expects it to.

The superiority of the complete view is a feature of many other Egyptian travel narratives. The experienced traveller and travel writer W. H. Bartlett, in his 1862 book *The Nile Boat*, finds his first view of Alexandria 'unimposing' because only fragments of it 'rise above the dead level of the sea'.[20] However, as soon as he has gained some distance from the city he finds that 'the view [of] the entire area of splendid and populous Alexandria, is so complete, that it requires little stretch of the imagination to recall vividly the many illustrious actors on this memorable theatre'.[21] Bartlett adds to Lane's grandeur a connection between the expanded vista and the imagination, as though the extent of the ocular horizon is temporal as well as spatial. Alfred Butler, in his 1887 travel book *Court Life in Egypt*, confirmed Bartlett's understanding of the complete view. From the top of a pyramid, Butler notes that 'surveying the wide horizon' led one to 'thinking how grand the pyramid of Chefren looked ... [and] about the building and the builders of the pyramid'.[22] One further reason for the complete view's privileged status was the difficulty in finding it. Edwards notes in *A Thousand Miles Up the Nile* how the statues at Abu Simbel rejected the traveller's all-encompassing gaze:

> Stupendous as they are, nothing is more difficult than to see the colossi properly. Standing between the rock and the river, one is too near; stationed on the island opposite, one is too far off; while from the sand-slope only a side view is obtainable. Hence, for want of a fitting standpoint, many travellers have seen nothing but deformity in the most perfect face handed down to us by Egyptian art.[23]

For Edwards the problematics of seeing give rise to artificial 'deformity' of the objects under visual scrutiny. Only a complete view, she suggests, would represent true observation.

Lane's enchanted view is also replicated in later travel narratives. While Lane artificially constructed his own enchantment, other travellers discover enchantment already existing at specific sites. As with Lane, however, these writers encounter the enchanted view only during the observation of the interior of Egypt's ancient monuments. Sarah Haight, on a visit to Egypt in 1840, discussed a particularly violent version of the enchanted view during an expedition to the interior of a pyramid.[24] Using the same 'uncertain light' drawn from a series of fires as Lane had done, Haight sees in the remnants of mummy's scattered around her 'a more sepulchral and demoniacal appearance than anything I had ever seen before'.[25] Her 'excited feelings' and 'surcharged vision' lead to an imaginative transformation of the scene that almost 'overcame all [her] remaining courage'.[26] Less pleasing, and more Gothic a view than Lane's, Haight certainly describes the enchantment of observing ancient Egyptian remains. For Norma Lorimer, writing her travel narrative while partly attached to a Thomas Cook tour of Egypt in 1909, the enchanted view of Egyptian monuments is one of delight: 'There is nothing', says Lorimer, 'to spoil your first enchantment' on seeing the temple's interior.[27] Lorimer brings the enchanted view closer to the complete view than other travel writers, arguing that it was 'the *ensemble* of the temple' that provided the enchanted perspective, allowing her to 'call up the *entourage* of the temple, the atmosphere of the time of its greatness'.[28] Edwards's definition of the enchanted view combines Haight's Gothic with Lorimer's sense of the enchanting presence of the past. At the temple at Denderah Edwards describes 'hurrying along by the light of a few glaring candles, one cannot but feel oppressed by the strangeness ... the very air tastes as it had been imprisoned here for centuries'.[29]

Egyptian travel narratives clearly construct a consensus on types of observation that collectively describe touristic ways of seeing ancient monuments. Both the complete view, in its similarity to the panoramic perspective that Western tourists would have been familiar with from popular entertainments, and the enchanted view, which imagines a Western-inspired Orientalist Egypt that is ancient, alien and pleasurably dangerous, seem nationally and culturally ideological and certainly subjective. They are surely very different, then, from the observational practices that professional Egyptologists like Petrie would have employed.

Certainly one of the primary functions of looking archaeologically is the very opposite of the preferred complete view of the traveller or tourist. The ability to look at and identify fragments of ancient objects is essential to archaeology in late nineteenth-century Egypt.[30] Petrie's assistant, F. L. Griffiths, discussed this in his inaugural lecture on the occasion of becoming a Reader in Egyptology at Oxford University in 1901. 'The skilled explorer', Griffiths noted, 'gathers history from the very potsherds which lie scattered beneath his feet'.[31] To identify them, Griffiths continued, required the archaeologist 'to cultivate the habit of observation and of reasoning from his observations that he is taught the truth by appearances and not deceived by them'.[32] Petrie exemplifies the practice of this fragmented view in one of his most important publications, 'The Discovery of Naukratis', which he announced in a short article for the *Journal of Hellenic Studies* in 1885.[33] Naukratis was found, Petrie reveals, from 'a few fragments of limestone columns' on one side of the site of the ancient town, and 'shells and bones ... and the streaks of stone dust' on another.[34] As Edwards noted a few years later in an article on Naville's work at Bubastis for the *Century Magazine*, from one excavation trench 'fragment after fragment was dragged out'.[35] Each of these 'fragment[s] of history', this 'slight and scattered data' helped the archaeologist to see an 'outline of the rise and fall of the temple'.[36] For Egyptologists the fragmented view is not only more common than the complete view but vital to the advance of archaeological knowledge.

Also necessary to the Egyptologist's desire to look into the past was an ability to reconstruct, often via the scientific imagination, those parts of objects and sites that were no longer materially available for observation. From a set of columns and a single pylon, for example, the Egyptologist needed to be able to reconstruct an entire temple. This reconstructive view is a further example of an archaeological way of seeing not shared by tourists and travellers. In the excavation report on work undertaken at Naukratis in 1885, for example, Petrie detailed the imaginative vision required to reconstruct one section of the town's temple:

> On this floor, at each branching of the cross passages from the main passage, and at each entrance to a chamber, there was a stone doorway built in the passage, the backing of the stones being visible in each remaining entrance against the mud-brick walling. These doorways of stone would hardly have been inserted without an object, and we can scarcely doubt but that wooden doors were fitted in them.[37]

Neither the stone doorways nor the doors remain at the site, yet Petrie is able to draw conclusions from the fragments of stone backing that he can see in order to reconstruct what he cannot. The scientific imagination plays a key role here; Petrie deploys his knowledge to imagine a temple that fits with archaeological possibility. This is an archaeological way of seeing that cannot easily be matched by tourists or travellers. It depends on extensive expertise that enables the observer, in Daston's phrase, 'to hold fast to facts, to keep the inventive imagination in check'.[38]

In a letter to Kate Bradbury in 1890 Petrie gave a more explicit example of the reconstructive view and its relationship to the imagination.[39] Discussing his observation of children drinking from water taps in the streets of Cairo, he exclaimed 'I can hardly believe that a little thirsty Cairene can enjoy a suck out of the brass mouthpieces at street corners without seeing any refreshment, unless he has the power of imagination enough to realize the whole waterworks reservoir as supplying it'.[40] For Petrie, it is the invisible structures for the provision of water, hidden and invisible to the drinker, that are most fascinating. These can only be accessed by using a combination of knowledge and imagination to provide a reconstructed view of the immense reservoir from which the water flows.

While the tourist or traveller in nineteenth-century Egypt sought to see completely when surveying Egyptian monuments and to look with enchantment at ancient Egyptian interiors, the Egyptologist concentrated on their fragmentary views and utilized their imagination in reconstructive observation. These ways of seeing seem to re-inscribe traditional boundaries between science and creativity: the former founded on knowledge, training and previous expert observation, the latter on notions of culturally appropriate pictorial scenes and the frissons of unchecked emotion. Indeed, Edwards argued for this difference succinctly in the opening chapter of *A Thousand Miles Up the Nile*, when she notes that 'one of our first excursions was, of course, to the Pyramids, which lie within an hour and a half's easy drive from the hotel door ... But it must be understood that we did not go to *see* the Pyramids. We went only to look at them.'[41] Edwards's differentiation between looking and seeing appears to reflect the gap between touristic and archaeological ways of seeing. This is confirmed in her subsequent explanation of the two terms:

> Later on (having meanwhile been up the Nile and back, and gone through months of training), we came again, not only with due leisure, but also with some practical understanding of the manifold phases through which the arts and architecture of Egypt had passed ... Then, only, we can be said to have seen the Pyramids.[42]

Edwards draws a decided line between the traveller's non-specialist, leisured looking and the knowledgeable seeing of the expert. Yet she also raises the possibility of crossing the boundary between these types of vision, of moving from sightseeing to scientific seeing. Several writers of travel narratives believed that their experiences in Egypt allowed them to make this ocular progression. Like Edwards, Norma Lorimer saw her travel book on Egypt as a combination of tourist activities and more knowledgeable study. As she notes in her preface, 'it was only when I was half way down the Nile, on my return journey ... that I began to see the sense of things in Egypt'.[43] For that reason, therefore, she describes her book as 'based on the diary which I wrote day by day, passed through the sieve of the study I have done subsequently'.[44] Equally, the interchange of travel writing and Egyptology was not all in one direction. Archaeologists commonly contributed to the numerous travel guides to Egypt, and E. A. Wallis Budge, the keeper of Egyptian antiquities at the British Museum, even wrote Thomas Cook's own guidebook to the country from the 1880s.[45] Budge went on to write a travel narrative of his own journeys through Egypt on behalf of the British Museum, combining once again the work of archaeology with the genre of the travelogue.[46]

These examples reveal how closely allied Egyptological work was to tourism, travelling and travel writing. Despite clear epistemological differences in observational practice, the two were otherwise cohabitants of the same geographic and cultural spaces, and their participants occupied complementary intellectual space, too. Nevertheless, while individuals like Budge, Edwards and Lorimer might have moved between Egyptian archaeology and travel writing, their examples suggest that the fields of activity remain separate and static, even when so clearly associated with one another. However, there were many other ways of seeing where there was significant overlap and continuity. Indeed, as the remainder of this chapter will show, Egyptology and travel writing exerted considerable influence on each other. The extent of this influence reveals that archaeology in Egypt was manifestly changed by its association with tourism and travel writing, just as the genre of Egyptian travel writing was shaped by archaeology.

Fictionalizing Egypt

One of the most prominent ways of seeing Egypt that was common to both tourism and archaeology was to look upon it as an animated version of fictional texts. Petrie's wife, Hilda, who worked with him at his excavatory sites in Egypt from 1896, wrote in her diary for 8 December 1897 that 'I never thought Egypt could be so Egyptian, or everything so rich and strange; it is a delicious medley of Biblical and Arabian Nights pictures'.[47] The simile of Egypt as an animated version of the fantastical stories of the Arabian Nights remained with Petrie, and she returned to it in her diary later in the same month: 'Time passes so rapidly here [in Egypt], and I find I spend all my time in doing nothing but taking in fresh impressions, and going over the new ground again and again, to make sure it isn't only an Arabian-Night's story'.[48] Hilda Petrie's advancement of the reality of Egypt at the expense of fiction should be seen as her recognition of and desire to maintain objective standards in her perception of the place of archaeological investigation. Yet her phrasing ('it isn't *only* an Arabian-Night's story') suggests the continuation of a specific way of looking at Egypt that is always to some extent rooted in previous fictional encounters with its culture and character.

Petrie's journals, the most extensive record of his archaeological work for the EEF, also reveal the influence of a fictional imagination on his vision of Egyptian antiquities. Looking at the mounds of Tahpanes in 1886 he writes, in similar terms to his wife, that 'this place marks a strange sort of fantasia to live in', and when he later discovers a store of *ushabti* statuary in a newly-dug tomb he describes them as 'a sort of solidified phantasy'.[49] Petrie also attributes this fantastical, imagined vision of Egypt to those around him. In a popular 1885 article for the *English Illustrated Magazine* detailing the life of an archaeological fieldworker he notes how

> continually you see heaps of ruins of the towns of former days, which seem only to need you to put the spade in to turn up almost anything you can dream of. No wonder that a people living in such a country have their heads full of treasure and jinns.[50]

This is clearly a rather ironic transference; it is Petrie whose head is filled with an imagined Egypt drawn from the popular tales of the East that were so popular in the 1880s after the publication of new translations of *The Thousand and One Nights*, which had first been disseminated in English by the travel writer Edward Lane in the 1840s and 1850s.[51]

The porosity of the boundary between scientific work on Egyptian antiquities and a fictionalized, fantastical Egypt is certainly due to the influence of imaginative narratives of Arabian history and culture (which Petrie and his wife Hilda clearly knew well) upon the visual imagination of the archaeological fieldworker. It is apparent from both Hilda's diaries and Flinders Petrie's journals that it was the sights of Egypt (ancient and contemporary) that provoked them to draw comparisons with fiction. For Petrie, however, such interactions were also in part a product of the unique combination of visuality and textuality that defined his Egyptological research. A significant proportion of Petrie's work was the discovery and subsequent reading of inscribed objects; objects such as *papyri* or tablets that were both archaeological artefacts and texts. Petrie was therefore always looking at ancient Egypt as both an artefactual and textual construction, which led him to make perceptual links to other texts-as-artefacts of which *The Arabian Nights* is the most common example. He noted this phenomenon, briefly, but profoundly, in his journal for March 1884, when he wrote 'I live in the remains of past ages actually and metaphorically'.[52]

Indeed some of the textual artefacts that Petrie discovered were themselves fictional narratives. A series of these stories were translated into English and collected together in 1895, with Petrie as editor, under the title *Egyptian Tales Translated from the Papyri*.[53] These stories once again reflect the influence of existing fictions on archaeological knowledge. Petrie organizes the stories by dynastic succession, but also links several together under a single heading. Three tales, focalized through three different narrators, are, for example, collected as 'Tales of the Magicians'. In organizing the stories in this way Petrie follows the pattern of *The Arabian Nights* which likewise interweaves separate tales by providing a framing narrative around them. In his introduction to the *Egyptian Tales*, Petrie directly addresses the issue of the fictionalizing imagination. Making a case for including illustrations to accompany the translated tales, Petrie argues that while 'it may seem presumptuous to intermingle translations of notable documents with fanciful illustrations' the extent of knowledge about Egyptian fashion and custom gives 'ample reason to provide such material for the reader's imagination'.[54] The relationship between archaeological knowledge and fiction becomes increasingly complex here. For Petrie, the fanciful may reasonably be used when it is based on existing knowledge drawn from visible artefacts. This is similar to Percival Lowell's acceptance that imagining Mars could be useful if founded on rational knowledge. The fiction of *The Arabian Nights*, as a translation of an existing ancient manuscript, would fall into the same category. Yet, as Hilda Petrie recognized, fictions cannot easily be stabilized as archaeological artefacts. Therefore, to continue to contextualize Egyptian archaeological fieldwork within fictional depictions of the Egyptian past still leaves an epistemological space for the imagination to occupy. Flinders Petrie's Egyptian archaeology, then, offers knowledge that is more than the accumulating, cataloguing, dating and mapping of artefacts and locations. It becomes an archaeology bound into the specific cultural practices of fiction and whose circulating knowledge of ancient Egypt is equally the circulation of an imagined Egyptian past.

The EEF's honorary secretary, Edwards, recognized that excavatory work provoked the imagination. In her popular articles and lectures, aimed at bringing the EEF's activities to a wider audience, she specifically drew attention to the relationships between Egyptian archaeology and fiction. In a lecture given in Boston in November 1889, entitled 'The Explorer in Egypt' Edwards compared the excavation of Egypt to the reading of books:

> It may be said of some very old places, as of some very old books, that they are destined to be forever new. The nearer we approach them, the more remote they seem; the more we study them, the more we have yet to learn. Time augments rather than diminishes their everlasting novelty ... This is true of many ancient lands, but of no place is it so true as of Egypt.[55]

That such old books might be fictions similar to *The Arabian Nights* is made more likely by a later metaphor, where Edwards compares the consumption of excavation reports and hieroglyphic dictionaries by insects to the 'novels and romances' that 'young ladies devour'.[56] At the same time as delivering these lectures, Edwards was also writing her article for the *Century Magazine* on Naville's discoveries at Bubastis, which opened with a direct comparison of Egyptian archaeology and fiction:

> Such a story as the one told in the following lines is a very uncommon one. It rarely happens that the pen of a novelist is inspired by archaeological facts, and withal the pen of a gifted and favourite author turned aside from romance, though it be only for a while, because she has found the Nile more enchanting and its soil full of tales more strange than fiction.[57]

For Edwards, linking Egyptian archaeology to fiction is a method of promoting it to her Boston audience and the readership of the *Century Magazine*. It is a technique of popularization, similar to that employed by writers of Martian fictions, that draws on fiction's popular cache as provider of the unique, the surprising and the marvellous. That is, Edwards's decision to draw a comparison between Egyptian excavations and fiction is to enable her more easily to agitate her listener's or reader's imagination and thereby gain consensus for the EEF's work as interesting and important.

Travellers and travel writers also viewed Egypt as a fiction; almost always as one fiction in particular, *The Arabian Nights*. Florence Nightingale, who visited Egypt in 1849–50, writes in a letter of her first days in Alexandria:

> We came into an enormous square hall lined with marble ... at the four corners were smaller halls, of immense height, with marble basins in each the floors slippery with water, – the whole like an Arabian Night's description. And when we came out again into that enchanted garden, it was still like an Arabian night.[58]

Charles Bell wrote of Cairo in 1888 as Nightingale had done of Alexandria. He recommends a visit to the bazaar 'if you want to be transported to the scenes of the enchanting stories of the "Arabian Nights", rub your "Aladdin's lamp", and go inside what still remains of old and picturesque Cairo, and your dreams will be realized, and your desires fulfilled'.[59] Bell's references to *The Arabian Nights* are much more self-consciously satirical than Nightingale's, perhaps indicating that the comparison had grown into cliché by the later 1880s. Still, E. A. Wallis Budge, in the 1921 edition of his *Cook's Handbook for Egypt*, continued to draw comparisons with *The Arabian Nights* in a discussion of Egyptian music.[60]

The Arabian Nights was clearly an integral part of the traveller's descriptive equipment from the mid-century, when Lane's translation into English was first available, and when Nightingale used it unselfconsciously to embellish her letter to her sister. The image of Egypt as a fiction circulates, therefore, within various public and private forms of travel writing as well as through popular writing on Egyptology and in the records of Egyptian archaeological fieldwork. Petrie's fictionalizing of the archaeological object can, in this structuring of the circulation of knowledge, be seen as an example of a colonial science which, as Tim Fulford, Debbie Lee and Peter Kitson argue, both 'gave Europe power to act from a distance' and 'convinced Europe that it already knew the cultures it was exploring'.[61] Certainly to define a newly-discovered artefact as a material example of an already-known 'phantasy' of Egypt's past suggests that the colonizing of that object began not with its return to centres of calculation in Britain's museums, but with an assumed knowledge of Egypt previously circulated through its own narratives, which were themselves colonized by English translation and publication.

The example of Petrie's excavations complicates Latour's understanding of the modes of operation of centres of calculation. For Latour, knowledge is accumulated at centres of calculation existing at the imperial centre. New scientific objects are first made mobile in being taken to such centres, and then made stable within the knowledge economy of that centre, which of course includes numerous other objects with which it might be compared.[62] Those artefacts which Petrie defined as fantasies are, by contrast, categorized immediately at the site of the archaeological excavation and in the context not of other objects within an existing imperial scientific structure but of fictions of Egyptian history. It may be more productive, therefore, to regard Petrie's practice as one example of what recent history of science scholarship has called the European Colonial Science Complex. Londa Schiebinger, in defining that phrase, makes central to it a recognition that 'indigenous European sciences changed as much as did non-indigenous European sciences' in the contact zone of colonial interchange, and that there 'was far more reciprocity ... than conventional models recognize'.[63]

In this new interpretation Petrie's (and Edwards's) fictionalizing of ancient Egyptian artefacts is illuminated as a form of cultural interaction and inter-influence, in which imaginative texts and their cultural role is central. The complex circulation of Egyptian artefactual knowledge begins with a non-indigenous fiction, colonized by translation, consumed as romance and used in travel writing to redefine its original location by recourse to a now imperial imagination. Thereafter, the archaeological fieldworker, also a reader of these travel narratives, transports this fictional imagination from the Egyptian present into the Egyptian past, and onto its artefacts, which popular archaeological writing then repackages for consumption once again at the imperial centre. There is no centre of calculation for the archaeological artefact in this process other than the centripetal force of the fictional text. This force makes its mark most obviously on forms of observation: whether that is the looking of the traveller or tourist, or the seeing of the archaeologist. The effect of vision that is mediated by a fictional text is to animate the object of vision according to fictional principles: to see Cairo and Alexandria as *The Arabian Nights*, or to see a series of *ushabti* statues as materialized fantasies. In this kind of observational praxis cultural and archaeological knowledge is also always fictional knowledge.

Egypt Alive

While a fictionalized ocularity animated Egypt and its artefacts as already known texts, there was another way of seeing which also brought ancient Egypt to life. In travel narratives, tourist guides and excavation records inanimate objects and sites of scientific and historical interest were often imaginatively reanimated and considered as a form of living Egypt. Alongside this act of mental revivification, contemporary Egyptians were commonly drawn into historical relationships with artefacts and locations which led them to be seen by observers as part of a living Egyptian past. Both travellers and archaeologists therefore arrived at an understanding of Egypt which was clearly influenced by their *in situ* observations of its people, places and cultures, as well as by their interactions with each other.

Virginia Zimmerman, in her analysis of the meanings of excavation, argues that its key activity is the discovery of traces of the past, which are rendered legible for 'empirical study' and 'imaginative interpretation'.[64] Central to this is the fact that 'the past is made to live again, figuratively through narrative, and literally through the removal of its trace into the material spaces of the present'.[65] The written records of the archaeologists working for the EEF exemplify such resurrections of the past, although they additionally reveal how revivification was a common product of excavatory observation. A part of the reconstructive vision that was a necessary skill in working out the topography of ancient sites, revivification was an imaginative process that enabled the archaeologist to see into the past with apparently greater clarity. As their narratives of revivification reveal, however, the past was always encumbered by the present; by ongoing relationships with contemporary Egyptian workers, or by the ever-present problems that beset large-scale excavations.

Petrie's excavations at Tanis in the early months of 1884 offer a useful case study of the revivification of ancient Egypt and the various influences that bear upon it. Petrie had found the initial organization of his excavatory site at Tanis difficult. Thomas Cook had failed to find him Nile transportation and continued to be erratic in forwarding mail and Egyptian currency. His domestic arrangements were unsatisfactory; he camped in a tomb for some time and the house he then chose to occupy was unweatherly and partly derelict. He was made uneasy by the appearance of Egyptians whom he believed to be either thieves or spies for unscrupulous collectors. His workers, and even his foreman Ali (also a collector of antiquities), he found suspicious and especially prone to avoiding the required digging work.[66] Petrie's bringing to life of aspects of ancient Egypt are coloured by these distractions in the present. One episode of reviving those who lived at Tanis in the age of Rameses II begins with the excavation of several burial sites, from which a number of skeletal remains are taken out along with the more usual artefacts. Petrie finds space for these in a small tomb he is using for domestic purposes, but which he finds 'full of antiquities of Ali's'.[67] Nevertheless, he 'stuck [his] jars in there for lack of space elsewhere' but spends the evening looking at 'a skull of some old Egyptian' who 'gazes at me from behind them as I stand in my other two tombs'.[68] Soon it is Rameses II himself who is being imagined in action. As Petrie moves into new accommodation, he notices that 'only one room is built with mud-bricks, the rest are mainly of broken limestone, limestone probably brought here by Rameses II, pillaged from his buildings ... and carried off'.[69] Once at the excavation site itself, the animation of Rameses II continues: 'here as elsewhere', Petrie writes in his journal, 'Rammy has smashed up his predecessors work, and we should have to search for their inscriptions'.[70]

This process of revivifying the ancient Egyptians captures the anxieties of Petrie's excavatory experience in the present, as much as they offer a new clarity on the past. Under surveillance from a skull, which is both spy and defender of artefacts, Petrie then imagines the ancient past as a period of artefactual destruction and removal. Rameses II, who has become the familiar 'Rammy' in Petrie's imagining of his acts, undertakes to desecrate the very same objects that Petrie fears being stolen or lost. While the archaeological evidence supports his reading of the actions of Rameses II, and is confirmed by other excavatory sites, Petrie's bringing to life of the ancient King is equally influenced by the contemporary moment.[71] His interpretation of the history of the excavatory objects, their 'coming into being' as Daston would term it, is reached only by looking through the lens of his dig, mottled as that is by individual suspicion, institutional failure and architectural decay.[72]

Travel narratives offer their readers remarkably similar ways of seeing ancient Egypt as reanimated within the contemporary world of the observer. Like Petrie's comments in his archaeological journal, these are often the product of scopic (and auditory) anxieties. For several travel writers, imagining the ancient Egyptians brought back to life was a way of leaving behind the Egyptian present and its continually stressful sights, noises and activities. For Lorimer, imagining at Abu-Simbel the 'hawk-headed Harmachis step forth from his niche above the door to greet the god as he mounts the bank of golden sand' it was the 'stillness of the dawn' combined with the 'majesty of the building enfolding you like the waters of a silent sea' that inspired her reanimation of the ancient Egyptian statues.[73] Similarly, R. Talbot Kelly, on a painting tour of Egypt in around 1900, came to the conclusion that the Sphinx was no 'mere monument of stone' because of the 'eternal dignity and calm' it brought him after spending time in Cairo's 'distracting noise and insistent detail'.[74]

It is Florence Nightingale who offers the most vivid analysis of the contrast between the Egyptian present and its past. At Abu-Simbel's main temple, Nightingale comments on the large statues, which she calls 'imperishable genii' who have 'seen three thousand years pass over their heads, and heed them not' before proceeding to imagine the active lives of those inhabiting the temple 3,000 years previously.[75] It is, however, Nightingale's later lament on leaving Abu-Simbel that is so illuminating:

> We shall never enjoy another place like Ipsamboul [*sic*]; the absolute solitude of it – the absence of a present, of any of one's fellow-creatures who contrast the past with that horrible Egyptian present. You look abroad and see no tokens of habitation; the power of leaving the boat and running up to the temple at any hour of the day or night, without a whole escort at your heels; the silence and stillness and freedom of it were what we shall never have again. At Luxor the present was such as to annihilate all pleasure in anything; and at Derr, where we stopped on the 13th for an hour, the cries and crowd were so insupportable that we saw the temple as quickly as we could.[76]

Nightingale's reanimation of the ancient Egyptians is an effort to capture the calm solitude of the past at the expense of the present's assault on all elements of sensory perception. This is only achievable by being able to limit perception to the single sense of sight. Nightingale's vocabulary for reanimation is entirely visual: she *looks* at the ancient statues, she *sees* no other signs of life, the statues themselves '*have seen*' 3,000 years of history. In contrast, the present requires, indeed demands, the attention of the full sensorium: the sounds and tactility of the crowd 'annihilate' the pleasure that might be had in *seeing* a new ancient monument.

For the travel writer, as it was also for the archaeologist, the reanimation of Egyptian artefacts was bound up with anxieties about observation, and in particular the limitations on observation set by the constant impress of the present. Travellers like Nightingale appear able to negotiate the present by taking the rare opportunities for scopic imagining unencumbered by the other senses. For Petrie, however, the continual negotiation of present and past that was part of the daily life of the excavatory site meant that no such opportunities were available. His archaeological reanimation therefore differs from that of the travel writers in continuing to be under the influence of present-day conditions. This does not make Petrie's reanimation of the ancient Egyptians any more subjective than the imaginative fiction of the travel writer; indeed his attempts to root his observations in available archaeological evidence makes it less so. Nevertheless, it does highlight how greatly that vision was imbricated within the present conditions of excavation.

The weaving together of past and present at the site of excavation is further evidenced by the unconscious (or at least uncommunicated) reading of present day Egyptians as themselves artefacts of the ancient Egyptian world. Thinking of contemporary Egyptians as a living version of the past was not uncommon in travel narratives and is clearly linked to the fictionalizing of contemporary Egypt as an example of historical texts. Many writers commented, often rather derogatorily, on the similarities or differences between present Egyptian subjects and the historic races that populated Egypt in the ancient world. For example, W. H. Bartlett, speaking of Alexandrians, noted that 'in the character of the population, at least, there remains a strong resemblance to the ancient city of the Ptolemies'.[77] But other travel writers took this connection further by drawing a direct comparison between present Egyptians and the artefacts left by their ancient ancestors. Lorimer compared her dragoman guide, Mohammed, to the many examples of ancient pictorial art or sculpture she had been taken to view:

> How splendid Mohammed looked in the temple ... Perhaps it was that his massive form was crowned with a head that is not altogether human in expression. He suggested some animal-headed god; what animal I cannot say, but something very mythological and Egyptian.[78]

Such a comparison is not making connections to an ancestral past. Rather it reconstructs the contemporary Egyptian as a living artefact of ancient Egypt, categorized not racially, but as part of a catalogue of artefactual discoveries, alongside statuary and papyri.

The excavatory practices of archaeologists working for the EEF also placed contemporary Egyptians – generally those working as diggers – within a broad matrix of artefact observation. The excavatory sites, their antiquities and the Egyptian workers who laboured on them, were all placed under the same regime of disciplining surveillance. This had the effect of denying the Egyptian workers any human agency and making them passive recipients of the archaeological gaze. The worker therefore became comparable to the antiquities they worked upon; transformed from active subjectivity into the same object-position as the artefact itself. F. L. Griffiths, in his lecture on Egyptology, reveals this in action at one of Petrie's excavations. 'At the outset' of any excavation, Griffiths explains, 'the excavator must be here, there, and everywhere amongst the diggers ... Here, he will carefully disengage from its sandy matrix some precious antiquity ... there, he will swiftly plan the positions of the objects in a larger find'.[79] The same all-encompassing gaze also allows the excavator to 'instruct his men to follow up a promising lead' or 'move them to another site'.[80] The observational regime of the excavator makes no significant allowance for the difference between artefact and worker: the artefact must be catalogued in a 'systematic and complete way' that allows for understanding while at the same time 'workmen [must be] disciplined' according to a 'system' that will transform their ignorance into knowledge.[81] Petrie's own excavatory journals support Griffiths's description. He details, for instance, a system of surveillance that he constructed while excavating at Tanis in 1884. Using a telescope strung up in a doorway, Petrie is able both to stay at home to conduct examinations of artefacts and still survey his workers at the distant excavation site. He notes, rather joyfully, on 3 April that 'I examine them with the telescope from my house, when they suppose me safely away'. This simultaneous examination of both antiquity and Egyptian worker illuminates how easily they come to have similar status for the archaeologist, requiring similar observational regimes of organization and classification.

In popular archaeology, too, the worker and the antiquity are co-terminally constructed as artefactual object. Edwards's essay for the *Century Magazine*, on Edouard Naville's discoveries at Bubastis, spends very little time discussing the relationship between the excavator, worker and artefact, but does illustrate their relationship in a series of plates giving visual details of the Bubastis site (see figures 5.1 and 5.2). In Figure 5.1, entitled 'Legs and Throne of Second Hyksôs Statue', the statue and an Egyptian digger both appear, although only the first is referenced within the title. The aesthetic of the line drawing does little to differentiate between the stone blocks of the statue and the living flesh of the digger.[82] Indeed the digger is placed so carefully against the edge of the stone block that it is difficult to see any separation between them. Figure 5.2 is a drawing taken after a photograph, entitled 'Fellah Woman and Head of Second Hyksôs Statue'.[83] On this occasion the title recognizes the inclusion of one of the site's workers in the illustration. However, this is done precisely so that a comparison can be drawn between the face depicted on the second Hyksôs statue and the face of the Fellah woman. The woman's face is drawn to mirror that of the statue; unsmiling, with lidded eyes and heavy brows, both the woman and the statue are solemnly and solidly composed. Together, they exemplify the descriptive term 'statuesque'. These two illustrations of EEF activities, in the context of the treatment of contemporary Egyptian workers at Petrie's excavatory sites, confirm that the contemporary Egyptian is continually being defined as a living artefact of an ancient past: an antiquity brought to life.

Figure 5.1: Ilustration entitled, 'Legs and Throne of Second Hyksôs Statue', for Amelia Edwards's Egyptological work, making clear the link between contemporary Egyptians and their historic sites. Source: A. B. Edwards, 'Bubastis: An Historical Study', *Century Magazine*, 39:3 (1890), p. 333, Egypt Exploration Fund Archive, Egypt Exploration Society; reproduced with permission of the Egypt Exploration Society.

Figure 5.2: Drawing taken after a photograph, entitled 'Fellah Woman and Head of Second Hyksôs Statue', highlighting the similarity between the present and the past. Source: A. B. Edwards, 'Bubastis: An Historical Study', *Century Magazine*, 39:3 (1890), p. 333, Egypt Exploration Fund Archive, Egypt Exploration Society; reproduced with permission of the Egypt Exploration Society.

Illustrating Egypt

Other forms of textual illustration also confirm the interplay between archaeological and fictional texts. Indeed, illustrations of Egypt and its excavation sites play a broader role in the visual culture of archaeology and travel writing. One of the most important points of connection between travel narratives and archaeological documents is their similar use of description and illustration. While travel writers attempted to be as scientific as possible in their efforts to depict Egyptian excavation, archaeological reports commonly described their ongoing activities using the literary approaches of travel writers. This interweaving of different genres and forms reveals the porousness of the activities of archaeology and travel writing, and their susceptibility to, and sometimes desire for, the influence of other methods of knowledge production.

Nigel Leask argues that travel writing in the period from 1770 to 1840 'ranges across modern disciplinary boundaries' and was therefore likely to try to deal with forms of knowledge that would today be defined as science.[84] One key technique employed by travel writers at this time was what Leask calls 'writing to the moment', a form of woolly objectivity that gained its authority from the authenticity of immediate personal observation, or 'autoptic witness'.[85] Later in the nineteenth-century, covering the period under investigation here, travel writing recognized the need to 'establish a more objective epistemological authority' in order to respond to the dominance of science, but at the same time was driven by commercial tourism to focus instead on 'entertaining reflections'.[86] Egyptian travel narratives in the later nineteenth century and early twentieth century, at the height of the EEF's excavatory activities there, combine autoptic witnessing with a more rigorous objectivity drawn from the example of archaeological reports.

Edwards pioneered this interweaving of generic traditions in *A Thousand Miles Up the Nile*. In her preface to the first edition, delayed because of Edwards's desire to be up to date in her Egyptology, she acknowledges that 'in answer to the repeated inquiries of those who looked for the publication of this volume a year ago' she could only respond that 'the writer who seeks to be accurate, has frequently to go for his facts ... to the pages of scientific journals and the transactions of learned societies'.[87] It would have been surprising, then, for the reader to find what seemed to be a very traditional travel narrative in the subsequent early chapters. There, Edwards follows travel writing's traditional generic tropes of writing to the moment; describing scenes touristically; and detailing not only the landscape and character of Egypt but also the domestic details of travelling and coming into contact with contemporary Egyptians that Leask defined as 'entertaining reflections'. However, as *A Thousand Miles* proceeds it becomes increasingly objective, eschewing the immediate response in favour of a more studied approach to Egypt and its antiquities. In discussion of a tomb at Abu Simbel, for example, Edwards writes: 'I must now return to the Speos [a mortuary chapel], and, as accurately as I can, describe it, not only from my notes made on the spot, but by the light of such observations as I afterward made among structures of that same style and period'. Here, Edwards is setting aside the 'naïve empiricism' of autoptic witnessing in favour of a more expert and objective form of observation based in the archaeological practice of comparative analysis. The same mixture of methods is used also by Lorimer in *By the Waters of Egypt* (1909). As Lorimer states in her preface, her work is based on a 'diary which I wrote day by day passed through the sieve of study I have done subsequently'.[88]

Edwards's approach to travel writing via scientific objectivity becomes increasingly clear in her chapter on a discovery made by members of her travelling party near Abu Simbel. As Edwards becomes more involved in the processes of excavation, so does her writing become increasingly subject to the formal structures of archaeological narrative (see figures 5.3 and 5.4). In Figure 5.3 Edwards gives detailed drawings of excavated friezes, including extra detail (the *ankh*) when further explanation is required. In Figure 5.4 she goes further, providing a plan of the excavation site, details the minutiae of measurements and directions. As Figure 5.3 shows, Edwards also incorporates scholarly footnotes directing readers to further information in the scientific literature of Egyptologists.[89] In these examples, Edwards's travel writing becomes difficult to differentiate from the official excavation reports published by the EEF which include identical formal elements. Indeed her own narrative includes the same amount of (and sometimes greater) illustrative detail as Petrie's own excavation journals (see Figure 5.5) where specific detail is reproduced visually as though in close focus.

Edwards's shift from autoptic witnessing and entertaining reflections to archaeological objectivity was a form of peripatetic genre-jumping that would have been familiar to readers of travel guides to Egypt. These narratives often did the same; offering their readers a combination of tourist advice, entertaining commentaries on Egyptian life, and cultural-scientific essays on Egypt's antiquities. In [Thomas] *Cook's Handbook for Egypt*, for example, Budge included an extensive article on the role of hieroglyphs, mummies and artefacts in ancient Egyptian culture. Remaining closer to the generic traits of the archaeological journal than to the tourist guide, Budge included illustration, footnotes and quotations from scholarly sources (see Figure 5.6).[90] Here Budge's closely focused hieroglyph characters reflect Petrie's sketches as well as Edwards's illustrative friezes. Budge's guidebook and Edwards's travel narrative provide, in their entirety, a combination of scientific information and entertainment and in doing so should be considered as examples of popular archaeology. Although their primary genre (and their commercial orientation) is as works of travel writing their contents resist easy demarcation, and provide further evidence of the complex and close connections between travel narratives and archaeological writing.

Popular archaeology on Egypt, whose primary function was to entertain and educate its readers in the new Egyptology, also paid little heed to generic boundaries. One of its common characterizations of the archaeologist was as the intrepid traveller or explorer who happened to interrupt their journeys with a period of excavation. In a popular lecture given in Boston, for instance, Edwards described Petrie as a kind of messianic 'explorer' who 'guided half by experience, half by instinct ... decides on a spot and calls up his workmen'.[91] This version of Petrie is both heroic and stoic, dropping into tombs on 'a flimsy twist of palm fibre' and continuing to excavate 'for as many next mornings as need be till the end is reached'.[92] Later, the explorer becomes the traveller, but is still awarded the honour of discovery: when Petrie arrived at Naukratis, Edwards describes him as 'the first European traveller who had set foot in that secluded hamlet'.[93]

340 *DISCOVERIES AT ABOU SIMBEL.*

beloved of Amen, but as Rameses beloved of Hathor, of Thoth, of Horus—in short, of each God depicted in the series.

These Gods excepted, the frieze is almost identical in design with the frieze in the first hall of the great Temple.

WEST WALL.[1]

The west, or principal wall, facing the entrance, is divided into two large subjects, each containing two figures the size of life. In the division to the right, Rameses II worships Ra; in the division to the left, he worships Amen-Ra; thus following the order observed in the other two temples, where the subjects relating to Amen-Ra occupy the left half, and the subjects relating to Ra occupy the right half, of each structure. An upright ensign surmounted by an exquisitely drawn and coloured head of Horus Aroëris separates these two subjects.[2] In the subject to the right, Rameses, wearing the red and white pschent, presents an offering of two small aryballos vases without handles. The vases are painted blue, and are probably intended to represent lapis lazuli; a substance much prized by the ancient Egyptians, and known to them by the name of *khesbet*. The King's necklace, armlets, and bracelets are also blue. Ra sits enthroned, holding in one hand the "Ankh," or crux ansata, emblem of life, ☥ and in the

STANDARD OF HORUS AROËRIS.

[1] I write of these walls, for convenience, as N., S., E., and W., as one is so accustomed to regard the position of buildings parallel with the river; but the present monument, as it is turned slightly southward round the angle of the rock, really stands S.E. by E., instead of east and west like the large Temple.

[2] Horus Aroëris.—"Celui-ci, qui semble avoir été frère d'Osiris, porte une tête d'épervier coiffée du pschent. Il est presque complètement identifié avec le soleil dans la plupart des lieux où il était adoré, et il en est de même très souvent pour Horus, fils d'Isis."—*Notice Sommaire des Monuments du Louvre*, 1873. DE ROUGÉ. In the present instance, this God seems to have been identified with Ra.

Figure 5.3: Amelia Edwards's detailed drawings of excavated friezes, including extra detail. Such drawings illustrate her employment of formal techniques similar to professional archaeologists in her travel writing. Source: A. B. Edwards, *A Thousand Miles Up the Nile* (London: George Routledge & Sons, 1890), p. 340; reproduced from the author's personal copy.

338 *DISCOVERIES AT ABOU SIMBEL.*

own measurements, and in part borrowed from the ground-plan drawn out by the Painter—may be accepted as tolerably correct. But with regard to the pylon, I can only say with

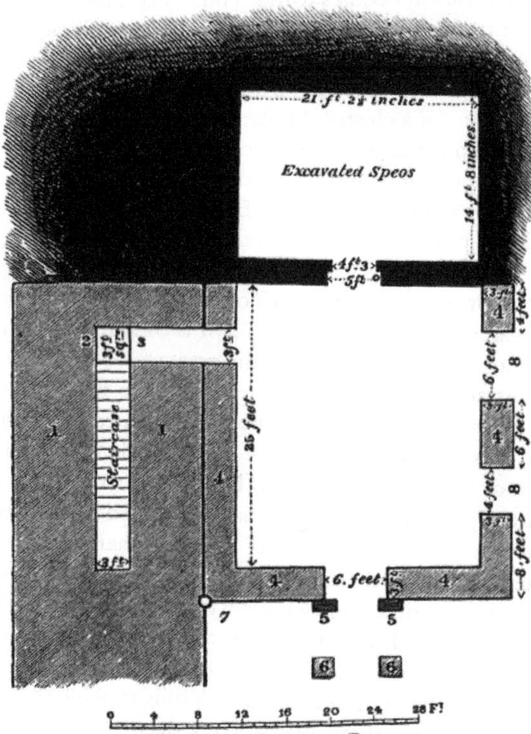

Scale ⅛ of an Inch to a Foot.

1. Wall of pylon.
2. Square landing.
3. Arched doorway and passage leading to vaulted hall.
4. Walls of outer hall or pronaos.
5. Door-jambs.
6. Stone hawks on pedestals.
7. Torus of pylon.
8. Arched entrances in north wall of pronaos.

certainty that the central staircase is three feet in width, and that the walls on each side of it are seven feet in thickness. So buried is it in débris and sand, that even to indicate where the building ends and the rubbish begins at the end next the Nile,

Figure 5.4: Amelia Edwards's plan of the excavation site, detailing the minutiae of measurements and directions. Such plans illustrate Edward's efforts at scientific accuracy extending even to the reproduction of detailed archaeological plans. Source: A. B. Edwards, *A Thousand Miles Up the Nile* (London: George Routledge & Sons, 1890), p. 338; reproduced from the author's personal copy.

1893–1897 91

the Triad. I hope we shall not find the statue, for it would weigh 10 tons or so.

Then Mr Wallis came over for a couple of days, and immediately we had visitors' luck, which always comes with friends. A complete stele of Ramesses III on a boulder of red granite, about 3 × 2 ½ft. very thick, weight about a ton. ... Then in a hole against the south wall of the temple area I saw a long block of limestone, lying deep down in some sand bed of foundations. I thought it was a piece of a clustered pillar. At last when it was cleared I found some carving and on turning it I saw it to be the legs of a statue of Khem (or Min as I should now write); and to my delight it was the primitive statue of the temple, perhaps the oldest statue in Egypt. This may see "a long pull and a strong pull and a pull all together" on the belief of my friends, but here are reasons. (1) It was the great statue of the temple at some time, and not a mere private object, as it must have been 15 or 16 feet high: the legs alone, from the fork to the ankles, are about 6 feet. (2) It is of soft limestone; whereas in any late age, or any historic age almost, the great statue would be of granite or hard stone. This seems to belong to the time when hard stone was not yet used, like the temple of Sneferu. (3) The work is extremely rough and plain, just a rounded mass, with a slight groove on each side  this section, and rude indication of knees. It could hardly be referred to any of the known ages of Egyptian art. (4) The attitude is not that sanctioned by regular use, no back pillar and one hand being at the side: differing therefore from all known figures of Khem, although there is enough of it to prove it to be of him. The hand at the side is very rude and elementary and pierced through to tie things on to the statue. (5) And on the surface on one side are several low relief carvings, most like the prehistoric rock figures of the white cliffs. ... The whole thing seems to have been a rude block only roughly shaped almost like a Greek island-figure and then occasionally decorated with stray carvings. From all these considerations I believe that I saw before me one of the primitive idols of Egypt, which may date from the days of the first settlement of the dynastic people. May we soon find the remainder of it.[1]

Next day we found the head of Khem; a round-topped lump, with big ears and whiskers but no face. Only a flattish smooth surface where face should be. All of it is very rudely fashioned on a shapeless lump; and I guess that finding no sufficient depth of stone in the top of the primitive monolith we carve a face, they flattened it and stuck one on, carved out of wood.

We found more of the Antef temple (I write awkwardly as I skinned my thumb tip with a stone). ... As a wide piece of that ground was in the hands of

[1] The head was not found. The figure is now in the Ashmolean Museum, Oxford.

Figure 5.5: Page from Flinders Petrie's field journal, showing the use of illustration as a means of providing visual education. Source: Egypt Exploration Fund Archive, Egypt Exploration Society; reproduced with permission of the Egypt Exploration Society.

678 SCARABS.

"I could stand. By the strength which was in my will "I laid the foundation [of things] in the form of the god "Shu and I created for them every attribute which they "have. I alone existed, for I had not as yet made Shu to "emanate from me, and I had not ejected the spittle which "became the god Tefnut; there existed none other to work "with me. By my own will I laid the foundations of all things, "and the evolutions of the things, and the evolutions which "took place from the evolutions of their births which took "place through the evolutions of their offspring, became "multiplied. My shadow was united with me, and I pro-"duced Shu and Tefnut from the emanations of my body, ". . . . . . thus from being one god I became three gods ". . . . . . I gathered together my members and wept over "them, and men and women sprang into existence from the "tears which fell from my eye."

Scarabs may be divided into three classes :—**1. Funereal scarabs; 2. Scarabs worn for ornament; 3. Historical scarabs.** Of **funereal scarabs** the greater number found measure from half an inch to two inches, and are made of steatite glazed green, or blue, or brown, granite, basalt, jasper, amethyst, lapis-lazuli, carnelian, and glass. The flat base of the scarabs was used by the Egyptians for engraving with name of gods, kings, priests, officials, private persons, monograms and devices. Scarabs were set in rings and worn on the fingers by the dead and living, and were wrapped up in the linen bandages with which the mummy was swathed, and placed over the heart. The best class of funereal scarabs was made of a fine, hard green basalt, which, when the instructions of the rubric concerning them in the *Book of the Dead* were carried out, was set in a gold border, and hung from the neck by a fine gold wire. Such scarabs are sometimes joined to a heart on which is inscribed the legend "life, stability, and well-being" .

Funereal scarabs were also set in pectorals, and were in this case ornamented with figures of the deceased adoring Osiris. Scarabs of all kinds were kept in stock by the Egyptian undertaker, and spaces were left blank in the inscriptions* to add the names of the persons for whom they were bought. **Scarabs worn for ornament** exist in many thousands. By an easy transition, the custom of placing scarabs on the

* The chapter usually inscribed upon these scarabs is No. 30 B.

Figure 5.6: Page from Budge's travel writing, showing the use of illustration to educate the amateur archaeologist/traveller. Source: E. A. W. Budge, *Cook's Handbook for Egypt and the Egyptian Sudan* (London: Thomas Cook, 1921), p. 678; reproduced from the author's personal copy.

Clearly it was advantageous in promoting archaeological work to construct the figure of the archaeologist as a glamorous adventurer, but to do so is to draw on the generic traits of travel writing. First, it employs a strategy of 'anti-conquest', identified first by Mary Pratt, in which tropes of heroism and adventure are used as 'strategies of representation whereby the European bourgeois subjects seek to secure their innocence in the same moment as they assert European hegemony'.[94] Second, it reflects the common trope in travel narratives of 'authorial "egotism"' in which representations of central characters (often the author, although not in this instance) are consciously overblown.[95] There is, then, a double shift from archaeology into travel writing here: the archaeologist is represented as the great traveller rather than the great scientist and the generic features of travel narratives are employed to do this work of representation.

Archaeological documents also slip from science writing into travel writing. Griffiths, for example, characterized the Egyptologist as an 'explorer' in his inaugural lecture in Oxford in 1901, even if it was 'systematic exploration' that was being undertaken.[96] Petrie takes this further. In his excavation journals he provides a series of linguistic illustrations of his excavatory work that are generically situated within travel writing. In January 1881, while excavating at Gizeh, Petrie wrote from inside a pyramid:

> I was lazy during the day, a hot sun and warm wind made it very delightful to sit in the cool bottom of the great hall, and look up the plain massive walls and shafts of pink granite to the deep blue sky shining above it all; there is neither ornament nor inscription, not even a moulding, simply beams and slabs of square cut granite ... The work is not very accurately done, but for effect it is more fascinating to me than anything else here, even the King's Chamber in Gt Δ; perhaps the Grand Gallery comes up to it for effect, but the cleanness of it all, and the contrast with the blue sky makes it charming.[97]

This vignette employs many more of the tropes of the travel narrative than it does of the archaeological report, for which Petrie's journals were an *aide memoir* and early draft.[98] The description privileges the position of the writer; the authorially egotistical self is located at the centre of activity. It also defines the writer as an autoptic witness writing in the moment of seeing. It responds emotionally to the environment by utilizing the traveller's gaze, which circumscribes the vista in an affective relationship with the writer. The end result is not knowledge but charm, a further form of affect. Shafts of archaeological insight do break through: the use of 'inscription' and 'moulding' hint at expert knowledge, and the comment on the lack of accuracy points towards prior experience of excavatory sites. Ultimately, however, it is the way of looking that clearly places this within the travel writing genre. Petrie observes touristically: his gaze is neither the fragmented nor the reconstructive view of the archaeologist but rather an effort to find the aesthetically 'charming' complete view of the traveller.

Ways of seeing are at the heart of genre-shifting. Petrie observes as though he were a travel writer, Griffiths and Edwards describe a vision of the archaeologist as explorer, and Budge and Edwards reproduce on the page a visual representation of the scientific article. Each of these are illustrations of Egypt that highlight the modes of visuality employed by the genres of archaeological and travel writing. They are also examples of the creative use of visual representations employed by those documenting Egypt and its antiquities – whether as archaeologists or travel writers.

Indeed, all of the examples in this chapter – Egypt illustrated, brought to life and fictionalized – show how archaeological writing, popular archaeology and travel narratives work in concert with one another to explore similar themes, access knowledge in comparable ways and reconstruct ancient Egypt from parallel perspectives. Travel writing has always been considered as a cultural and literary product, reflecting and refining historical, political and social understandings of place through rhetorical techniques developed in fictional narrative. Yet travel writing can also be scientific writing, as the above examples testify. Egyptian travel narratives, by Edwards or Lorimer, explicitly claim the status of a scientific document by emulating the practices of scientific writing. They do so, in part, because of an inherent understanding of the close proximity of travel writing to archaeological reports and, moreover, of the traveller or tourist to the archaeologist. Equally influenced by cultural ways of seeing, such as through the fiction of *The Arabian Nights*, travel writers recognized their epistemological similarities to archaeologists and were emboldened to articulate their own perspectives on antique artefacts. Archaeological writing, commonly held to represent the enactment of objectivity, is in fact a great deal richer than this. More credibly regarded as narrative, as the archaeological historian C. M. Hinsley has argued, Egyptological documents are an interweaving of the objective with the subjective.[99] Their representations of ancient Egypt pool scientific experience, instinct and expertise (defining features of objectivity) with travel narrative, historic fiction and imaginative links between past and present.[100] This does not diminish their status as works of science. Rather they are enhanced by the texture which cultural involvement gives to their writing. The alternative ocular horizons that travellers, tourists and particularly their narratives laid bare for the archaeologist gives their investigations of Egyptian antiquity a place within the world not accessible when only objective systems and practices are allowed to structure their knowledge and its dissemination.

6 EGYPTIAN ARCHAEOLOGY AND FICTION: THE ARTEFACT AS THING

On Christmas Day in 1892 Flinders Petrie wrote to Miss Bradbury, Amelia Edwards's closest friend and companion on her Egyptian travels, of his growing frustration at the excavatory practices of the Egypt Exploration Fund (EEF) archaeologist Edouard Naville, who, he claimed, was destroying important Egyptian antiquities.[1] A week later, in a letter to Edward Maunde Thompson, the director of the British Museum, he returned to the subject, concluding that he was concerned 'solely with the subject of the destruction of archaeological material, which is equally to be deprecated whether done in the name of science or by a plundering Arab'.[2] He proposed that the EEF send as a replacement a more capable man, Howard Carter, 'who understands what he sees' even if he had not undertaken the 'formal education for the best work in excavating'.[3] To be anxious about the loss of precious artefacts is hardly surprising, and it should be even less so in the context of the EEF's stated intention to save Egyptian antiquities from destruction. Yet Petrie's letters do still break certain boundaries of British propriety: of professional courtesy, certainly, but also of personal politeness, by explicitly regarding Naville as having fallen to the level of the Arab antiquities thief. Yet what is most interesting about Petrie's attack on Naville is his excessive anxiety about the loss of artefactual objects: the 'things' of Egyptological research. Indeed Petrie is so anxious about the loss of things that he suggests Naville be replaced by Carter, whose limited archaeological knowledge and expertise is disregarded in favour of his superior seeing.[4] Why is it that Petrie privileges the observation of things? What is it about these things that make them such fascinating objects of perception?

This chapter will explore the Egyptian artefact as a thing within the context of vision; specifically, how Egyptian artefacts are defined by the different ways they are seen. Both history and philosophy of science scholarship have become increasingly interested in scientific objects as things. This work has emerged from greater attention to science's material culture: its components, technologies and museums, for example. This scholarship has argued that objects reveal not only the matter but also the meaning of science. Such interpenetration between material existence and epistemological significance has led Lorraine Daston to argue that scientific objects are 'things that talk', and Ken Alder to regard them as 'thick things' which are assemblages of science, ethics, aesthetics and politics.[5] A further area of inquiry, particularly pertinent in discussions of the scientific museum, considers the role of the scientific object in human relationships.[6] In this scholarship the object is anthropomorphically charged, given life, and is scrutinized through its personal biography. While each of these approaches has been enormously valuable in extending an understanding of science and its role in society, very rarely do they allow for a particular focus on visual culture. Yet many of science's things are encountered first by the eye, and not by an objective eye but by an eye embodied in an observer who is seeing that thing at the same time as seeing many other things. To understand more of the observation of scientific things, then, is to give further nuance to how these objects are defined as things within the numerous other assemblages of the world which invite attention and influence thought and action.

It is important, therefore, to address the thing *in situ*; that is, to examine carefully the contexts within which Petrie and the other archaeologists working for the EEF brought Egyptian artefacts out of the ground. Yet it is also vital to understand the thing in other contexts. As Bruno Latour has argued, the things of science need to be considered in an 'extended number of assembling practices' which together give the thing its thickness, confirm its multi-stability and describe its relationships. The most significant networks for Egyptological things, as the previous chapter revealed, were a number of different narratives which gave antiquities definition within human society. These included both imaginative fiction and travel writing which attempted to understand artefacts in relation to the contemporary world, to draw into alignment the present with the ancient past, and ultimately to comment on the continuing life of the ancient artefact in late nineteenth- and early twentieth-century cultures. These textual products offer insight into the texture of the Egyptological artefact that are as equally valuable as the practices and writings of the archaeologist. Indeed, as cultural historian W. J. T. Mitchell argues, things always draw upon and mix multiple 'genres of discourse'.[7] It is only through interrogating as many of these discourses as possible that the 'elusive inflection' of things can be given 'specificity and salience'.[8]

Petrie's archaeological practice, published works, field notebooks and letters all exemplify his desire to understand the things he uncovered in the Egyptian soil. Through action and in narrative Petrie gives the Egyptian artefact definition. In an article on ancient tools written late in his career, for example, he concluded that a more complete knowledge of such things would become 'an important aid in tracing the growth and decay of civilizations, the natural history of man'.[9] Travel writers of the late nineteenth and early twentieth century also defined the scientific object as a thing offering access to a much broader history. In *A Thousand Miles Up the Nile* and other later works that promoted the work of the EEF, Amelia Edwards commented continually on the importance of understanding the nature of things. In a lecture given in 1890, for example, she argued that recently discovered Egyptian 'fragments' were 'not only a series of valuable finds, but an "object-lesson" of the highest interest' for cultural and historical study.[10] Imaginative fictions dealing with Egypt, such as H. Rider Haggard's *She* (1887) and Bram Stoker's *The Jewel of Seven Stars* (1903) were equally immersed in Egyptian archaeology, often via their authors' interest in the work of organizations like the EEF.[11] Indeed Rider Haggard was a keen amateur archaeologist and later a subscriber to the EEF, giving £2. 2*s* annually.[12] Central to both these fictions are Egyptian artefacts. The entire narrative of *She* is given impetus by the discovery of an ancient potsherd, called 'The Sherd of Amenartas', and Rider Haggard sets aside over a dozen pages to describe this single artefact, including two illustrations and numerous transcriptions of the writing that appears on it.[13] Stoker's novel, too, is concerned almost exclusively with the meaning of things: material artefacts with manifest lives of their own, 'enough things to evoke the curiosity of any man'.[14]

Together, each of these ways of looking at an Egyptian artefact contributes to its assemblage as a thing of science. Clearly, such a statement implicitly accepts that narratives outside the sphere of scientific archaeology might still be involved in the making of scientific knowledge. This does not mean that these narratives are scientific in any objective sense; they do not follow specifically coded procedures for procuring knowledge nor do they exhibit the kinds of expertise and experience necessary for the attainment of objectivity. What they do manage, as Ken Alder eloquently writes, is to represent the artefact 'in ways that at least partially and temporarily coordinate the diverse sets of human agents who design, make, and use them'.[15] At the same time, archaeological definitions of the Egyptian artefact are not free from challenges to their own objectivity. The Egyptologist, especially when working in the field upon an excavatory site, is not only an archaeologist but also a workforce manager, a traveller's guide, a cook, a reader, a political agent and a commercial actor. These other aspects of the *in situ* work of the Egyptologist have a decided influence upon their perception of the artefact and thereby upon their characterization of it. To consider the artefact as a thing of science across the disciplines of Egyptology, travel writing and fiction is not only to illuminate science in other cultural practices but also to shed light on its recurve: the cultural practices already within science.

Resistant Things

The alien nature of Egypt and its artefacts to the archaeologist, travel writer and novelist, problematized representation. What language should, or indeed could, be used to describe and define an Egyptian temple, or interior, or even a piece of statuary or hieroglyph-covered carving? The ontological deficit felt by those who looked at such things for the first time is captured perfectly by Edwards in her description of the temple at Karnak. 'To describe it, in the sense of building up a recognisable image by means of words, is impossible', wrote Edwards. 'The scale is too vast; the effect too tremendous; the sense of one's own dumbness and littleness, and incapacity, too complete and crushing. It is a place that strikes you into silence; that empties you, as it were, not only of words but of ideas.'[16] For Edwards, the very thingness of the Egyptian artefact breaks the connection between observation and perception. To perceive is to give thoughtful reflection and organization to the images captured on the retina. Yet this is exactly what Edwards finds impossible. Although Karnak can be given visual attention and made subject to the gaze of the traveller, this process of looking admits an anxiety about organizing vision into a form of representation that in turn occludes perception. Edwards was not alone among observers to feel the resistance of the artefact as thing. Norma Lorimer, whose travels in Egypt were begun in the early twentieth century said of Karnak that 'you are thankful for the simple and the understandable, thankful for the object which well tells its own story without the need of mental reconstruction'.[17] Lorimer grasps that what is at stake in the visual interaction between human and thing is the power to give the thing definition. Karnak's artefacts immediately resist categorization and definition, and the representational forms of travel writing appear to offer little to combat that resistance.

Archaeological observation suffered in similar ways. While it may seem logical to assume that the expert vocabulary of Egyptologists would provide a much more powerful representational force when confronted with the resisting artefact, there were many moments when archaeological ways of seeing were also resisted by things. Petrie, in his field notebook for 1889, spoke of looking at a collection of statuary in his excavation tent, without being able to find any language appropriate to categorize them: 'I gazed on the great stacks of ushabtis in my tent, as a sort of solidified phantasy'.[18] Petrie can find no reliable reality from which to give the statues definition, and they must therefore remain beyond representation as a form of the fantastic. Griffiths, who assisted Petrie's at several EEF excavations, warned in a lecture of the deceptive quality of artefacts, arguing that 'it is no new observation that appearances are deceitful', but that the archaeological observer should make every effort 'to cultivate the habit of observation and reasoning' so as not to be 'deceived by' things.[19] Here in archaeological discourse, both *in situ* at the site of excavations and in reflection afterwards, the problematics of defining resistant things emerges, even when, as Griffiths wrote, the Egyptologist understands how to 'interrogate their material, mode of manufacture, purpose and date'.[20] As Petrie and Griffiths make clear, it is in the visual relationship between observer and thing that the difficulty of representation first erupts. Just as Edwards and Lorimer attested to an occluded perception of things, so too do Griffiths and Petrie recognize that despite the tools of expertise that they might bring to bear on artefacts, the first moments of seeing often produce perceptual anxiety.

Stoker's *The Jewel of Seven Stars* both imagines and interrogates the ocular resistance that interaction with the Egyptian artefact illuminates. The novel is a Gothic tale of Egyptological research and antiquity-finding, with an experiment to revivify a mummy as its centre-piece. Stoker's narrator, the lawyer Malcolm Ross, finds himself confronting a collection of antiquities early in the novel while keeping vigil on an ill friend. 'The room and all in it gave ground for strange thoughts', writes Ross. He continues:

> There were so many strange relics that unconsciously one was taken back to strange lands and strange times ... More than once as I thought, the multitudinous presence of the dead and the past took such hold on me that I caught myself looking round fearfully as though some strange personality or influence was present.[21]

While the fictional genre allows Stoker some freedom to imagine artefacts as potentially alive, and certainly active, Ross's experiences repeat those of travel writers and archaeologists. The artefacts he observes give rise not to understanding but to strangeness (the word 'strange' is repeated as an incantation and reflects the 'phantasy' Petrie discusses), and rather than being defined by him they take possession of him, denying him the opportunity to exert his own perception. The narrator of Rider Haggard's novel *She*, Horace Holly, experiences the same failure of categorization as Ross. Encountering the artefacts of the fictional ancient city of Kôr, a reimagined version of Karnac, Holly, like Edwards and Lorimer, notes that 'it is quite impossible for me to describe its grim grandeur' and that 'it almost awed me by the intensity of its lonesome and most solemn greatness'.[22] Once again it is the human observer who finds himself under the influence of the artefact and without a representational vocabulary suitable to capture its qualities.

The Egyptian artefact's resistance to definition seems, in these examples, to be part of its undefinability. Yet the very fact that the artefact can be shown to offer resistance to the observer tends to give it meaning. Specifically, it is this quality of resistance to definition that places it in the category of a thing of science. Daston has made a similar point in another context, in a discussion of scientific observation based on the example of cloud patterns. For Daston, clouds which were by their very material structure 'too mutable to yield regularities' but which were nevertheless made subject to scientific scrutiny, are examples of the 'ways in which observation generates new scientific objects'.[23] Cultural theorist Bill Brown would argue that such objects are actually 'things', exemplifying their thingness (as opposed to their objectness) in the complex textures that emerge when they are observed.[24] The Egyptian artefact, therefore, becomes a new type of scientific object, a thing of science, at the very moment it begins to show the 'specific unspecificity that "things" denotes'.[25] Indeed the Egyptian artefact exemplifies what Brown calls the index of things: 'a certain limit or liminality, to hover over the threshold between the nameable and the unnameable, the figurable and the unfigurable, the identifiable and the unidentifiable'.[26] The Egyptian artefact, which may be imagined to be a particularly good example of a simple scientific object created by the science of archaeology, is rather better defined as a thing of science, brought back into human history by the archaeologist but resisting the 'thinning process' of archaeology's 'synthesising explanatory power' and instead reflecting the thick assemblage of meanings it gathers from the 'diverse set of human agents' who interact with it.[27] To uncover what meanings might lay alongside resistance and undefinability it is important to return to the thing *in situ* and consider how further observations begin to tease new definitions from beneath its opaque surface.

Domestic Things

Not dismayed by her initial inability to offer definition to the artefacts at Karnac, Edwards responded by developing a comparative approach that drew illustrative connections to familiar European objects. The Great Hall at Karnac, Edwards reiterates, is impossible to define, but 'it covers four times the area occupied by the Cathedral of Notre Dame in Paris' and might also be seen as comparable to 'St. Peter's' Basilica in Rome. While the Hall of Pillars at Karnac has majesty, 'the Colosseum covers more ground' and 'the Parthenon is more beautiful'.[28] To make such comparisons is an effort to educate the eye of the observer (or the mind's eye of the reader) by analogy. Yet what it achieves is a characterization of the Egyptian artefact as a thing that is similar to European things. Indeed the artefact is made homely by regarding it as one part of a familiar set of images that are already institutionalized in the common visual vernacular of European populations. Edwards was not idiosyncratic in identifying Egyptian antiquities as part of a European tradition. Budge, in his handbook on Egypt for Thomas Cook's travellers, wrote that 'it has been calculated that modern Paris could stand on [the] space' taken up by the great city of Thebes, before going on to describe the skins of mummified Egyptian bodies as similar to 'the skeletons of the monks preserved in the crypt beneath the Capuchin convent at Floriana in Malta'.[29] Mary Louise Pratt, in her work on explorers and travellers, would categorize such analogies as representative of the imperial eye of foreign writers.[30] It is true that such comparisons work to enclose observation within colonial practices, yet they also begin to define the artefact as a national domestic thing rather than as an unidentifiable alien object.

Stoker and Rider Haggard also align Egypt with national domestic sites through the process of analogy. Rider Haggard's narrator, Holly, illustrates the extent of the ruins of Kôr by noting that they 'stretch away for an almost immeasurable distance, like the gas lights of an empty London street', while one of Stoker's characters, bewildered by the range of antiquities in the private collection of Abel Trelawny, wonders 'whether I am in a private house or the British Museum'.[31] Both novels define the artefact by their similarity to familiar national spaces, and they do so to impose some form of categorization on the things they have seen. Stoker's alliance of the Egyptian thing with the British Museum is particularly interesting, for it seems self-referential. The artefacts are compared to other artefacts. There is, though, a key difference: the artefacts housed in the museum are already subject to the categorization of the museum's curators and are therefore objects whose meanings have been calculated according to the knowledge economy of the colonial centre and now remain static.

The interconnection between a colonial politics of observation and a national domestic mode of seeing remains apparent in archaeological fieldwork, too. Petrie's field notebooks, journals and diaries consistently switch between discussions of artefacts and domestic concerns, giving insight into the ongoing European character of his periods of excavation in Egypt: from church-going and letters written to friends and colleagues in Britain, to guiding visiting British tourists around archaeological sites. These moments of national domesticity interrupt and influence his work with newly discovered artefacts and play a part in his efforts to understand them. Often such influences are subtly portrayed in the character of his narratives. In a letter to his wife from a dig at Tahpanes, written before their marriage, Petrie writes:

> This place makes a strange sort of fantasia to live in; here on one side of my tent, is a group doing the der-wishes, all in the dark, howling and groaning ... and the mounds of Tahpanes on the other, seen in dark starlight, with the glimpse of a fire here and there, under the bushes.[32]

Although not obviously a discussion of Egyptian things, such written testimony reveals the continual interpenetration of the strangeness of Egypt with the European psyche. Here a Christian ethic underpins the description of the religious rites (the der-wishes) of his Arab workers, demonstrated by the brief allusion to the burning bush that reflects Petrie's own anxiety about doing the work of God. The scene that Petrie observes appears as a 'fantasia', almost impossible to categorize, yet he is able to give it some form of identity by overlaying it with a biblical narrative he knows intimately. The mounds of Tahpanes – the things Petrie wishes to understand – therefore become imbricated within a particular set of discourses and practices, Islamic and Christian (but also Egyptian and British), before he has had an opportunity to bring his archaeological expertise to bear upon them.

A more explicit example of the national domestic influence upon the observation of things is Petrie's discussion of the different ways of seeing ancient Egyptian art. The focus of his analysis is the lack of understanding of pictorial objects by his native workforce:

> As to sight, it is said that generally they cannot recognise a picture, but it may be European mannerisms of representation that confound them, as they decorate their own homes with pictures after their fashion, which are quite intelligible. But they are brought up (at Gizeh at least) in thorough familiarity with ancient Egyptian drawing, which is simply form without shadow or perspective to puzzle them, and yet if a figure is damaged, or fragmentary, they do not catch at all readily what it has been; and until the connection of parts is pointed out to them they are all at sea.[33]

This example, from Petrie's field journal for 1884, shows a sympathetic recognition of cultural difference while at the same time highlighting the primitivism of his native diggers. Although he is providing this illustrative example in order to speak about the observational paucity and archaeological ignorance of the native Egyptian it also reveals how European visual culture influences his analysis of Egyptian representational art. It is only by comparison with European art traditions (on the use of perspective, say) that he can look at Egyptian art as simple, or claim that it is easily analysed, even when in fragments.[34]

The national domestic vision of travellers, novelists and Egyptologists works to bring some form of definition to the previously unknowable Egyptian artefact. In giving the artefact this additional texture they also consolidate its existence as a thing of science. However, these textures are not always one and the same. While some analogies recognize the thing as internationally significant and historically and culturally valuable (as large as Paris or four times the size of Notre Dame) others reduce it to primitive icon or commonplace object (in simplistic forms or akin to London gas lamps).

The domestication of artefactual discoveries also worked on a much smaller scale than pan-national analogies. The everyday organization of domestic arrangements – eating, cleaning and cooking, for example – were so interwoven with the discovery and observation of antiquities (for both the traveller and the archaeologist) that any separation between the two activities often broke down. As they overlapped, so the artefact began to be defined in relationship to domestic objects and actions. Stoker's novel depends entirely upon the recognition of this relationship. Set for the most part in a London bedroom filled with Egyptian antiquities, the narrative continually explores the unique interface between domestic Britain and alien Egypt. Often the Egyptologist himself, Abel Trelawny, caught in the suspended animation of a coma, is used as a vehicle for Stoker to investigate the artefact as domestic object. Trelawny is 'as white as a marble monument', has 'all the pathos of a great ruin' and is 'wrapped in impenetrable sleep'.[35] By describing the Egyptologist as monument, ruin, and finally mummy, Stoker directs attention to the temporal history of Egyptological things, first living, then falling into ruin and finally dead. Implicit in this continuum is the thing's afterlife: archaeological efforts to bring it back into circulation as a thing of science (just as the aim of the narrator is to bring Trelawny back to life from his coma). Stoker imaginatively suggests that Egyptological excavation will always end with the thing being reconstructed in the image of the Egyptologist and thereby at least partly domesticated.

By contrast, Rider Haggard makes comedy from recognizing the close association of everyday domesticity and the alien nature of Egyptian artefacts. When his narrator first observes the supernatural figure of Queen Ayesha, beginning to emerge from behind a curtain, he asks 'Who could be behind it? – some naked savage queen, a languishing oriental beauty, or a nineteenth-century young lady, drinking afternoon tea?'[36] He extends this link between Egyptian things and everyday social practice later in the novel, when Holly finds himself in the darkness of one of Kôr's ancient tombs: 'I got out my box of Bryant and May's wax matches, and they struck as merrily, there, in that awful place, as they could have done in a London drawing-room'.[37] While these comparisons are intended to introduce an element of farce to the novel, the interpenetration of Egypt by banal events in the daily calendar of British domestic life manages to make the artefact on the one hand laughably ordinary and on the other increasingly unknowable and frightening. In these moments of description the thing is held in suspension between very different organizational categories, it becomes, in Mitchell's memorable phrase 'evanescent [and] multistable', ephemeral and yet also enduring.[38]

Both Stoker and Rider Haggard are concerned with the temporality of Egyptian things, a key element of what Brown calls their 'thingness'.[39] For Brown, thingness emerges in the consideration of 'the before and after of the object', when the object is 'temporalized'.[40] The recognition of the temporality of the Egyptian artefact was one of the key motivations for the EEF's excavations. As Petrie wrote in an undated letter to his wife early in his career with the EEF: 'Egypt is exhausting itself, the natives are ceaselessly digging, and unless we look to it pretty quickly the history of the country will have perished before our eyes'.[41] It was not only writers of fiction who examined the temporality of artefacts by linking them to contemporary domestic objects. Egyptologists, too, bound together the domestic items they saw around them everyday with the antiquities they dug from the earth. Petrie makes this connection in his journal for 1884 when he reminisces about the visit of a colleague, Professor Amos:

> Amos works a little at hieroglyphs; so after a meal, before washing up, out would come dictionaries, lists of kings etc. and we would be thick in the discussion of a dynasty when conscience would whisper 'that tapioca pan will be hardening if you do not wash it soon.' So four or five days passed in a fascinating jumble of history, antiquities [and] cooking.[42]

The jumble of antiquities with everyday domestic chores gave such artefacts additional thickness. This was promoted by the fragmentation of the archaeologist's observational arena where artefacts and domestic items were thrown together in the same spaces so that visual attention was continually shifting between them. The setting up of excavatory sites was often haphazard. It was driven mainly by the availability of resources and funds rather than by any specific scientific need. Petrie and other EEF archaeologists found themselves housed in tents or tombs, the remains of ancient dwellings or ramshackle modern buildings. Once beyond the reach of Thomas Cook's tours, they were beholden to local amenities and lines of trade for their provisions and travel. Finding space for the study and storage of artefacts was done responsively, and sharing space with domestic items was common. For Petrie this often meant stacking his new discoveries in the same tomb as he cooked, washed and slept, so that 'a skull of some old Egyptian' was always staring at him 'from behind [my jars]'.[43]

A visual dialectic between archaeologist and artefact/domestic object was therefore continually in play throughout the excavatory season. Petrie writes specifically of how his vision oscillates between, and ultimately combines, these different things: 'I have to sit watching the pot boiling for breakfast while I mark skeletons, and divide my attention between vertebrae and eggs, ribs and coffee'.[44] Although Petrie suggests a division of observation, his language betrays the combination of his domestic stores with Egyptian artefacts; and the artefact accrues further definition as a thing *of* domesticity, an item of produce akin to coffee and eggs which may be given value and a place in the market. In fact, as Petrie's notebooks reveal, the Egyptian thing was indeed defined in such a way, set alongside domestic purchases in the numerous shopping lists that the notebooks include: 'meat 8½, sugar 3¾, coffee 4, basalt column 20'.[45]

The definition of Egyptian things as domestic objects became so very thoroughly ingrained in the excavatory experience that there were even examples of reversal, when domestic objects were mistaken for artefacts. In a moment of conscious self-parody, Petrie tells the story of one such mistake with the fluency and comic timing of an experienced writer of imaginative narratives:

> Yesterday I was much puzzled at seeing something in the plain below the house, as I stood in the temple. Yes, it was certainly the capital of a column, yet I never saw it there before; but it can't be anything else; it's polished, for I see a glint from it in the sunshine; but how is it that I have overlooked it? It is certainly very strange. I know there are some blocks further on, but I never observed this before. I must go and see what it is. My bath; bowled down the hill some two hundred feet by the wind and standing upside down.[46]

Observational anxiety pervades this short narrative. It is anxiety produced by the failure of the excavator's vision to identify and categorize the artefact. But it is also a retrospective narrative that denotes the failure to distinguish between the domestic object and the artefactual thing. This is not to suggest that Petrie or his EEF colleagues found it impossible to separate the thing from its domestic cohabitants. Rather, such examples highlight how the Egyptian artefact gathers to itself an excess of meanings, and that the archaeologist is one of those knowledge-makers who contributes to this excess. Brown argues that the thing is always 'what is excessive in objects'; that things 'exceed their mere materialization as objects or their mere utilization as object' and become instead 'a metaphysical presence' of 'values, fetishes, idols and totems'.[47] The Egyptian artefact becomes such a thing through its participation in the excavatory process. It is therefore a thing of science, and a thing *created by* science, even when aspects of its thingness seem to be outside the realm of scientific knowledge.

Gothic Things

As well as becoming a thing through its unidentifiability and domestication, the artefact also gains further texture by being identified as Gothic. Alex Warwick has argued that excavated artefacts accrue significance as Gothic objects primarily in fictional narratives.[48] While it is certainly true that fiction plays an important role in the signification of the artefact as Gothic, this process of making the artefact a Gothic thing begins while the thing is still *in situ* at the excavatory site. Indeed travellers and Egyptologists who looked at artefacts while in Egypt very readily constructed what they saw in Gothic terms, most especially by using a descriptive vocabulary common to the genre of Gothic writing. When Gothic writing emerged as a genre of discourse in the late eighteenth century it did so because those who consciously constructed their work as Gothic drew on common linguistic structures and similar intellectual epistemologies to describe the world. At the heart of the genre was an interest in describing the effects of the philosophical construct of the sublime, a recognition of the thin dividing line between the known and the unknown (which Freud later called the uncanny), and a consistent effort to explore the relationship between life and death.[49] Many Gothic texts therefore considered the ontology of landscape and its influence on the individual, as well as exploring those places considered as borderlands between life and death, such as tombs or sepulchres.

Travellers to Egypt, who thereafter wrote about their experiences, often constructed Egyptian artefacts using these principles of Gothic writing. The experienced travel writer, W. H. Bartlett, for example, wrote in *The Nile Boat* of his first encounter with the pyramids near Gizeh. 'After passing the fork of the Delta ... we first caught sight of the mighty pyramids', writes Bartlett, and 'how familiar and yet how strange they appeared – hovering afar in the dusky grandeur upon the edge of the yellow Libyan desert'.[50] Bartlett's assertion of the grand artefact as simultaneously familiar and strange is a precise recognition of their status as uncanny objects and therefore also an assertion of their Gothic nature. Stoker's *The Jewel of Seven Stars* similarly privileges the uncanny. The novel stresses its attachment to the Gothic genre in the opening sentence, where the narrator, Ross, tells the reader that his subsequent encounters with Egyptian things 'all seemed so real that I could hardly imagine that it had ever occurred before; and yet each episode came, not as a fresh step in the logic of things, but as something expected'.[51]

For other travel writers the artefact was linked to the larger landscapes of Egypt, which were themselves sites of Gothic unease. Budge, for example, defined artefacts within the context of the Egyptian desert as a sublime space. On first seeing the desert in 1886 Budge believed himself to be entering 'a new world' which 'filled me with a certain fear' because 'everything was strange, everything wonderful to me'.[52] This commingling of beauty and fear persists in Budge's remembrance of the Egyptian landscape and 'returned whenever I have looked upon the desert'.[53] Budge's description of the desert as both wonderful and dreadful specifically constructs it as a landscape of the Gothic sublime, continually productive of terror whilst also recognizable as exquisite. As Budge intimates, this oscillation between two opposing poles ultimately means that the landscape is indefinable and therefore unknowable, and such an epistemological *aporia* is the space of the Gothic. Norma Lorimer, another traveller to Egypt in the early twentieth century, was able to define this Gothic mode more succinctly in narrating her visit to the temple at Denderah. 'From the moment that you pass under the outer pylon you are embraced by Mystery', writes Lorimer.[54] 'Mystery walks hand in hand with you as you go from pylon to dromos; from dromos to hypostyle hall, from hypostyle hall to sanctuary; Mystery stands with you while you gaze upon the shrine'.[55] The capitalization of 'Mystery' and Lorimer's anthropomorphism of it into an active yet ghostly travel guide captures the powerful influence of the Gothic sublime felt by those observing Egypt's artefacts. Rider Haggard's *She* places the Gothic sublime at the very centre of the narrative, and also depicts this at work within a living artefact, the immortal Queen Ayesha. The narrator, Holly, feels 'inward terrors' upon seeing her 'swathed mummy-like form', but also finds her 'instinct with beauty in every part'.[56]

Just as the extensive landscapes of Egypt and the larger of its artefacts are depicted as Gothic territories, so too are the more intimate spaces of excavatory sites. Amelia Edwards, in fact, draws an explicit connection between Egyptian landscapes as expansive Gothic panoramas and the smaller interior spaces of archaeological investigation, and in doing so shifts attention from the uncanny and sublime to the relationship between the living and the dead. In her lecture on the explorer in Egypt, based almost exclusively on her work with the EEF, Edwards writes: 'It has been said that all Egypt is but the façade of an immense sepulchre'.[57] This, she argues, is 'literally true; for the terraced cliffs that hem in the Nile to east and west, and the rocky bed of the desert beneath our feet, are everywhere honey-combed with tombs'.[58] Later in the same lecture Edwards returns to this image of the Egyptian landscape as a holding ground for the dead:

> If you but stamp your foot upon the sands, you know that it probably awakens an echo in some dark vault or corridor, untrodden of man for three or four thousand years. The mummified generations are everywhere – in the bowels of the mountains, in the faces of the cliffs, in the rock-cut labyrinths which underlie the surface of the desert.[59]

This is an exemplary Gothic narrative, weaving together the living traveller, the potentially awakened mummy and the labyrinthine spaces of the dead that are only just hidden from view. It characterizes the Egyptian artefact – the mummy in this instance – as a particularly Gothic thing by regarding it as part of an encounter between the living and the dead. It becomes, therefore, an intermediary between two poles, replicating Gothic writing's consistently operating binaries of known and unknown, terror and beauty.

Although encounters with artefacts of the dead were very much a common experience for practising Egyptologists, their narrative representation of these artefacts still reveal the characteristics of the genre of the Gothic. Petrie's extensive experience of excavating in Egypt might well have led him to regard tombs, sepulchres and mummies as the banal and ordinary objects of archaeological investigation. Yet in his depiction of such objects he continually uses both a Gothic register and generic tropes of the Gothic genre to represent his encounter with them. In his field journal for 1888, for example, Petrie describes his first exploration of the tomb of Amenenhat III: 'The whole of the walls are pitch black, owing to some deposit or growth when the water has filled the chamber. So it is very dark, and the candle only just shows you when you collide with floating coffins or some skulls that go bobbing about.'[60] The artefacts that are about to become objects of archaeological study are here represented as part of a Gothic scene. The dark and oppressive tomb can only be glimpsed in fragments; perception and therefore understanding are balanced uncannily between the known and unknown. Petrie characterizes himself as the nervous explorer, unnerved by his inability to anticipate the encounters with coffins and skulls, and yet in the next moment discovering an amulet which he describes rapturously in the phrase 'what gorgeousness!'[61] The skulls, too, have an active existence of their own, they 'go bobbing about' uncontrollably as Petrie's movements disturb the water in which they float. Linguistically this short narrative of the discovery of artefacts reveals a provocatively fecund and fluid space of natural growth and organic matter, a space of life which is protecting the objects of death. In only a few words and phrases Petrie represents the things he uncovers as particularly Gothic: uncanny, sublime and actively existing on the boundary between life and death.

In all of these encounters with things, the Gothic both emerges from observation and simultaneously seeks to construct that observation in particular ways. This is, in part, because Egyptological artefacts so prominently display themselves as Gothic objects. They are indefinable, hitherto unseen, objects of the dead that immediately offer themselves as subjects for Gothic analysis. However, it is also because the Gothic mode was founded on a particular epistemology of vision closely related to scientific observation. The Gothic arose in the late eighteenth century as a coda to Enlightenment empiricism and its championing of objectivity emerging from keen and rational observation. The Gothic asked questions of empiricism's objective vision, offering a series of alternative ways of seeing that instead suggested potential weaknesses in observation. As Andrew Smith argues, the Gothic explored 'the role that the apparently irrational could play in critiquing quasi-rationalistic accounts of experience'.[62] Gothic observation was characterized by occluded vision and an ever-present optic unease about the truths apparently made plain by the eye. It gave attention to things only partially seen or not seen at all; it dealt also with the effects of secrecy, of surveillance, and of the power of the gaze to disturb objectivity. The Gothic therefore created a subjective Gothic vision which concentrated on transgressive vision, vision's ambivalence and the relation between vision and the imagination.

In Egyptological excavation the very act of discovery often placed the archaeologist in a position of only having a Gothic vision available to them. The partially uncovered artefact, not yet identifiable, nor seen as a whole, or the partly cleared space of the tomb or burial chamber, accessible only from a small crack in the desert floor and still predominantly in darkness, all invited the archaeologist to look with a Gothic sensibility. Petrie highlights this common occurrence in his journal when discussing the discovery of a new store of mummy cases at Illahun in 1889:

> One case I found apparently almost crushed under fallen pieces of rock, but as *on peeping* in at a broken bit of the lid – *I saw* the inner carved case with a painted face in good state, we cleared it carefully. On getting the lid boards off the inner coffin was found to be much rotted, so I began to break away the lower part so as to remove the head and shoulders alone. *I soon saw* some beads showing at the feet, so as I wished to note the pattern of the network for re-threading it I cleared it delicately. Next I found on the legs a band of hieroglyphs executed in beadwork, and *then spied* the corner of a pectoral showing.[63] [my emphasis]

While this is clearly a Gothic scene in its depiction of the archaeologist as scrabbling around in a space of the dead, deciding upon the correct way to break a corpse into pieces, it is the language of vision that is most interesting. Petrie describes his vision as entirely splintered: he sees a painted face, then beads at the feet and on the legs, and finally he sees the corner of a further artefact on the corpse's chest. His language and actions become increasingly Gothic as he considers this fragmented body. He begins with a broken coffin and then finds much that is 'rotten' before beginning to dismember the remains.

Nevertheless it is the language of observation that is most striking. Petrie's ways of looking at this artefact – his peeping, seeing and spying – are very much the product of a Gothic occlusion of objective vision where the mummy remains largely unknown and as yet undefined and uncategorized. Moreover, the vocabulary of peeping and spying with which Petrie's connotes his role as observing excavator suggests other Gothic ways of seeing; the discovery of things hidden through a covert surveillance which makes them subject to a penetrating gaze. That is, Petrie does not simply describe the mummy as a Gothic object, he imposes Gothic categories upon it by employing Gothic ways of seeing when first looking at it. The scientific thing (here, the mummy) is not defined solely by its own inhabitation of a particular set of characteristics, nor even by being defined in the context of such characteristics, but additionally by the characteristics of the excavator himself, which influence and inform his perception. To put this more schematically, the mummy becomes Gothic because the excavator already is. Theorists of artefacts and things have come closest to making this argument in their recognition of the importance of the relationships between humans and things. For Chris Gosden and Yvonne Marshall, for example, the archaeological object should be defined through a 'new focus' in cultural archaeology on 'the way human and object histories inform each other'.[64] Brown takes this innovative new practice into his theory of things, arguing that 'the story of objects asserting themselves as things, then, is really the story of a changed relation to the human subject and thus the story of how the thing really names less an object than a particular subject-object relation'.[65]

Public engagement with Egyptology in the late nineteenth and early twentieth centuries reveals a clear understanding of the important role of the subject (the excavating archaeologist) in its relationship to the object (the Egyptian antiquity). Very often archaeologists were challenged on their excavatory practice. In particular the discovery and removal of mummies from their tombs and burial grounds was open to very different interpretation. Budge discusses this at length in his popular science book and travel guide of 1920, *By Nile and Tigris*. 'My endeavour to make the British Museum collection of mummies as representative and as complete as possible has brought upon me much criticism', writes Budge. 'I have been called "sacrilegious", "inhuman", "brutal", "wicked" and "diabolical"; and the epithets "goul" and "body snatcher" have been frequently applied to me'.[66] What Budge describes here is the transfer of the Gothic tropes often associated with the artefact onto the archaeologist. Budge is, in fact, made to represent the accumulated characteristics of Gothic villainy: he is religiously perverse, evil, demonic, transgressive and barbaric.[67] For certain public groups, then, it was a Gothic Egyptology that defined the things of ancient Egypt rather than the inherent Gothic nature of those things. Both at a distance from the site of excavation, as here with Budge, or *in situ*, as exemplified by Petrie's Gothic gaze, the archaeologist and his excavatory practice was as influential on the definition of antiquities as were those characteristics they embodied within their own material existence.

Whether in Egypt or in Britain it is the nature of observation that dominates the construction of the Egyptologist as a Gothic figure. Budge continues his discussion of public opinion on the work of the Egyptologist by giving details of a series of letters written to him at the British Museum. In one, his correspondent finds it upsetting to know that the Egyptologist is breaking open tombs, dragging out the bodies of mummies and then bringing 'them to England to become gazing-stocks for irreverent crowds in the British Museum'. The only thing worse than this, he continues, is that 'you strip the dead of their wrappings ... and then you leave them naked'.[68] Another letter-writer notes that by 'placing the body of the Neolithic Egyptian on a board in a case in the First Egyptian Room' it would simply 'be stared at by a gaping mob'. This, he continues, is 'degrading alike to the living and the dead'.[69] What both these letter-writers find problematic is the degrading nature of the archaeological gaze, which extends from the initial Gothic observations and actions of the Egyptologist to the viewing public in the museum. The writers implicitly argue that it is in the Gothic nature of the first visual encounter with the artefact that definition takes hold. The artefact then carries those ways of seeing with it as it makes its journey from the excavatory site to the museum cabinet. The original Gothic observations are therefore transferred to, and made manifest in and on, the thing itself. These public comments therefore confirm that the epistemological significance of the Egyptian artefact is tied very closely to the mode of perception of the archaeologist who first sees it. As Johannes Grave has argued, what leads things of science to accrue meaning is 'a specific kind of encounter' between the thing and 'a gaze that looks ... in a new way'.[70] Egyptian things, even when taken out of their context, continue to be Gothic because their meaning has been partly crystallized in the very first moment of their ocular interaction with the scientist.

The observer who encountered Egyptian things was also susceptible to the Gothic mode because, in its many different cultural forms, it gave a free rein to the imagination. In literature, art and music, the Gothic of the late eighteenth and nineteenth centuries promoted the use of the imagination as a method for explaining 'the complexity of human experience'.[71] The Gothic imagination was therefore easily called upon to offer a set of defining principles whenever the viewer was unsettled by a scene or object that was unfamiliar or unknowable. Sarah Haight, on her visit to Egypt in the mid-nineteenth century, gave a clear picture of the Gothic imagination at work upon the things she saw yet found difficult to understand. Her narrative begins in the middle of a guided trip to locations of scientific interest. At one location Haight looks out from her tent at a group of native Egyptians who have set up a camp fire near a collection of mummies:

> While gazing at these hideous creatures, my imagination transformed the hooded females who flitted by the blaze into Hecates and witches, the swarthy myrmidons into devils incarnate, and half-consumed mummy-fuel into some victim they were tormenting ... [This] gave the whole a more sepulchral and demoniacal appearance than anything I had ever seen before in real life, or in the mock horrors of *Der Freischutz*.[72]

Haight's narrative is clearly the product of the Gothic imagination. Indeed it includes so many of the tropes of the genre as to be almost parodic. It begins with observation, and finds inspiration in the internationally successful German Gothic opera, *Der Freischutz*, written by Carl von Webber and first performed in the 1820s. The narrative uses the Gothic imagination to turn objects of scientific inquiry into living things, to link the living to the dead and to inter-animate the present with the past. It is further evidence of how scientific things acquire definition and texture from outside the usual limits of scientific investigation.

It is as revivified or reanimated objects that Egyptian artefacts are most commonly given meaning as Gothic things. Archaeologists, travel writers and novelists all allow the work of the imagination to reconstruct as living things the inanimate artefacts they look at and write about. Although Amelia Edwards wrote *A Thousand Miles Up the Nile* after considerable study of the science of archaeology, the dominant motif of the Egyptian artefact is as a thing caught in suspended animation. At the enormous statues of Abu Simbel she describes how she comes 'to believe that there must sooner or later come some one sunrise when the ancient charm would snap asunder, and the giants must rise and speak'.[73] Later, in the interior of a tomb at the same location, Edwards felt that the ancient Egyptians depicted on wall-paintings were so 'instinct with supernatural life' that she 'should scarcely have been surprised to hear them speak – to see them rise from their painted thrones and come down from the walls'.[74]

Yet Edwards's imagined animation of ancient artefacts was not as far from scientific objectivity as it may appear. Petrie specifically discussed the employment of the imagination as one tactic of reanimating the past in order to better understand it. In a letter to his close colleague and fellow archaeologist Flaxman Spurrell in 1881, he wrote that he was 'trying as far as possible to get into the minds of the old people; for that is the only way to realise what they intended, and how they looked at things'.[75] The word 'things' is important to contextualize here, for Petrie was not meaning this in a general sense, but very specifically in relation to ancient Egyptian workmanship. The thing, then, is the artefact. Elsewhere in his Egyptological writings Petrie tries to put this stated desire into his practice. In a discussion of a series of stone carvings of the men of Ashkelon, in a popular essay in *Harper's New Monthly Magazine* in 1888, Petrie describes the carving in neutral terms as showing the men 'standing on top of their fort, with outstretched hands in supplication'.[76] Yet he continues 'Probably they were feeling much as did the unhappy people of Megiddo when 103 prisoners yielded themselves to the Egyptians, "starved out by the enemy"'.[77] There is shift in register from an objective account of what Petrie observes to an entirely subjective effort to 'get into the minds' of the men of Ashkelon, to bring them back to life as thinking and feeling human beings.

The artefacts themselves also offered some support to this use of the Gothic imagination, especially those artefacts that provided hieroglyphic accounts of ancient Egyptian storytelling. Edouard Naville's account of his excavations at Abydos, for example, begin with a very Gothic account of ancient revivification, taken from a series of ancient histories. Abydos 'became famous', writes Naville, 'as a place for the worship of Osiris'.[78] He then interweaves the story of Osiris, who 'had been cut into pieces by his rival, Set, or Typhon; but his son Horus had brought him back to life by reconstructing his body'.[79] Naville's readings of the artefacts at Abydos are therefore contextualized within a Gothic story of reanimation and in turn his own story of rediscovering Abydos as a site of archaeological importance becomes a narrative of similar piecing together and revivification. The process of archaeological discovery and Gothic fiction are here intertwined; and in becoming so offer to a wider public new meanings of the artefact as an active reanimated object and the archaeologist as a god-like figure of Gothic potential.

Writers of late nineteenth-century Gothic fiction depict the Egyptian artefact using similar imaginative tropes to the travel writers and archaeologists who respond to it *in situ*. Both Rider Haggard's *She* and Stoker's *The Jewel of Seven Stars* focus on the power the artefact gains in the process of reanimation. In *She* the revivified ancient queen Ayesha holds unquestioned authority over all others due to her Gothic ability to keep herself alive by magical means. Yet she is still an ancient object, as the narrator Holly makes clear when Ayesha's magic fails and she ages swiftly, becoming 'like an old piece of withered parchment' or a 'badly-preserved Egyptian mummy'.[80] Similarly Stoker presents the mummified remains of the Egyptian queen Tera as always potentially alive. She exists in a state that seemed 'not like death at all' but rather like a 'living person'.[81] From this state of Gothic liminality, precariously suspended between death and life, Tera controls the narrative action, attacking the archaeologist Abel Trelawny, and so overpowering the mind of his daughter Margaret that the narrator Malcolm Ross asks whether 'the personality present was my Margaret ... or the other new Margaret, whom I hardly understood'. Indeed Ross ultimately comes to understand that Margaret may well be under the control of Tera's 'active intelligence' and that 'whatever power of necromancy the Sorceress [Tera] had might have been exercised' over Margaret since the day of her birth, was leading to the inevitable conclusion that Margaret was 'not an individual at all, but simply a phase of Queen Tera herself'.[82]

Both novels, too, interrogate the role played by the excavator in the definition of the artefact. Rider Haggard invites the reader to consider this relationship by linking his heroic explorer Leo Vincey to Ayesha's ancient lover, Kallikrates, who is his physical double. Ayesha is entirely defined by her relationship to Kallikrates, and hence also by her relationship to Leo Vincey. Rider Haggard's aim in allowing the reader to believe that Vincey is a reincarnation of the ancient Egyptian Kallikrates is to show how the discoverer of the artefact is also responsible for defining it, and that such definition may not be controllable but rather emerge out of the encounter itself and the complex range of interactions that take place in that moment. Stoker plays with a similar idea of reincarnation and connection across history. By linking Margaret to the mummified form of Queen Tera, and making Tera's continued animation dependent upon Margaret's existence, Stoker suggests that the archaeologist Trelawny is father to both. He is, therefore, the producer of the mummy as an active thing and responsible for how its life might be defined in the present. As Rider Haggard has Ayesha thoughtfully note as she considers whether to raise the dead Kallikrates back to life, 'The life in thee would be *my* life, and not *thy* life'.[83]

In each of these fictions, too, the discomforting Gothic gaze plays a prominent role. Rider Haggard's archaeological explorers Holly and Vincey find themselves in the same ocular relationship with the mummy as Budge described. Like Budge's correspondents, they are made both anxious and Gothic by this process of looking. When Holly first recognizes that Ayesha's clothes, which he describes as 'wrappings', 'were so thin that one could distinctly see the gleam of the pink flesh beneath', he becomes 'more frightened than ever at this ghost-like apparition' but also notices that he 'was in the presence of something that was not canny'.[84] That is, Holly is filled by a Gothic anxiety produced by looking. Malcolm Ross similarly describes the emergence of an ocular anxiety as he watches Trelawny and his archaeological assistant unwrap the mummified form of queen Tera: 'As [Trelawny] stood back and the whole glorious beauty of the Queen was revealed, I felt a rush of shame sweep over me. It was not right that we should be there, gazing with irreverent eyes on such unclad beauty: it was indecent, it was almost sacrilegious!'[85]

These fictions of Egyptology offer interesting parallels to the narratives of excavation and travel, but they do not simply replicate their relationship to artefacts. Both Rider Haggard and Stoker investigate and reveal the power of the Egyptian artefact as it becomes a thing under surveillance by science and a wider public culture. As they place the artefact in a series of imaginative relationships with archaeological experts, explorers and laypersons, they illuminate both the thing itself and the thing in interplay with human culture. They confirm Latour's polemical conclusion: 'consider things, and you will have humans'.[86] More than this, these fictions suggest that the imagination, whether it be the scientific, traveller's or fictional imagination, might offer access to truths about the thing. As Rider Haggard's narrator Holly muses, at the conclusion of a daydream of ancient Egyptian life, 'who shall say what proportion of fact, past, present, or to come, may lie in the imagination? What is imagination? Perhaps it is the shadow of the intangible truth.'[87]

Vision Assembles the Thing

With each additional strata of definition the Egyptian artefact shifts its status from object pursued by science to thing of science. Where the thing gains thickness from its resistance to meaning, its domestic texture and its Gothic animation, the solidity and multiplicity it acquires comes from its being observed. And it is in the detailed examination of the observation of scientific things that they emerge as more than assemblages of ethics, aesthetics and politics, as Alder argues. Yet, Alder also recognizes that scientific things might incorporate 'entire worlds' while also 'insisting on their particularity' and that in doing so they 'can help break down distracting dichotomies like science/technology, idea/thing, and especially developed/underdeveloped'.[88] To consider the thing from the position of the observer – to unpick something of the phenomenology of vision within a broad spectrum of writings on, and the practices of, archaeology – reveals how things of science are also things of the everyday, of the imagination, and of fiction. Seeing the thickness of things developing in these directions also unravels and complicates further unhelpful dichotomies, like science/literature, objective/subjective and fact/imagination.

The archaeological things looked at, for example, by a travel writer who founded the EEF, by a fiction writer who subscribed to it, and by the archaeologists whom the Fund employed were all observed by individuals using their own eyes. They did not look in accordance with an abstract observational ideal but rather with an embodied perception whose ocular horizon was populated with personal experience. The eyes that looked at Egyptian things were the same eyes that had been, or were about to be, involved in reading hieroglyphs, cooking dinner, writing journals, washing tapioca pans, planning excursions, watching Gothic opera or plunging into the darkness of a tomb. Sometimes they were eyes that looked blankly, unable to come to any reasoned perception at all; eyes that registered only an ab-sense.[89]

Understanding the complexity of things of science is, perversely, to understand what is not scientific about them. It is to recognize that scientific observation is also bound up with the observation of other aspects of culture, from everyday tasks to popular fiction. The viewer is looking as an archaeologist, but also as a writer, reader, cook or traveller. As Victoria Carroll has argued in her work on museum display, 'the act of looking is never free from prejudice' and to recognize the different 'viewing conventions' of an object is to understand 'how different people made these objects speak'.[90] The various ways of seeing that defined the Egyptological artefact reveals, moreover, how these different people (the writer, the traveller, the archaeologist) interact with and influence one another to the extent that their own viewing conventions multiply and overlap. In that sense even the individual observer becomes 'different people', looking from the multiple perspectives that human identity enjoys.

PART IV: FUTURE

7 ARTHUR CONAN DOYLE AND HARRY HOUDINI: OPTICS, OPHTHALMOLOGY AND MAGICAL PERFORMANCE

Having spent earlier chapters considering how scientific instruments gave rise to new ways of seeing and how contexts for vision affected the observer, the following two chapters of this final section examine the eye itself. The eye is both an optical instrument and a conduit for acts of perception. In this, it is very like either the microscope or the telescope, both of which 'see' by focusing light through a lens and seem to have an active influence on what is observed. But the eye also sees from within the body of an observing human subject and is therefore always looking from specific contexts. The study of the eye has a long history before the nineteenth century. Yet as a distinct medical practice, as ophthalmology, it took hold during the nineteenth century. Indeed in theory and particularly in practice the second half of the century is the period of greatest activity; when Hermann von Helmholtz began to publish his *Treatise on Physiological Optics* while developing the ophthalmoscope, and when Edmund Landolt and Jonathan Hutchinson made optical disease the subject of popular lectures for the citizens of Paris and London.[1]

What emerges from much of the published work on optics and ophthalmology, written either for a general audience or the specialist, is both a fascination with the eye's fallibility and an admiration for the complex adaptation that made it such a successful organ of sight. Countless textbooks, courses of lectures, short introductions and technical compendia spend pages of description and analysis on lens distortion, muscle damage, corneal irritation and nervous dysfunction. Landolt began one of his 1879 introductory lectures on the eye (later published as an introduction to ophthalmology) by telling his audience that 'all the tissues of the body are represented in the eye and its accessories' and that the eye was therefore potentially subject to any kind of disease and defect known to human medicine.[2] Similarly, Helmholtz made reference throughout his *Treatise* to the 'subjective phenomena' of unhealthy vision that might effect normal forms of observation.[3] The eye, in these examples, is a fragile organ prone to myriad forms of visual occlusion or failure. Helmholtz's analysis, though, concluded that the eye was a hugely successful visual tool that allowed the human subject to orient itself in the world.

Much of the contemporary scholarship on visual culture regards the eye as an organ wielding this extraordinary power, even when it is shifted into metaphor as the lens of other media instruments.[4] Victorian oculists would have recognized this aggrandizement of the eye but also have thought of it as misleading. They often regarded it as an organ of mischance, whose visual capacity was often overwhelmed by a modern habitat for which it was unsuited. Jabez Hogg, surgeon at the Royal Westminster Eye Hospital from the 1850s, and commentator also on microscopy as we have seen in Part I, notes the propensity of new professions such as engineering to raise the occurrence of eye disease. His statistical calculations, possible by the mid-century because of the increase in reliable data from the numerous new eye hospitals across Britain, revealed that modern 'profession[s], diet, general health both physical and mental' all take their toll on the eye.[5] Optical investigations showed ophthalmologists an eye struggling to cope with new ways of seeing.

Popular entertainment also responded to the increase in optical investigations. As several scholars have noted, optical illusions began to play a central role in public spectacle from the early nineteenth century, and especially so in the mid-century.[6] Of the many examples that have been recovered in historical accounts, the ghost illusion of Henry Dircks and John Henry Pepper stands in the foreground. Germinated in geometric optics, but given impetus by physiological studies of the eye, Pepper's Ghost was an optical illusion that employed the algebraic certainties of catoptrics to produce a ghostly stage presence with which to amaze audiences. While Pepper and Dircks displayed their optical illusion in the context of the popular scientific lecture other performers, especially magical performers, soon made optical tricks a feature of their work. After 1850, Victorian illusionists almost always applied the principles of optics to one or more of the illusions that contributed to their stage act. Hermann the Great performed the illusion 'Gone!' from the 1870s. This 'clever adaptation of the "Pepper Ghost"', as Victorian magical historian Albert Hopkins called it, employed a change in lighting and a reflecting mirror to make the 'fair occupant' of a suspended chair 'disappear' when the magician 'discharges a pistol'.[7] Similarly, 'Vanishing in Thin Air', another illusion based on the rules of reflection, was described by Harry Houdini and attributed to the late Victorian magician

Chung Ling Soo (whose real name was Billy Robinson) who performed it in the 1890s. This illusion required speed and timing to fool an audience: 'With proper construction, this illusion should work with perfect precision, and its action is so fast that there is no chance of the spectators' realising what takes place', wrote Houdini.[8] As these examples show, a working knowledge of optics went hand in hand with a keen understanding of the physiological responses in the eye of the observer.

From the early 1890s, and continuing into the early years of the twentieth century, two particular magical performers caught the public attention more than most. One of these was an illusionist: Harry Houdini. The other was an amateur detective who existed in the fiction of Arthur Conan Doyle: Sherlock Holmes. Houdini and Holmes seem to make a rather incongruous pairing. Even the several works of popular history that debate the spiritualist conflicts between Houdini and Conan Doyle pay very little attention to Conan Doyle's fiction, and often equally little to Houdini's early career.[9] Nevertheless, Houdini's illusion practice and Conan Doyle's Sherlock Holmes fiction share a fascination with vision as mediated by ophthalmic and optical debate. In their sustaining of this interest across the boundary of the nineteenth and twentieth centuries they provide a valuable point of departure for a sustained investigation of Victorian and modern visual cultures.

Conan Doyle and Ophthalmic Practice

Conan Doyle was a trained ophthalmologist. Having graduated from the University of Edinburgh's Medical School in 1881, he first discovered an interest in 'eye doctoring' while a general physician in Southsea, near Portsmouth, where he practised for eight years from July 1882 until December 1890. His colleague Arthur Vernon Ford was chief eye surgeon at the Portsmouth Eye Hospital, and by March 1889 Conan Doyle was working there 'three hours a day'.[10] By this date Conan Doyle had plans to specialize in ophthalmology if his fiction enabled a certain financial independence. He told his friend Margaret Ryan that 'should there be a prospect of my being able to depend entirely on literature I should sell my practice here and start in London as an ophthalmic surgeon and oculist'.[11] Less than a year later Conan Doyle set this plan in motion: he gave up his Southsea practice in the winter of 1890 and left for Vienna to study the eye with the leading ophthalmological lecturers in Europe.

It is common in biographies of Conan Doyle to downplay this, admittedly brief, excursion to Austria, and to lay little emphasis on his ophthalmological enthusiasms.[12] Until recently, it has been taken for granted that what Conan Doyle wrote of this period in his autobiography, *Memories and Adventures* (1924), represented the historical reality:

> I attended the eye lectures at the Krakenhaus [general hospital], but could certainly have learned far more in London, for even if one has a fair knowledge of conversational German it is very different from following accurately a rapid lecture filled with technical terms ... Therefore, so far as eye work goes, my winter was wasted.[13]

Yet Conan Doyle's letters and papers from this period tell a different story, and one that has only come to light with the release of these documents into the public domain. While Conan Doyle later downplayed his studies in Vienna it is clear that he spent a valuable portion of each day of his two-month stay either in lectures or practical laboratory sessions. At the beginning of his stay, in January 1891, he wrote to his mother Mary Doyle with the details of his daily itinerary:

> I have after Monday to be at a class at 8 in the morning, but it is only a couple of hundred yards off. I will have a cup of coffee ... with Touie [his wife, Louisa] when I come back. Then five days in the week I write from breakfast until dinner at 1.30 ... After dinner we have a walk and do a little sight seeing and afterwards work until 5 when I have another class. At 7 we have supper, and afterwards I do an hour or two more writing.[14]

The morning class was a practical session on surgical technique overseen by Austrian eye surgeon Oliver Bergmeister, in his early forties and at the peak of his professional career. There is no remaining evidence to offer the same exactitude about the evening class, but it is very likely that Conan Doyle would have been attracted to the English-speaking lectures of Ernst Fuchs, Vienna's most celebrated ophthalmologist. Indeed Fuchs's *Text-Book*, a collection of his lectures, was first translated into English in 1892, with Fuchs's help, and probably provides a very close approximation to the lectures heard by Conan Doyle a year earlier.[15]

Conan Doyle appeared to be spending two hours a day on the study of ophthalmology and six hours on his fiction. While this appears rather imbalanced for a physician aiming to specialize in the eye we should recall that Conan Doyle always aimed to become an oculist as a form of retainer for his professional writing life, a lucrative fall-back if the literature failed to sell. However much he was misjudging the competitive ophthalmology market we should see his concentration on fiction in this context, rather than believing that he was remiss in his ophthalmological studies or simply unable to cope with technical German. In fact, the obvious interweaving of Conan Doyle's writing life with his professional medical practice is an important fact to glean from his time in Vienna.

Conan Doyle left Vienna for Paris in mid-March. His aim was to spend two weeks studying with one of the most influential of nineteenth-century ophthalmologists, Edmund Landolt. This had been planned while in Vienna: the holiday he had first envisioned set aside in favour of further oculism. He wrote to his mother in February: 'I shall not go to Buda now' but instead spend 'a fortnight at Paris to be able to say that I have studied under Landolt. I can gain a fair idea of his practise in two weeks.'[16] Conan Doyle's views on this period of study with Landolt are unknown, but he must certainly have found parallels to his Viennese experiences. By March 1891 Landolt had completed some of his most important work. His large series of published lectures under the title *A Manual of the Examination of the Eyes* had first appeared in 1879 and been continually revised through numerous editions in the 1880s. He had also, in 1886, published *The Refraction and Accommodation of the Eye and Their Anomalies*.[17] Both of these were works of real significance in bringing together theory and practice in diagnostic techniques. Landolt believed that analysis of the eye as an 'optical instrument' needed to be married to 'clinical investigation'; physiological optics and ophthalmology were essential to each other and in being knit together would ensure a 'systematic method' that would 'overlook nothing and attain our ends with certainty'.[18] Although he had only latterly arrived at the originating source of such a view, Conan Doyle would already have been familiar with Landolt's argument. Ernst Fuchs had made systematic observation the subject of the first page of his *Text-Book*'s first lecture.

Returning to London late in March Conan Doyle opened his ophthalmological practice in Upper Wimpole Street, as close to Harley Street as was affordable. He aimed to work on the correction of astigmatism: one of those defects of vision consistently discussed by ophthalmologists which had contributed to more general anxieties about the eye's fitness and vision's weakness. These consulting rooms are often discussed in the scholarship because it was here that Conan Doyle wrote the first of the Sherlock Holmes stories, apparently while their author waited in vain for any medical patients. Again, however, the reality of Conan Doyle's engagement with ophthalmology is different. Conan Doyle wrote to his friend and fellow physician Reginald Ratcliff Hoare around Easter 1891 to alert him to his new location, and in the hope of some advice on the general medical scene. 'I had good opportunities at Vienna and Paris', Conan Doyle notes, 'and as I have now hooked on at the Westminster Ophthalmic I shall keep up to date'.[19] The Royal Westminster Ophthalmic Hospital was the second London medical establishment, after Moorfields, to be dedicated to diseases of the eye. It had been opened in 1816, under the presidency of George James Guthrie, and had been the first hospital to offer a course of lectures on ophthalmology in 1817. By 1891 it was central to contemporary research in ophthalmology. After Vienna and Paris, Conan Doyle now found himself at the heart of a third European powerhouse of ophthalmic practice.

The first of the Sherlock Holmes short stories, 'A Scandal in Bohemia' (1891) attests to the influence of vision on narrative content.[20] In the first instance the geographical location in the story's title was inspired by Conan Doyle's ophthalmic sabbatical in Vienna, Bohemia's historic capital under Hapsburg rule. The story itself is suffused with visual and optical referents: disguises and misdirections, lenses, lighting and silhouette, modes of observation and perception, and visual evidence. Holmes is described in the first paragraph as an 'observing machine',[21] a phrase with clear echoes of Landolt's cautionary description of the eye as an 'optical instrument'. Watson extends the parallel by comparing Holmes's susceptibility to emotion with mechanical and optical failure: 'Grit in a sensitive instrument, or a crack in one of his own high-power lenses, would not be more disturbing than a strong emotion in a nature such as his' (p. 161). Holmes is, in fact, the eye itself; a perfect specimen both of ocular power and its potential for failure. The narrative action is woven with specific instances of optical analysis, each of which place the eye (and therefore Holmes) at the centre of the mystery's solution. Watson discerns Holmes's mood from reading the patterns his body makes in silhouette; Holmes uncovers the social and geographical location of his client from illuminating in bright light a watermark on a piece of paper; the central object of the story is a photograph, a visual representation explicitly contrasted with other, less powerful pieces of evidence.

The central scene, however, is an elaborate illusion designed to force Irene Adler, the 'criminal' of the story, to reveal the hiding-place of the photographic evidence which Holmes has been charged to uncover. Constructed as a form of magical performance, the scene enacts a complex negotiation of observation, ocular trickery and illusion practice. Holmes, disguised as a clergyman, gains access to Adler's home, where he enables Watson to throw in a smoke-bomb and raise a cry of fire. Adler, believing her possessions to be at risk, retrieves the photograph from its hiding place and thereby allows Holmes to discover its location. The entire scenario is made up of a series of visual manipulations: Holmes's disguise hides from Adler's eye his true identity, his contrived entry into Adler's home allows him to place his own eye in exactly the right position to observe the photograph's hiding place, and the smoke-bomb presents the illusion of fire where none exists. Holmes controls these events like a stage magician, employing the techniques of misdirection and sleight of hand in a grand *coup de theatre*. Watson admiringly calls Holmes 'a fine actor', while Holmes modestly waves away such praise, claiming his illusion as only 'an old trick' (p. 172).

As 'A Scandal in Bohemia' testifies, illusion practices are ideal fictional material for the study of scopic culture. Illusion practices demand the performance of vision and necessitate active observation. They are also often confrontational; a site of conflict between the performer and the observers of performance. To that extent they are superbly suited to detective fiction with its generic antagonism between detective and criminal that is very often situated as a struggle between discovery and concealment. Imagining Holmes as the magical performer of staged illusions, or as the observer of illusions organized for criminal purposes, allows Conan Doyle to connect visual trickery and optical expertise to wider cultural concerns about the role of the eye in modernity, and especially to the dialectic between the eye's power and fragility which defines the later part of the nineteenth century.

Harry Houdini and Magical Performance

Harry Houdini's career was on a similar trajectory to Doyle's, a trajectory that would see them correspond from the beginning of 1920 and meet in April that year (see Figure 7.1). Although Houdini was born fifteen years after Conan Doyle, his career as a stage magician mirrors, at least chronologically, Conan Doyle's writing of the Sherlock Holmes stories. *A Study in Scarlet* was published in December 1887, only a few months before Houdini's first public performance, early in 1888. Neither gained much public notice. The first Sherlock Holmes short story was published in the *Strand Magazine* in July 1891. Three months before this, on 3 April, Houdini left his job to concentrate on a career as a magician. It took longer for Houdini to gain the kind of public recognition that Conan Doyle achieved. Indeed Conan Doyle had 'killed off' Sherlock Holmes before Houdini's name became familiar. Yet by the time Holmes returned in *The Hound of the Baskervilles* in 1901, Houdini was one of the most prominent magical performers in Europe and the United States of America. Thereafter Houdini and Conan Doyle remained at the forefront of popular culture. Houdini's final magical performance came in October 1926, a week before his death. At the same time Conan Doyle was working on the last of the Sherlock Holmes stories, whose long series was brought to an end early in 1927.

In some degree the temporal correspondence of their careers is unremarkable. Significantly, though, their work occupied the same imaginative territory and revealed a similar interrogation of vision. Houdini's magical act had its foundation in the performance of criminality and detection. All of his important illusions drew on his escapology. This demanded that Houdini take on the role of the incarcerated criminal: tied, bound and handcuffed Houdini would be placed into 'cells' of varying types (packing cases, boxes, large caskets, sometimes actual prison cells) from which he would escape. Audiences would be invited to observe each moment of his incarceration in order to act as detectives for any trickery or deception, and thereby to confirm that Houdini's escapes were genuine rather than acts of magic.

While outside the theatre Houdini cultivated his image as an exposer of criminal trickery (a kind of Sherlock Holmes in the social sphere), inside the theatre he relied on the concealment of these same tricks. Although it is undoubtedly true that a proportion of Houdini's escapes were performed through physical effort, the vast majority were illusions dependent upon the management and control of audience observation. While Houdini's magical performances can appear spectacularly superficial, as some writers have observed, they were actually carefully crafted examples of illusion practice.

Houdini's stage performances were layered investigations of vision. He did not depend upon a single optical device, however extraordinary its effects, but

SIR ARTHUR CONAN DOYLE AND HOUDINI PHOTOGRAPHED AT THE AUTO CLUB, LONDON, ENGLAND

Figure 7.1: A rare photograph of Houdini with Conan Doyle. Source: H. Houdini, *A Magician Among the Spirits* (New York: Harper, 1924), frontispiece; reproduced from the author's personal copy.

on constant visual manipulation. Houdini's biographers often implicitly characterize him as a man of little education and limited intellect. This is contrasted with his physical strength and active stage performance. It is certainly clear from Houdini's own writing that he had poor written English, including quite shocking spelling, and was ignorant of many areas of knowledge that contribute to intellectual refinement. Yet he understood dramatic performance acutely and was hugely knowledgeable on all aspects of magical history, enough to make him one of the most important magical historians (and collectors) of the early twentieth century. Above all, though, Houdini had a practical understanding of the physiology of vision and perception. He read on this subject widely; not often in the original scientific literature but in summaries and popular articles, and in the magical literature that found inspiration in the facts of optical and ophthalmological knowledge. Houdini's library contained countless articles on the psychology of vision that embodied the arguments of, say, David Brewster or James Sully on the eye's susceptibility to illusion. It also included runs of periodicals such as *Science and Invention*, where optical illusions were explained from the standpoint of geometric optics and optical gadgetry was described enthusiastically from the perspective of the amateur observer.[22] Houdini therefore understood the primacy of vision to the successful performance of illusions and developed a range of techniques to manipulate audience sense perception. Little of this knowledge can be traced directly into the scientific sphere, but it remains the case that Houdini applied popular science in a practical context both consciously and with a clear idea of the effects he might achieve.[23]

Houdini's magical performance, and especially the escapology that was the backbone of his stage acts, exemplifies his multifaceted interrogation of visual perception. To successfully perform his escapes Houdini relied on disguise and distraction (as does Sherlock Holmes); magical theory denotes this as misdirection. This optical hoodwinking enabled Houdini to conceal handcuff keys (known as 'fakes') and other necessary accoutrements of escapology from the audience. Houdini's magical performances also based their visual manipulations on the physiological characteristics of the eye. Walter B. Gibson, who collated and published Houdini's working notes after his death, gives several examples of Houdini's exploitation of the physical changes in the eye when it is subject to bright light and darkness in quick succession, when, for example, Houdini invited audience members to examine the interior of his escape cabinets.

Houdini's illusion practice embodies what is a common element of all optical tricks: it hides from the audience the true optical landscape by overlaying it with one that is illusory. Crucially, it allows the audience to verify that illusory landscape for themselves. Houdini's performance demands that the audience conceptualize the theatre exactly as he would wish them to: as an authentic space, fully visible and without the magical apparatus that might defy observation.

Conan Doyle's detective fiction and Houdini's magical performance should be seen as parallel experiments in an evolving visual culture, a culture clearly in motion in response to emerging knowledge of the eye and perception from physiological optics and ophthalmology. There are similarities in their approaches to vision, as we have seen, but how vision is represented and what it is made to mean, varies substantially. Although it is one of the implicit arguments of this chapter (and of the book) that vision can only be understood historically and specifically, this does not of course mean that the historical actors will develop similar representations of vision at contemporaneous moments.[24] Indeed Conan Doyle and Houdini rather prove the reverse: their analyses of vision meet, diverge and come into conflict. To gain an appropriate understanding of vision it is important to investigate the full variety of their visual representations; revealing and exploring the layers and textures that give them their depth and subtlety.

Criminal Visions

Anxieties concerning the fallibility of the human eye were expressed both by the individual and across the public sphere. As ophthalmology and physiological optics revealed more of the eye's weakness and susceptibility to sight-endangering diseases and activities, the individual's concern for their own vision increased. Brewster's *Letters on Natural Magic* (1832) had indicated that failures of visual perception were entirely the fault of the individual, whose 'fancy' often constructed illusionary and even hallucinatory images.[25] Later, by the 1870s, Helmholtz's work on the physiological responses to light of the organ of the eye and its associated muscles and nerves confirmed the observer's poor visual capacity.[26] The nervous response to light, for example, was shown to be extraordinarily slow, especially by comparison with the speed of light itself, which was another of the physical characteristics of the natural world to be understood and given definition in this period. While light sped through the pupil of the eye and hit the retina at approximately 300 million metres per second, the optic nerve transported this light to the brain at a meagre 50–100 metres per second. As Timothy Lenoir shows, this left Helmholtz 'incredulous' and suggested greater barriers to perception than had been previously imagined.[27] Other popular works, such as the publisher Charles Knight's series of books on *How To Observe*, from 1835, seemed to suggest that a general lack of observational expertise was the norm, and across Europe the increase in specialized ophthalmological clinics, surgeries and teaching schools reinforced social awareness of the eye's common disabilities.[28]

Conan Doyle and Houdini reflected this visual unease in their fiction and performance by hitching perceptual failure to another social anxiety: crime. The criminal behaviour represented in Conan Doyle's detective fiction is rooted in visual trickery and often relies upon the creation of optical illusions. In several stories criminal acts are accompanied by, and made possible because of staged illusions. In 'A Case of Identity' (1891) the criminal James Windibank successfully concludes his immoral plan by vanishing from a cab:

> when the four-wheeler drove up we waited for him to step out, but he never did, and when the cabman got down from the box and looked there was no one there! The cabman said he could not imagine what had become of him, for he had seen him get in with his own eyes. (p. 195)

'The Golden Pince-Nez' (1904), a story that even in its title suggests the importance of vision, opens with a disappearance that also turns out to be an illusion. On this occasion the criminal, Professor Coram, employs the disguised hiding place common to magical performance, which is eventually found to be a concealed room behind 'a high bookcase in the corner of the room' (p. 619). Jonas Oldacre, the criminal figure in 'The Adventure of the Norwood Builder' (1903), uses exactly the same optical illusion, and is discovered behind 'a door ... [in] what appeared to be solid wall at the end of the corridor' (p. 508). In two of these stories the optical trickery of the criminal is offset by the weak eyesight of their victims. The concealed woman in 'The Golden Pince-Nez' has lost her glasses and is essentially blind when she is made to 'vanish'. Miss Mary Sutherland, the victim of Windibank in 'A Case of Identity', has such 'short sight' that the criminal is 'doubly secure' (p. 200) of the success of his illusions.

Further optical illusions are reinforced by the use of magical apparatus, commonly employed in visual tricks where disguise of an object or person is necessary for the illusion to hold. In 'The Speckled Band' (1891) Grimesby Roylott uses 'a dummy bell-rope' (p. 267) as a disguise for the rope used by the poisonous snake that commits murder on his behalf. 'The Red-Headed League' (1891) offers a slightly different version of disguise by making the obtrusive hair colour of the criminal an identifier of his allegiance to a (fake) secret society and so misdirecting the victim. James Windibank gives a more traditional performance of disguise when he uses a variety of magical apparatus to become 'Hosmer Angel' and thereby conceal his own identity from Mary Sutherland. One early Holmes narrative is based entirely on disguise as a form of ocular misdirection: 'The Man with the Twisted Lip' (1891) tells the story of middle-class city businessman Neville St Clair, who disguises himself as a disfigured beggar to dupe observers into providing him with money. The ocular vocabulary of these stories extends the contrast between the weak eyesight of the victims of crime and the visual acuity of the criminal class. Often this is done by representing good eyesight as a metaphor for perceptual ability. James Windibank is described as having 'keen eyes' (p. 200), while Neville St Clair has 'a pair of very penetrating dark eyes' (p. 235). Grimesby Roylott's vision is more complex. He is described as having the appearance of a 'fierce old bird of prey', which certainly characterizes his eyesight as acute, but at the same time his eyes are 'deep-set [and] bile-shot' (p. 264). As one of the most morally debauched of Conan Doyle's criminals his eyesight is both powerful and corrupt. Clearly the dialectic of powerful vision and ocular weakness is continuing in such examples.

Houdini's escape act, which gave visual representation to criminal incarceration, provided him with a platform from which to speak publicly of crime as the performance of illusions. In his popular 1906 book *The Right Way to Do Wrong* Houdini used his position as a handcuff expert and jail-breaker to reveal the inherent optical trickery of criminal activity.[29] A series of short chapters detail the various illusions employed by professional criminals to commit burglary, theft and swindles. His aim, as he argues in the preface, is to expose the 'various tricks' of the 'criminal classes' and 'explain the adroit methods by which they seek to defraud'.[30] However, Houdini also argues for the necessity of his book on the evidence of the public's visual fallibility and their ignorance of the ocular deceptions taking place around them: 'of the thousands – yea, tens of thousands – of undiscovered crimes and unpunished criminals, you know but little'.[31] Like Conan Doyle's detective fiction, Houdini connects criminal activity to observational failure, suggesting a causal link between limited perception and the rise of crime.

In fact *The Right Way to Do Wrong* might easily be read as a practical companion piece to Conan Doyle's imaginative narratives. One of the optical tricks he reveals is the professional criminal's clever disguise as a beggar, which includes an example that parallels almost exactly Conan Doyle's 'The Man With the Twisted Lip'. Houdini tells the story of Cecil Brown who 'every morning went into the city, and every evening came home'. Yet Brown, like Neville St Clair is not a city businessman but a beggar, who can be seen 'shuffling painfully along' with 'one arm hung helpless by his side, his head hung with the weakness of paralysis'.[32] When his double identity is discovered his criminal practice proves similar to St Clair's:

> the lame man dragged himself to an adjacent tobacconist's shop, where he changed his silver and coppers into bills. Here, too, he left the cigar-box and the matches until the morrow, and then he boarded a car to a cheap lodging-house, and by the time he had arrived there his lameness had disappeared and he went up the steps two at a time. Finally, he went home to the smart little villa.[33]

Brown's wife, like St Clair's, representative of the general public ignorance, was entirely oblivious to her husband's criminal career and 'was surprised to hear' of the 'goings on'.[34]

Whether or not *The Right Way to Do Wrong* was directly influenced by Conan Doyle's fiction, as this example suggests, it is certainly true that Houdini regarded it as a response to late Victorian detective fictions. In writing of a particularly clever criminal trick – a jewel theft from a travelling train – Houdini exclaims: 'Here was a case for a Sherlock Holmes or a Martin Hewitt'. But

Houdini was not satisfied with remembering his adept predecessors, he would rather replace them: 'but either these gentlemen were not forthcoming, or they totally failed to solve what is, perhaps, the most mysterious railway robbery of recent days. Let me lift the veil and show how the little job was done.'[35] Houdini's standing in for Sherlock Holmes reveals a self-aggrandizing confidence in his ability to perceive the optical tricks of the criminal class. More than this, though, it shows that criminal activity might be curtailed by someone equally expert in illusion practice. Anxieties over increasing social insecurity occasioned by ocular failure could be reduced by employing as powerful an optical knowledge to the solving of crime as that brought to bear by its perpetrators.

Houdini's characterization of Sherlock Holmes as the failed observer is not reflected in Conan Doyle's fiction. In fact Holmes embodies the role for which Houdini auditions. His expert vision is characterized by its similarity to the visual power enjoyed by the contemporary ophthalmologist and physiologist. While optical research had revealed the many different ways in which the eye's ability to see might be reduced, this had only been achieved by increasingly powerful optical instruments and by ever more astute methodologies for understanding how vision worked. Optical and ophthalmological practices already had their own cure in the very research they undertook, and fears about the eye's malfunctioning could be countered with the many new processes designed to conserve and maximize its visual power.

Holmes's techniques of observation echo the systematic study of the eye that Viennese ophthalmologists in particular saw as the only proper method of examination. Ernst Fuchs began his course of lectures on ophthalmological practice by stating that 'too much stress can not be laid upon the necessity of proceeding systematically, since otherwise important matters can very easily be overlooked'.[36] One should begin, Fuchs continued, by 'establishing the history of the case'.[37] Holmes's detection works from the same principle. In 'A Scandal in Bohemia' he warns an over-eager Watson that the process of detection has a starting point in the collection of data: 'It is a capital mistake to theorise before one has data. Insensibly one begins to twist facts to suit theories, instead of theories to suit facts' (p. 163). Similarly, when investigating the criminal John Clay in 'The Red-Headed League' Holmes warns Watson against speculation: 'this is a time for observation, not for talk' (p. 184). Holmes begins, as does Fuchs, with the establishment of the case's historical narrative: 'pray give us the essential facts from the commencement', Holmes asks of John Openshaw in 'The Five Orange Pips' (1892), and of Helen Stoner, the victim in 'The Adventure of the Speckled Band', he demands 'that you will lay before us everything that may help us in forming an opinion upon the matter' (p. 259).

While this provides Holmes with a properly organized starting point for his observations, he has also to understand the data of the visual field laid before him. This he does by a process of eliminating all that is normal within his scopic landscape and focusing on the remaining abnormalities. The maxim Holmes uses to describe this technique is one often repeated across the Holmes canon, and which first appears in *The Sign of Four* (1887): 'when you have eliminated the impossible, whatever remains, *however improbable*, must be the truth' (p. 111). This is recognizably the practice of both ophthalmological diagnosis and physiological research. To simplify complex calculations, physiologists of vision created what they called the 'reduced eye', a simplified ocular model that allowed for easier notation of difference in optical experiments.[38] Similarly, ophthalmologists based their examination of the eye on continuous comparisons with the so-called 'normal eye', a statistically stable, ideal eye against which difference might be calculated. As Fuchs states, 'every observation sifts the statistical identity of normal eyes to find any deformation or mutation'.[39] Holmes determines the visual evidence of crime on this basis, seeking out the 'mutation' of that evidence from the numerous visual clues left by normal human interaction with a given landscape or object.

Although such expert observation sets Holmes apart from the unsystematic looking of those around him, it is his perception of the things he observes that makes him unique. As Watson points out in 'The Boscombe Valley Mystery' (1891), 'I could plainly see the traces which had been left by the fall of the stricken man. To Holmes, as I could see by his eager face and peering eyes, very many other things were to be read upon the trampled grass' (p. 212). One of Holmes's greatest abilities is his connection of visual traces to prior events as though he were looking at the events themselves. For example, Holmes can determine a person's gait from the traces they leave behind as footprints. Metaphorically, Holmes is able to 'see' what Helmholtz argues cannot be seen. As Helmholtz argues it,

> we are exceedingly well trained in finding out by our sensations the objective nature of the objects around us [but we] are completely unskilled in observing the sensations per se; and that the practice of associating them with things outside of us actually prevents us from being directly conscious of the pure sensations.[40]

These pure sensations that cannot be seen but are unconsciously participating in the organization of vision can be seen (fictionally) by Holmes.

Holmes needs these fantastical ocular skills to be able to perceive the optical illusions of those criminals he wishes to bring to justice. In many of the stories we witness Holmes simply looking, seeking the optical clues that will uncover visual trickery in action. 'The Speckled Band' is exemplary: 'Holmes drew one of the chairs into a corner and sat silent, while his eyes travelled round and round and up and down, taking in every detail of the apartment' (p. 267). Holmes's method of systematic observation discovers the bell-rope, and his perceptual ability 'sees' its potential function. While others fall victim to their own perceptual failure, Holmes's employment of ophthalmological diagnostic methodology and his ocular physiological adeptness situates him at the pinnacle of the visual hierarchy Conan Doyle's fiction has created. In Holmes the reader finds a scopic hero who saves the fictional victims of crime from their own ocular weakness but also allays the anxieties of visual deficiency in a nervous public.

Houdini's escapology performs a similar function; his expert knowledge of optical illusions, gained over years of magical performance, is put to use in uncovering those criminal behaviours that rely upon the same tricks. Houdini's escape acts are often regarded by his biographers as exemplary physical feats rather than illusions. Houdini did require strength, endurance and suppleness for his escapes but they were still essentially optical tricks, demanding adept misdirection, sleight of hand and magical apparatus. As Neville Maskelyne and David Devant note in their 1911 work on the theory of illusions, *Our Magic*, 'all physical magic is from the same category, both optical illusion and escapology, both are visual magic, illusion practices'.[41] This was not, of course, what Houdini's spectators saw, as one report makes clear:

> Last evening ten men were on the stage, by invitation, and all were supplied with an assortment of handcuffs. The magician [Houdini] was bound with five pairs of handcuffs and a set of leg irons, the chain of the leg irons crossing the links of the handcuffs, so that the champion 'jail-breaker' was bent double. Curtains were drawn about him and in less than ten minutes Houdini emerged with the open shackles dangling from him.[42]

However astonishing such an escape might have appeared, its explanation was much more mundane. Houdini puts it succinctly in his notes on magical performance: 'you enter your cabinet and, obtaining possession of your duplicate key, you simply unlock the cuff and again conceal key'.[43] While this illuminates the method of Houdini's escapes it does not give a full picture of the skills needed to perfect the sleight of hand techniques required to hide, bring out into use, and hide again lock-picks and skeleton keys, especially when the escape was performed in full view of the audience, as Houdini's often were. Nor does the simple explanation of the duplicate key give a true picture of the performance required to conceal from the audience (and the inspecting committee invited on stage) the magical apparatus needed for the successful escape illusion. Especially when Houdini performed for a specialist audience, such as the exhibitions he gave at police stations or prisons, his skills of misdirection and disguise needed to be exceptional in order to bring off a successful escape under the observation of experts in incarceration. Although not as theoretically proficient as Sherlock Holmes, Houdini's own career gave him expertise in practical optics and especially in the physiological relationship between the human eye and the external world. More than this, of course, his own performance provided him with a store of knowledge on how optical illusions might be created and presented.

Houdini often employed his optical knowledge for what he regarded as the public good, revealing frauds and con artists, and exposing crime. Clearly *The Right Way to Do Wrong* is the best example of this. By exposing the optical tricks so prevalent in crime, Houdini hoped to 'so familiarize the public with the methods of the criminal classes that it will enable law-abiding citizens to protect themselves from the snares of evil-doers'.[44] Houdini's relationship with an ocularly-ignorant public is more instrumental that Holmes's. He does not only seek to be the hero of his own narrative but also to provide the public with their own visual expertise. As the preface states, 'I wish to put the public on its guard, so that honest folks may be able to detect and protect themselves from the dishonest'.[45]

However, there is an assumption here that 'the public' do not have the optical nous necessary to protect themselves, an assumption that has parallels with the stress on ocular weakness with which Conan Doyle characterized Holmes's clients. Houdini reiterates this in other works. In his 1910 book *Handcuff Secrets*, for example, he takes to task those magicians who do not perform genuine escape acts: 'those who evade the critical tests (which might crop up in such challenges) by fraud and colossal lying, those are the individuals who may have something to say against the divulging of their various methods of hoodwinking a credulous public'.[46] Public observers are again characterized as victims of their own fallible eyesight. But there is also a quiet exasperation in Houdini's view of public vision, a sense of public disinclination to engage in better ways of seeing that is ultimately an abnegation of social duty. As Houdini says in *The Right Way to Do Wrong*, 'You who live your life in placid respectability know but little of the real life of the denizens of this world [of crime and illusion]'.[47]

There was certainly a sense in the work of ophthalmologists and physiologists that individuals (and by extension the public body) were culpable for their own failing vision as much as victims of it. Leading ophthalmologists across Europe – including Hutchinson, Hogg, Landolt, Fuchs and Laurance – gathered extensive statistical and clinical data that proved how greatly an individual's choices affected their ability to see clearly. The use of 'alcohol and drugs'[48] figured highly in lists of potential sight-affecting activities, but nearly every ophthalmological textbook from the 1850s onwards charted many different effects of individual action on eyesight. Lionel Laurance, summarizing nineteenth-century research in his textbook for students at the London School of Optics in 1920, notes the following as potentially disruptive of perfect sight: 'fatigue ... a sudden change from brightness to obscurity, or vice versa ... environment ... occupation and food ... education and alertness ... the brain or intelligence behind the eye ... mental torpidity and illiteracy'.[49] Laurance is not arguing something unique about the twentieth century here. Much of what he says was commonly expressed in earlier treatises, such as Charles Wheatstone's translation of J. E. Purkinje's work on vision from the 1820s.[50] In many respects this is a list of Victorian social problems for which no individual could be blamed. However, as we have seen in discussions of disease in the first two chapters, this logic did not necessarily follow, and individuals were often blamed for personal failings that might easily have been regarded as the responsibility of the state.

One aspect of the ocular anxiety of the period was therefore a sense of guilt that individual action had exacerbated visual instability, and while Holmes and Houdini might well stand out as individuals with exceptional visual acuity their exposure of criminal illusion practices highlighted a general ocular deficiency leading to social fragmentation, at the same time as they were supposed to counterbalance that deficiency and contribute to the making safe of society. Yet both Houdini's performance as the shackled criminal and Holmes's mastery of criminal trickery do more than offer up a heroic example of the visual knowledge that should be the baseline of social responsibility. Their positioning of criminal vision actually attests to a broader reading of the observational culture of the late nineteenth and early twentieth centuries where failures of vision are the fault of institutions rather than individuals.

Conan Doyle sets Holmes's visual power against a series of institutions, all of whom fail to match his perception. The police are the most common dupes of criminal illusions, never once seeing the optical trickery that is the route to successful detection. In several stories, such as 'The Boscombe Valley Mystery' or 'The Adventure of the Norwood Builder', the official police force come very close to condemning an innocent individual for an act committed by another, often because of their failure to read visual evidence properly. Areas of state government are also revealed as ocularly incompetent: in 'The Naval Treaty' (1893) and 'The Adventure of the Second Stain' (1904) Whitehall departments lose sight of documents central to Britain's position on the world stage. In 'The Adventure of the Priory School' (1903) a late Cabinet minister loses his own son. Even monarchical government is not spared: 'A Scandal in Bohemia' shows the stupidity of a king who enabled visual evidence to be gathered against him, thus undermining the stability of his nation. Sherlock Holmes may make up for the visual weaknesses in individual victims, but he is far more often saving institutions from their own poor sight.

Houdini's escape acts depict comparable institutional folly, although it would be easy to see them as a glamorization of the criminal classes. After all, on stage he appears as the shackled figure, bound and handcuffed with regulation items of institutional restraint. Houdini actually goes far to embody the criminal role: he notes how some of his 'lockpickers or keys' were 'taken from a French murderer',[51] and he uses himself as the final example of criminal trickery in *The Right Way to Do Wrong*.[52] By allowing himself to be photographed in prison cells and jails, in chains and behind bars, Houdini seems to revel in his association with criminals usually found in such places and positions. However, his escapes are an act of transformation from base criminal to masterful illusionist. John Kasson argues that Houdini's 'handcuff escapes gave him the opportunity to create a new kind of metamorphosis: from impotent victim of authorities to manly victor over them'.[53] It also produced a similar change in his audience, as one report of his escape act perceptively notes:

> They are with him in his work, but they are as ready to be against him. Watch a house when he has an exceptionally difficult proposition to tackle. If it seems that some one has at last confronted him with something that is beyond him applause comes heavily for the captor. Let Houdini finally free himself and the ovation is for him. To keep himself the hero always is the difficult task.[54]

Houdini walks a tightrope between his twinned roles as unexceptional criminal and heroic magician with acute illusionary skill. To succeed in his escape is both to banish the criminal role and to uncover the impotence of repressive institutions. For the spectating public this transformation is evidence of the individual's triumph over institutional power. Houdini never failed to make certain that his audience knew that the handcuffs were police issue or that the prison guards who had bound him were satisfied that escape was impossible. It did not matter that the audience had also failed to recognize his escape as an optical illusion. It was the material objects of institutional surveillance that had been overcome, and the representatives of these institutions that had revealed their observational weaknesses.

By displaying institutional ocular failure Conan Doyle's and Houdini's representation of scopic culture is at odds with the general thesis of vision often applied to the late nineteenth century and early twentieth century. Originating with Foucault, this thesis argues that the period witnessed an increasingly institutionalized scopic regime, with visual power linked to surveillance and control. The evidence of Conan Doyle's and Houdini's creative responses to the period's visual culture offers a different perspective. Visual anxiety was not a product of institutional control of the ocular landscape but a much more complex reaction to the duality of visual power and ocular fallibility that exerted an influence both individually and institutionally.

Performance

This chapter has so far considered Holmes and Houdini as heroic manipulators of optical knowledge and read optical illusions as a mode of trickery used in criminal deceptions. However, both Holmes and Houdini also perform optical illusions. How should we read their magical performance? A closer consideration of the presentation, structure, performance and conclusions to the optical illusions they perform should tell us a great deal of the modes of operation of vision. Magical performance, like text, offers an opportunity for representational analysis. Maskelyne and Devant, writing at the height of Houdini's career in 1911, argued that 'as grammar is to literature, or versification to poetry, so are sleights and fakes to magic'.[55] If that is the case then an attention to the forms of magical performance will illuminate the structural meanings of visuality in Conan Doyle's and Houdini's optical illusions.

Holmes often brings detection to a close by presenting a magical performance. In 'The Second Stain', for example, he performs what magicians would recognize as the box trick, first described by the French conjurer, Robert Houdin, from whom Houdini took his own stage name. In this trick a letter, written by the magician's interlocutor, is destroyed or made to vanish, a locked box is retrieved and opened, and the letter is found inside and intact. 'The Second Stain' tells the story of a letter stolen from the British European Secretary, which Holmes is charged with recovering. Having done so, Holmes replaces it without the Secretary's knowledge in the box from which it was originally taken. He then invites the Secretary to open the box again: 'This is a fanciful waste of time ... I have always had the key on my watch-chain ... Good heavens, what is this? ... Yes, it is it – and the letter is intact! ... Mr. Holmes, you are a wizard, a sorcerer!' (p. 666). This box trick is only one of many that Doyle creates for his short fiction. Holmes also reinterprets a late Victorian vanishing illusion based on Rider Haggard's *She* in 'The Norwood Builder' and re-enacts the three cup trick with breakfast trays in 'The Naval Treaty'.

Holmes brings to his performance of these tricks all the skills required for successful illusion practice. In the first instance he is a consummate actor, a quality Houdin argued was essential for any magical performer. In his 1868 book on magical theory and practice, *Les Secrets de la Prestidigitation de la Magic*, he argues that 'a prestidigitator ... is an actor playing the part of a magician'.[56] Holmes's acting abilities are often on display when in disguise, but also at other key moments in his investigations. 'The Reigate Puzzle' (1893) offers one example, as Watson narrates: 'My poor friend's face had suddenly assumed the most dreadful expression. His eyes rolled upwards, his features writhed in agony, and with a suppressed groan he dropped on his face on the ground' (p. 403). Watson recognizes this as a symptom of a recent illness, but as Holmes later explains, it was simply a ruse to alter the direction of a conversation. 'Speaking professionally', says Watson, 'it was admirably done' (p. 409). While this moment of dramatic performance goes undetected, more commonly Holmes's role as actor, and as the actor playing the part of a magician, is explicitly drawn to our attention. Watson concludes his description of another of Holmes's tricks (pulling a ruby from inside a plaster bust as though he were drawing a rabbit from a hat) by noting that 'he bowed to us like the master dramatist who received the homage of his audience. It was at such moments that for an instant he ceased to be a reasoning machine, and betrayed his human love for admiration and applause' (p. 594)/ During a similar feat in 'The Norwood Builder' Watson notes that 'Holmes stood before us with the air of a conjurer who is performing a trick' (p. 508).

Holmes also reveals his knowledge and dexterous application of misdirection, the art of 'distraction', as Maskelyne and Devant describe it, which is 'mainly derived from some action on the part of the magician himself' whenever he 'has anything "magical" to do'.[57] Holmes's method of misdirection is similar across several stories and normally involves knocking over an object to 'distract' everyone nearby. Watson describes one such moment in 'The Six Napoleons' (1904):

> Near the foot of the bed stood a dish of oranges and a carafe of water. As we passed it Holmes, to my unutterable astonishment, leaned over in front of me and deliberately knocked the whole thing over ... I stooped in some confusion, and began to pick up the fruit ... the others did the same ... Holmes had disappeared. (p. 405)

Holmes's disappearance on magical business follows the misdirection of his audience and leads directly to the discovery of a piece of evidence necessary to proving the guilt of the suspected criminal.

One magical performance that depends entirely on misdirection is the three cup trick, much beloved of sideshow swindlers, who invite spectators to guess under which cup a ball is to be found. It exemplifies Houdini's definition of misdirection as 'prompt[ing] the thought that a thing is really not what it seems nor where it appears to be'.[58] Holmes reprises this trick for the conclusion to 'The Naval Treaty'. Apparently unsuccessful in retrieving a lost document of state, Holmes, Watson and the document's owner, Phelps, sit down to breakfast:

> [Mrs Hudson] brought in three covers, and we all drew up to the table, Holmes ravenous, I curious, and Phelps in the gloomiest state of depression ... 'Mrs. Hudson has risen to the occasion,' said Holmes, uncovering a dish of curried chicken ... 'what are you going to take, Mr. Phelps ... try the dish before you.' Phelps raised the cover, and as he did so he uttered a scream and sat there staring with a face as white as the plate upon which he looked. Across the centre of it was lying a little cylinder of blue-gray paper ... 'it was too bad to spring it on you like this, but Watson here will tell you that I never can resist a touch of the dramatic.' (p. 466)

Phelps's scream reveals the success of Holmes's misdirection: the breakfast dish was not what it appeared to be and the document beneath it entirely out of place, dislocated from its perceived location elsewhere.

Illusions involving misdirection fooled the eye of the observer. Phelps's 'staring' at the uncovered document betrays a moment of ocular confusion, a dislocation between the image on his retina and the meaning made of that image in his mind. Conan Doyle would have been aware of the many different eye defects that produced similar fragmentations of vision, and conscious of the parallels between ophthalmic experiment and the performance of optical illusions. There is a clear connection between the manipulation of the eye in magical performance and the effects produced by experimental practices in clinical ophthalmology. Landolt, for example, describes an experiment to determine the extent of 'false projection' in 1913, the effect of the paralysis of eye muscles in locating objects in the external world:

> If the patient usually uses the non-paralyzed eye for fixation, and if this eye be covered so as to oblige him to use the paralyzed eye for fixation, instead of locating the fixed object in its real position he projects it towards the side to which the paralyzed muscle should normally rotate the eye. He will estimate the situation of the object according to the amount of nervous energy which he requires to expend to direct the deviated eye towards that object. The more pronounced the paralysis, the greater is the expenditure of innervation. In reality the degree of false projection usually is in relation to the degree of paralysis.[59]

Landolt's experiment induces the patient to believe that objects are not where they appear to be. The patient's dysfunctional muscle sends to the brain signals of eye movement that do not tally with the actual movement of the eye. The eye's ocular activity prompts the brain to locate an object in one place when it actually exists in another. This is misdirection prompted by the eye's physiological breakdown. Optical illusions, such as the one Holmes performs on Phelps, are likewise experiments in ocular fallibility, turned from clinical investigation into creative performance. Like ophthalmic research optical illusions explore the nature of visual dislocation. Holmes's own optical illusions take this further, for they are designed to be revelatory, to bring to light causes and to provide knowledge.

Jeremy Brooker, in an otherwise excellent article on Pepper's Ghost, claims that 'magical illusions were based on a desire to deceive'.[60] This is too reductive. Optical illusions are experimental performances that investigate vision. Their desire is not to be deceitful but rather to illuminate visual weakness by showing the extent of human ocular deficiency. Conan Doyle certainly employs Holmes's magical performances to these ends. His optical illusions are not examples of deception but acts of discovery. The cup trick he performs on Phelps is a prelude to revealing that he has discovered the missing document, and similarly the box trick that so amazed the European Secretary was a dramatic method of revealing the missing letter. Holmes summarizes his illusion practice as a way of reaching for the truth: at the conclusion of 'The Naval Treaty, and in the immediate aftermath of his three cup trick, he explains to Watson that he aimed to show how 'what was vital was overlaid and hidden by what was irrelevant' (p. 467).

Conan Doyle therefore provides a clear distinction between Holmes's optical illusions and the illusion practices performed by the criminals he investigates. Criminal illusions instigate mystery and maintain secrecy, while Holmes's provide solutions and offer transparency. For contemporary theorists of magical performance only Holmes's illusion practice was genuine. Maskelyne and

Devant, in giving definition to professional magic in *Our Magic*, argue that 'a legitimate magician never deludes his audience as to the character of his performance. He makes no claims to the possession of powers beyond the scope of physical science.'[61] Holmes's optical illusions do not prey upon ocular anxiety, as do those of his adversaries. Rather they offer interpretations of the physiological weakness of ocular structure. They do this within a context of joyful discovery – of the mystery at the heart of the narrative and the visual deceptions that enabled it. They exploit, ultimately, the power of the eye rather than its fragility, coming, in doing so, to the same conclusions as Helmholtz and the other leading optical researchers of the period.

It is in this light that we should read Houdini's exposure of deceitful magicians, as well as his exposés of spiritualist performance, which is the subject of the following chapter. Houdini's distaste for dishonest magicians was a constant of his career: he wrote an entire book, *Miracle-Mongers and Their Methods* (1920), on historical cases of conjuring that had been exploited as supernatural or miraculous acts.[62] He also harangued, usually in the daily press, magicians who copied his own act, or aspects of it, or the acts of his closest associates. In his library he kept a series of playbills advertising his imitators, often rudely annotated, in order to build a general picture of the extent, as he saw it, of magical deception. They all make similar claims, often suggesting their own creation of tricks that Houdini had successfully performed for many years:

> Newton Lumar Mentalist and Escape Artist.
>
> Newton Lumar Presenting His Sensational Death-Defying Escape from a Pine Box which will be locked, strapped and Roped.
>
> The Great Mail Bag Mystery
>
> Escaping from 32 feet of Chains in full view of the audience.[63]

Some performers were even more gauche. Robert Doige, a representative of the Society for American Magicians (SAM), sent Houdini a packet of pamphlets in 1909 advertising the performances of 'Oudini, the Handcuff King'. Doige writes in an accompanying letter to Houdini:

> I came into the possession of these handbooks by purchase, in order to keep them from general circulation and out of the hands of those having no genuine interest in magic. All magicians will, of course, understand that OUDINI, the author, is not Harry Houdini, the President of SAM, and chief exponent of the handcuff act. The author understood his subject however, although it is to be regretted that he adopted the name he uses.[64]

What is interesting here is the status of that complex word 'genuine'. It is understandable that professional magicians wished to protect their own magical commodities from theft and imitation. Yet Doige and Houdini appear to regard themselves as defenders of a magical tradition that upholds illusion practice as essentially genuine. But what meanings adhere to the term, and how far can we see Houdini's work as genuine, when magical performance relied upon sleights of hand, misdirection, disguise and concealment? Houdini's performances are excellent examples of these deceptive techniques. In his escapes Houdini's audiences witnessed his strength, suppleness, even agility. They marvelled at the abilities of his body: his toes could untie ropes and his teeth unbind buckles. Yet Houdini employed the techniques of magical performance to allow him to use various tools and implements out of sight of his observers. *Handcuff Secrets* lists the many different lock-picks and keys that he could hide on his person or clothes. In the notes he left at his death, he tells of how he released himself from buckles: 'This is done by having a secret pocket in my clothes, containing a small, sharp knife and a duplicate strap for my wrists'.[65] His working notes on cabinet escapes show the various methods by which an assistant might be concealed in order to help him out of his chains, or how various other devices could be hidden inside cabinet curtains or even the steel rods holding the sides together.[66]

Even the most celebrated aspect of Houdini's performances – the challenge escapes – were carefully constructed pieces of theatre. On arrival in a new city where he was engaged to perform Houdini would invite challengers to approach him with any piece of restraining equipment they preferred. These were often handcuffs or chains, brought to him by the local police or prison warders. Sometimes they were more complex, such as the 'old and obsolete padded cell suit' which the Cardiff army challenged him to escape from when he performed in that city in January 1913.[67] Leaflets explaining these challenges were distributed to the public, and the challenge items were placed on show in the theatre's entrance. However, the challenges were themselves fake: Houdini approached the challengers himself, the playbills were written by Houdini and his assistants, and the restraints were often substituted for Houdini's own, complete with the magical apparatus that allowed for easy escape. We can build this picture of the challenges from a great deal of historical evidence. Houdini, for example, carried 'challenge items' as part of his general properties for each performance, the challenge leaflets (written apparently in different countries, even on different continents) are printed with the same design details and contain the same linguistic constructions.[68] Walter Gibson, who had access to Houdini's performance notes after the magician died, writes that 'most challenges were planned beforehand' and for those that were most difficult 'Houdini smuggled tools or implements in [to the cabinet] with him in order to effect a release. The music of the orchestra covered the sounds that came from the cabinet.'[69]

Since the audience were unaware of the tools Houdini used, or the magical apparatus of the cabinet, or the planning of the challenge escapes, how far can his illusions be regarded as genuine magical performances? To be genuine, though, was to reveal to the audience the character of the performance, not its content. Just as Holmes's acts of magic revealed deception rather than dramatized it, Houdini's optical illusions were presented in the conjuring tradition as examples of drama that combined 'art and science'[70] to create 'mental impressions' that achieved their goal 'by misdirection of the senses'.[71] Houdini never pretended that his achievements were the result of supernatural intervention or of faked presentation. His performances were not 'the offering of puzzles for solution'[72] but optical illusions that were to be enjoyed for their ability to deceive the eye. Houdini's performances were examinations of the different ways in which the ocular landscape might be manipulated. While they did not offer explanations of the laws of physiological optics that were the foundation of the illusions they revealed by practical example the limitations of human perception that optics and ophthalmology had uncovered.

Houdini had a keen understanding of how the eye worked physiologically, and how the eye in combination with the brain arrived at perception. One of his most important works on magical performance was an article written in response to the French psychologist Alfred Binet's earlier essay on magic and psychology in the *Revue Des Deux Mondes* in 1894.[73] Binet had written extensively in his early career on visual memory and suggestibility. Houdini's article, 'The Psychology of Prestidigitation' is his most complete theoretical examination of illusion practice and the most astute analysis of optics he provided at any point in his career. Houdini opens the article with a stark statement of his understanding of illusion as 'deception of the optic nerve', which he links with delusion as 'deception of the mind'.[74] In agreement with Helmholtz's analysis of the relationship between retinal vision and mental analysis, Houdini goes on to connect illusion with delusion by seeing both as examples of perception, although his understanding of the process by which light reaches the brain is rather less accurate than it might be: 'the optic nerves through the eye, receive vibrations, produced by light, color and form and these vibrations are interpreted by the mind'.[75] Despite misunderstanding the physical properties of light waves, the remainder of Houdini's analysis of optical illusions shows a subtle knowledge of the role of vision.

In particular Houdini highlights how optical illusions, in his own performance practice as well as in general theory, exploit the gap between sight and perception. It had been recognized for over a century that the retinal image of the external world is not complete. The area where the optic nerve attaches to the retina is devoid of light-sensitive cones, and is essentially blind. This is colloquially known as the blind spot. Any image of the external world is therefore only ever partial as some aspect of it is lost to the blind spot. David Brewster discusses this in *Letters on Natural Magic*, providing his readers with a 'try this at home' experiment that both reveals the existence of the blind spot and shows how we are perceptually unaware of it in normal observation.[76] The mind, he proves, fills in the dark space of the blind spot and provides a complete image. Houdini argues that optical illusions often work to take advantage of the desire for completeness in human perception. Illusion practices often involve incomplete actions (such as the pretence of throwing away something which is actually retained). Done with appropriate prestidigitatory skill, audience perception will complete the action despite the eye's failure to see it. For Houdini, who criticizes Binet's failure to note this, such manipulation of perception is permanent:

> A common error with Binet ... is that he says most of the illusions are of but momentary endurance; that is not so ... the height of the illusion as presented by the magician, is, where the illusionary effect is permanent and perpetual so long as the visual rays from the audience are directed on them.[77]

The constant attention of the audience's vision also allows the illusionist to celebrate the perceptual gap, by 'calling particular attention to the details, the impossibility and impracticality of doing certain things'.[78] Again this is successful only when the illusionist 'draws particular attention to the very thing that is deceiving the eye'.[79] Moreover, this drawing in of the focus of the eye onto the actual objects of the illusion, and to the illusionist himself, serves to heighten perception and increase the illusion's potential effects. Helmholtz, in his lecture on the 'Facts in Perception' argued that the spectral illusions produced when there is a dsyfunction in the eye are at their most intense when there was a 'transference of the state of excitation from the part of the brain that is active to the visual apparatus'.[80] Houdini's technique of continually bringing the audience's perception of an illusion back to what they were observing on stage embodies this transference of stimuli in his performative practice.

In the performance of a magician such as Houdini optical illusions become dramatic spectacles of human vision. They reveal the imperfections of the ocular structures that enable sight and illuminate the shortcomings of perception. Provoking an audience's optical consciousness by making real the eye's defects might be said to heighten ocular anxiety rather than to alleviate it. Instead, however, performers of optical illusions transform any emerging optic nervousness into pleasurable enjoyment of the dangers of ocular failure.

Houdini learned how to achieve this from his extensive study of Robert Houdin. Whatever he thought of Houdin's writing on escapology – and in his own copy of one of Houdin's works he scrawled 'Rot!' in the margin – Houdini followed his teaching of performance practice very closely. Houdin, in an extensive piece on the importance of the magician's own eyes to successful illusions, advises the illusionist to take charge of the audience's eyesight immediately on entering the stage. Houdini's own performances show how he continued to follow this advice throughout his career and he eventually paraphrased Houdin in a short undated work of his own, 'Addressing an Audience'.[81] Houdin's original is worth considering unabridged:

> In conjuring parlance, 'to have a good eye' is the phrase used to denote a particular stage qualification whereby the sympathy of the spectators is attracted to the artist, and which moreover has the faculty of greatly heightening the effect of the tricks performed. You have doubtless, reader, sometimes found yourself in the company of persons whose glance you felt as if you could not meet with comfort, and whose eyes seemed in like manner to avoid meeting yours. Such a state of things is embarrassing to both parties, and the pleasure of the conversation is pretty sure to suffer in consequence. The reason of this is that the eyes of your interlocutor are timid, wandering, uncertain; that he cannot support your straightforward look; it arises, in a word, from the fact that he has not 'a good eye'. This kind of ocular nervousness, this feeling of uneasiness and embarrassment, is extremely catching, and where a public performance is afflicted with the malady it is not unusual to find it spread through the whole of the company. In such a case the spectators are unsympathetic, and often even unfriendly. 'To have a good eye' is to possess the quality which is the antithesis of the defect I have just mentioned. Note the advance to the footlights of yonder artist, whose keen, intelligent, self-reliant glance goes straight to meet the eyes of the company. A relation of an almost mesmeric character is instantly established between all parties. The spectators are at their ease with the performer, they at once catch his eye, they listen to him with indulgence, and from this double relation there speedily arises a feeling of sympathy [and] ... conviction.[82]

The eye of the magician brings the audience into sympathetic alignment with the performance of the optical illusion, banishing 'ocular nervousness' and replacing it with 'ease', 'indulgence' and 'conviction'. This was also what Houdini aimed to do, so that as his magical performances brought to the audience's attention their weak and unruly vision, they also created an atmosphere in which there was no anxiety in the knowledge of their own ocular inadequacy.

Both Holmes's optical illusions and Houdini's magical performances reveal contemporary cultural concerns of increasing visual disability. Yet they do so without initiating anxiety. Indeed they show how the public's disquiet about degenerative visuality can be made safe, even instructive, and ultimately turned into entertainment. This is a very different visual culture to the more common historical view of visual fragmentation engendered by the shock of the new in the early twentieth century. The frisson in the work of Houdini and Conan Doyle is not the developing sense of visual blindness confronting the modern subject, but rather that in the moment of recognizing visual blindness comes the knowledge that there are those with the visual power (as optical and ophthalmological researchers had suggested) to turn ocular opacity into clear sight.

The Role of the Observer

Conan Doyle's detective fiction and Houdini's magical performances contend that optic nervousness was not simply a product of an institutionalized surveillance society nor always productive of a scopic culture of fear and bewilderment. Ocular instability was as likely to limit the gaze of repressive institutional hierarchies as it was individuals, and visual dysfunction could be celebrated as entertaining and revelatory spectacle. However, whether collectively or individually, through enjoyment or anxiety, observers were subject to visual manipulation. What forms these manipulations took, and what they represented, must be understood to achieve a complete picture of the ocular interactions on which optical illusions depend.

Optical illusions were often investigated by physiologists. Brewster wrote an influential paper on one such illusion in 1826, and returned to the topic for one of his *Letters on Natural Magic* in the 1830s.[83] His paper, which dealt with 'the optical illusion of the conversion of cameos into intaglios', gave further impetus to one particular way of understanding sight that had been under scrutiny from the mid-eighteenth century and was known as the judgemental theory. Brewster claimed that an observer viewing the raised contours of a cameo would, if the angle of light was altered so that the shadows fell in a different direction, recognize the cameo as an incised *intaglio*. He concluded that 'the illusion, therefore, under our consideration is the result of an operation of our minds, whereby we judge of the forms of bodies by the knowledge we have acquired of light and shadow. Hence the illusion depends on the accuracy and extent of our knowledge on the subject.'[84] As Jutta Schickore has argued, 'Brewster's experiments demonstrated that the element of mental activity was a general feature of sense perception and he showed how precarious was the act of perception because of this element of mental activity'.[85]

By the 1850s and 1860s Brewster's view that judgement was central to sight had been taken on by Helmholtz, whose empirical theory of vision had its roots in earlier judgemental principles. In the third part of his *Treatise*, published in 1867, Helmholtz argued that what we regard as an 'illusion of the senses' is actually an 'illusion in the judgement of the material presented to the senses, resulting in a false idea of it'.[86] For R. Steven Turner, this view of perception was 'a methodologically legitimate explanation' of illusions, highlighting how 'unusual patterns of stimuli "tricked" the perceptual process into an inappropriate application' of inference from things seen.[87] For Helmholtz, however, it was not merely mental inference that was at play in judging the images appearing on the retina, but also memory. In 'Facts in Perception' Helmholtz had first introduced elements of his empirical theory by arguing that 'the memory vestiges of previous experiences also play a further and highly influential role in the observation of our visual field'.[88] For Helmholtz, then, just as 'the manner of action of light [is] conditioned by the way in which the nervous apparatus reacts'[89] so too is an understanding of the retinal image dependent upon mental acts of recollection and knowledgeable reflection. Judgemental and empirical theories essentially argued that our understanding of what we see depends greatly on our comparison of an image with previous similar images and with our knowledge of the usual pictures produced by similar combinations of light, tone, depth and colour (known in physiological optics as the 'local signs' of vision).

Performers of stage illusions clearly understood such modes of perception, at least as they were practised by observers. Nineteenth-century works on magic often stressed the combined importance of attention and misdirection to potential conjurers; a clear attempt to draw on the knowledge of the observing subject while at the same time presenting them with those 'unusual patterns of stimuli' that would trick them into ocular misprision. Maskelyne and Devant, for example, stressed the importance of the audience's 'mental adaptation'[90] to the performer's illusion practice, and Houdini discusses how the audience's visual attention is central to any illusion. As Gibson notes in his editorial to the published version of Houdini's private notebooks, the 'holding of interest'[91] was key to Houdini's work.

Conan Doyle also considers the connection between the visual apparatus and mental activity. Indeed the key difference between Sherlock Holmes (the performer) and John Watson (the observer) is their varying ability to connect visual stimulus to mental action. Holmes has this in abundance, as seen by his method of linking visual evidence to prior actions but also in his knowledge on subjects such as cigar ash, and his memory for historical criminal activities. Watson, by contrast, is unable to make the necessary links between what he sees and what that vision leads him to understand. Holmes puts this to Watson succinctly: 'you see, but you do not observe' (p. 162). Indeed, this line subtly confirms Helmholtz's explication of pure sensations, which are not 'seen' but unconsciously taken up by observers for their eye to function correctly. Conan Doyle's detective fiction provides many more examples of the visual failure of the observer, such as Colonel Ross's inability to translate what he has seen into any meaningful narrative of events. Ross, whose horse Silver Blaze has been stolen, has access to the same visual evidence as Holmes – including the 'cataract knife' (p. 342) which is representative of the ocular dysfunction from which Ross suffers – but is incapable of bringing to bear on it any act of mental organization. When Holmes reveals to Ross the meanings of the evidence, Ross can only cry, 'I have been blind!' (p. 349). Holmes often despairs at the failure of observation he encounters. In 'The Copper Beeches' (1892), he complains to Watson of the disregard the public have for observational expertise: 'what do the public, the great unobservant public ... care about the finer shades of analysis and deduction' (p. 317). Holmes uses a range of optical vocabulary here; his 'finer shades' recalling particularly the cameo/intaglio illusion and Brewster's experimental work on light and shadow.

If the mind, and not the eye, was emerging (at least to some physiologists) as the most important part of the complete ocular apparatus then the mind's susceptibility to fanciful visions of its own creation required analysis. As Schickore argues, Brewster's 1826 paper on the conversion illusion of cameo into intaglio was little short of a 'demonstration of the "power of fancy" and its dangers'.[92] In Brewster's materialist philosophy, with its anti-Romantic ideology, the imagination was a weak link in the optical system, transforming retinal light into illusory, even delusionary, images. Clinical ophthalmology appeared to support Brewster's view. Hogg, in his 1858 book on the ophthalmoscope, represents the views of many ophthalmologists when he claims that spectral illusion caused by motes in the eye were often made worse by patients suffering from diseases associated with imaginative excess:

> Floating bodies, muscae, are not unfrequently the cause of very troublesome spectral illusions. A young lady suffering from muscae often imagines she sees persons or animals moving about her room, which is particularly troublesome towards evening, or in a dull light, when the pupil becomes dilated. This condition will explain many curious illusions of which we hear, and find associated with particular temperaments; hysteria, hypochondriasis, febrile and other affections.[93]

Like Brewster and the ophthalmologists Conan Doyle also warns of the dangers of 'fancy' in the interpretation of visual evidence. Holmes's methodical approach to detection does not provide a role for the imagination, and in fact he distrusts it so extremely that he is disgruntled even by Watson's use of it in giving narrative cohesion to his solving of crime. In *The Sign of Four*, Holmes dismisses Watson's work: 'Detection is, or ought to be, an exact science and should be treated in the same cold and unemotional manner. You have attempted to tinge it with romanticism' (p. 90). Across the range of Conan Doyle's detective stories it is always the observer who falls foul of their own imagination; sometimes to comic effect. The official police are most often subject to the vagaries of fancy:

> 'Ha! I have a theory. These flashes come upon me at times ... What do you think of this, Holmes? Sholto was, on his own confession, with his brother last night. The brother died in a fit, on which Sholto walked off with the treasure? How's that?'
>
> 'On which the dead man very considerately got up and locked the door on the inside.'
>
> 'Hum! There's a flaw there.' (p. 113)

The ontological failure that imagination engenders is further reinforcement of the problematic relationship between ocular vision and mental interpretation. Such a relationship is also recognized by Houdini. His 1924 short story, 'Imprisoned with the Pharaohs', ghost-written by H. P. Lovecraft and published in the popular periodical *Weird Tales*, is a narrative that investigates visual uncertainty.[94] Opening the story, Houdini (in the guise of a first-person narrator) tells the reader:

> What I saw – or thought I saw – certainly did not take place; but is rather to be viewed as a result of my then recent readings in Egyptology ... These imaginative stimuli, magnified by the excitement of an actual event terrible enough in itself, undoubtedly gave rise to the culminating horror of that night.[95]

Of course Houdini, the heroic illusionist, does not remain in thrall to his imagination for long: 'I am quite sure that I preserved a logical consciousness ... that I did not add any fullgrown phantoms of imagination to a picture ... explicable by a type of cerebral illusion'.[96] As Hogg and Brewster explain, the imagination is often at fault for fancifully enlarging what is an easily identifiable optical illusion. By confirming this view Houdini also represents the imagination as a dangerous influence, reimagined in this short story as a barrier to survival and an opponent of rational thought.

Despite articulating a sense of anxiety about the effects of the imagination on accurate observation, physiologists were confident that these could be controlled. Schickore concludes her discussion of a range of optical illusion experiments by attesting that, 'In the investigations and experiments on popular optical deceptions ... the possibility of acquiring visual knowledge as such was never at stake'.[97] Hogg, too, was unconcerned by spectral illusions, calmly noting that they were easily 'curable by judicious medical treatment'.[98] New optical knowledge was the key to weakening the influence of the imagination. Developments in physiological optics and advances in clinical techniques in ophthalmology gave cause for optimism in the ongoing attempts to secure visual acuity. The recognition that optical knowledge advanced accurate vision was also taken up in the public sphere. Increasingly, popular scientific lectures combined optical entertainment with educational dissemination, providing spectators with a knowledge of optics that enabled them to understand, and thereby see more properly, those ocular deceptions that had previously baffled them.[99]

The London Polytechnic Institute was a significant example of the drive towards visual learning. Helen Groth, writing about Henry Pepper's ghost illusion, argued that Pepper's accompanying lectures on the optics of the ghost 'allowed Polytechnic audiences to rationalize the mystery of Pepper's Ghost via the lens of past knowledge'.[100] This did not undermine the effect of the illusion, Groth maintains, but rather refined it: 'Pepper's explanation of his own spectacular illusion suggests that an emerging and rapidly expanding visual vernacular fostered sceptical curiosity, which worked in concert with, rather than against, visual and textual illusions'.[101] Indeed the 'puzzled and uncertain'[102] audience that had not been witness to the explanatory lecture, were transformed into an expert collective appreciative of Pepper's skill in manipulating optical laws that they too understood.

However, the optical knowledge that inoculated spectators against the infecting potential of ocular trickery is only one aspect of a broader observational dynamic that occurs when optical illusion are performed and witnessed. Optical illusions were still understood by observers even when they were fooled by them. Their understanding might be different from the fact-based knowledge of optical laws, but it was nevertheless a *coming to know* the meanings of visual manipulation that proved to be as useful a form of cultural knowledge as an understanding of optical principles was useful scientific knowledge. In fact, an instrumental scientific knowledge – a simple 'knowing how it works' – tended to undermine the broader representational structures that were the most instructive in optical illusions. As David Abbott argues in his 1908 book on magic as a form of wonder, 'Were the public at large to become thoroughly instructed in the means by which magicians perform their effects, the noble art of magic would disappear. It can exist only by their being suitable subjects, upon whose minds the performer can produce his illusions.'[103]

So dominant has the perception of the Victorian period as one of progress through science become that it is difficult to recover from history those moments where science impedes knowledge. The relationship of optical illusions to visual culture is one of these moments; where an observer's understanding of optics promotes factual knowledge over an intellectual engagement with the social and political implications of vision's fallibility. Maskelyne and Devant highlight the important distinction between hoodwinking the sense of sight and deceiving the intellect:

> The misdirection which forms the groundwork of magic does not consist in telling lies, with the object of deceiving the spectator's *intelligence*. It consists, admittedly, in misleading the spectator's senses, in order to screen from detection certain details for which secrecy is required. It militates against the spectator's faculties of observation, not against his understanding.[104]

In spite of physiologists' fears that optical illusions make uncertain the mind's analysis of the eye's visual landscape, and thereby disable attempts to understand the external world, performers of optical illusions maintain that alternative forms of understanding remain available. Moreover, optical illusions also offer other, perhaps more creative, ways of understanding vision's weaknesses, both its failures of judgement and its susceptibility to imagination.

Audiences witnessing magical performances rarely exhibit the straightforward puzzlement and uncertainty that Groth regards as characteristic of observers at Pepper's Polytechnic entertainments. Indeed, such spectators bring to their experience of optical illusions a different method of understanding: revelling in their own deception and enjoying the speculative discussions that generally followed the conclusion of a successful stage performance. Numerous reports on magical performance from the late nineteenth century through the early twentieth century highlight the sense of wonder experienced by audiences. A report on the celebrated illusionist Harry Kellar from 1897 notes that 'the magician's audiences apparently like to be deceived if expressions of dumbfounded amazement, wonder and pleasure count for anything'.[105] Even if audiences did appear 'nonplussed', as another report on Kellar states, they were also conscious of the effects produced on them: 'It was not unnatural that the spectators should conclude they were being treated to a clever optical illusion'.[106] Houdini's own performances provoke similar responses, but also leave space for contemplation on the nature of vision. A report of his performance at Keith's Theatre on 3 April 1906 highlights the stimulating effect of optical illusion on the intellect:

> Houdini not only gives the spectators at Keith's something unusual to see, but material for comment and speculation long afterward ... With Houdini the interest remains keen for several days ... Last evening Houdini slipped out of one or more pairs of especially binding handcuffs, escaped from his prison cell and from his barrel. He was even quicker than usual. The spectators watched intently as ever and at the end of the act the house hummed with the speculations of the audience. Never was a vaudeville turn the cause of more wonder.[107]

Optical illusions do not, in these instances, bring the mind to a halt. Rather the enervation of the eye activates an intellectual engagement founded in the pleasure of deception. Wonder initiates a process of speculative curiosity that allows the spectator space to analyse but also to celebrate their recognition of the eye's frailty.

The observers of Holmes's optical tricks respond with similar blithe exuberance, and crucially do so before Holmes offers an explanation of the illusion. When Holmes produces the 'famous black pearl of the Borgias' (p. 594) from inside a plaster bust, Lestrade and Watson are both spectators: 'Lestrade and I sat silent for a moment, and then, with a spontaneous impulse, we both broke out clapping, as at the well-wrought crisis of a play' (p. 594). Likewise, when Holmes employs his own illusion to reveal the hiding place of the apparently 'vanished' Jonas Oldacre, Watson regards the trick as 'amazing' (p. 508) while Lestrade believes it to be 'the brightest thing that you have done yet', even though 'it is a mystery ... how you did it' (p. 509). Percy Phelps, the unwitting spectator to the optical illusion Holmes employs to return the naval treaty, offers the most extreme example of celebration as he 'danced madly about the room, pressing [the treaty] to his bosom and shrieking out with delight' (p. 466). Of course the ultimate observer in these detective fictions is the reader, who is interpolated into the position of the narrative observer and invited to celebrate with them in finding Holmes's revelations wondrous.

What did spectators come to understand from observing optical illusions, and why did they respond so positively to their deceptions? For one thing, optical illusions downplayed individual anxieties about deteriorating vision by hoodwinking everyone. For the single observer (in a crowded audience at one of Houdini's performances, or as a reader of the *Strand*) there was satisfaction in this flattening of ocular hierarchies, especially as ophthalmological research continued to suggest that visual acuity was related to social status. R. Steven Turner notes that 'the scientific study of vision assumed its modern outlines between 1838 and 1868',[108] and it was also in this period of ocular reformation that the relationship between sight and social power was solidified. Statistical evidence of a scale useful to make judgements only became available to ophthalmologists after the emergence of the larger eye hospitals in Britain and across Europe in the first thirty years of the nineteenth century. Even then, diagnosis of eye disease was difficult until the modern ophthalmoscope was invented and made clinically available in the 1850s. From the 1850s through to the 1920s the data of ophthalmological research was often interpreted through a politically conservative ideology that made assumptions about the relationship of the eye to other structures of social and cultural power.

Hogg's statistical evidence of the preponderance of eye disease in various professions, for example, reveals that 'Labourers' suffer most from dysfunctional vision (some 305 cases) while artists suffer least (only 4 cases).[109] Far and away the greatest number fell in the category of 'Married women with household duties' (634 cases), although of course this was not included as a profession.[110] No account was taken of the relative number of men who were, in London or across England, employed as labourers or artists. Nor, for example, did Hogg mention the relative position of the Westminster Eye Hospital to London's East End, one of the poorest areas of England, and one with a high incidence of disease generally.

From the 1850s ophthalmological textbooks, study guides and introductory courses of lectures perpetuated a belief that visual weakness increased in parallel with conventional social hierarchies: the further one descended down through the different classes of Victorian Britain the worse eyesight became. To some extent there is a recirculation of this knowledge between these different books. Nearly all of them reference those that have come before, and draw on, or simply repeat, their illustrative cases. Yet in 1920 Lionel Laurance, director of the School of Optics in London, could still write that 'what would be taken as average normal visual acuity in the out-wards of an East End hospital could not be so regarded in the consulting-room of a West end oculist'.[111]

Optical illusions recreate the subjective conditions of eye disease and ocular dysfunction. Observers are made to see the spectral illusions associated with the hysteric suffering from problematic *muscae* on the cornea. Or, clever misdirection provides the observer with the sensation of muscular paralysis that leads to false projections of the location of objects. Yet in pathologizing the individual observer, optical illusions also pathologize every observer. Every reader of the *Strand*, every witness to Holmes's tricks and every spectator at Houdini's performances were subject to the same ocular instability. Optical illusions offered an egalitarian world of shared visual fallibility. They also engendered a carnivalesque enjoyment of communal failure: one could laugh at the ocular weakness of the police inspector, take pleasure in the disbelieving eye of the prison guard, or celebrate the exposure of the illusion practice of the criminal.

In particular, however, the optical illusions of Houdini and Conan Doyle disrupt, even flatten, traditional social hierarchies. In Conan Doyle's detective fiction, every character, regardless of their social status or cultural role, finds it difficult to maintain an accurate observation of what Watson calls 'the ever-changing kaleidoscope of life as it ebbs and flows' (p. 424). Holmes acts for powerless daughters, governesses and clerks but also for government ministers, lords and kings. Each of them has either been betrayed by their eyesight, or is tricked by Holmes's optical illusions. They all represent the same visual failure. Houdini's magical performances also exercise a disestablishment of social privilege. His stage committees, always groups and never individuals, bring together sailors and physicians, the first because they are likely to be 'familiar with knots', the second because they can attest that 'there is no way of contracting the bones and muscles so as to slip out of the knots'.[112] When Houdini unties and slips these knots both are revealed as equally visually inept. In his challenge escapes Houdini illuminates once again the equally poor visual capacity of, on the one hand, employees of a mail bag manufacturing company, and on the other, the chief of police. For the observer of Houdini's optical illusions ocular weakness is neither a corollary of social status nor embodied in any specific individual.

In a very real experiential sense observers of such optical illusions and experiments still felt a sense of the wonder that such encounters offered. In direct contrast to Daston's and Park's argument that wonder was mostly lost, and when it was not it was sidelined and emasculated within specific forms of entertainment, wonder appears in these examples as powerfully egalitarian, popular and democratic. Indeed the democracy of the emotion of wonder is significant here as it allows optical illusions to be experienced and understood across social divides. Wonder, then, takes on a new role in divesting observation of social and class status and bringing a new democracy to previously privileged knowledge.

Houdini's and Conan Doyle's optical illusions illustrate the weakness of human eyesight, certainly, but even as they fragment vision they offer a collective experience of that fragmentation. For observers, the optical principles of illusions did not need to be understood for their meanings to be clear; factual scientific knowledge actually impeded the experience of being fooled by optics, and it is in the moments of experience that the meaning of Houdini's and Conan Doyle's illusions were felt. As the observer marvelled at the effect of an illusion they recognized the similar emotive response of their co-spectators. In such a way the individual becomes connected to a larger community of observers, equally perplexed and similarly speculative. By speculating together observers also become social together. The staging of an optical illusion constructs an egalitarian observing community, positively disenfranchised from the inequalities outside the text or the theatre. Social authority is discarded the moment the illusion takes hold: it is overshadowed by visual anxiety that is transformed into a celebration of the observer's likeness to every other observer. For the spectator, the heroism of the illusionist – whether detective or magician – is not in the performance alone but rather in the performance's binding together of the collective body who witness it. The illusion of perception is not, as the psychologist of vision James Sully would have it, a 'deviation' from 'collective experience' that 'breaks the chain of intellectual solidarity' and leaves the observer in solitary despair for their ocular frailty.[113] Instead it is, in Houdini's and Conan Doyle's hands, an affirmation of the universality of optic nervousness; a quietly, yet importantly political, recognition that perceptual disruption is commonplace but only threatens when individuals are left to see for themselves.

8 ARTHUR CONAN DOYLE AND HARRY HOUDINI: SENSATION, SPECTACLE AND SPIRITUALISM

While Houdini's magical performances and Conan Doyle's fictionalized optical illusions reveal a sympathetic synergy in their articulation of a scopic democracy – and in the process undermine modernity's assumed fragmentation of optical space – their very different relationships to spiritualism can be read as a fracture of that sympathy. Moreover, their fierce opposition to one another on the subject of spiritualism illuminates the possible divergent paths of science and vision under pressure from early twentieth-century commodity capitalism and its stress on conformity to spectacular society. This final chapter will consider Houdini's and Conan Doyle's responses to the key site of spiritualist practice – the séance – as a starting point for a more extensive investigation of the role of vision as it crosses from the nineteenth century into the twentieth, or to put it differently, as it shifts from Victorianism to modernity.[1]

Spiritualism's role in British and American culture has been broadly examined by historians. Its gender, racial and class politics, its role as a new religious movement and its relationships to the 'dominant ideology of the era', as Molly McGrath puts it, have all been fertile ground for a consideration of the place of the séance in nineteenth- and twentieth-century societies.[2] Spiritualism's relationship to science has also been investigated, most profitably by Richard Noakes, and partly as a result of this work, and partly due to the turn towards the periphery in history of science scholarship generally, the séance has gained some cache as an interesting site of scientific contest.[3] The role of magicians – one of the most influential groups who stood almost unanimously in opposition to spiritualism from the mid-nineteenth century – has been examined only tangentially. Janet Oppenheim, for example, in her important early study of British spiritualism, briefly considers those 'professional conjurers' who aimed to 'expose the tricks of dishonest mediums' but only to introduce a broader argument about spiritualism's relationship to 'man's age-old belief in witchcraft and spirit possession'.[4] Yet from the time of spiritualism's emergence as a transatlantic phenomenon séance performances always attracted the attentions of spiritualists, scientists and magicians.

Conan Doyle's relationship with Houdini is the best known of the many interactions between spiritualists and magicians. Both of them encountered spiritualism early in their careers, before either had become figures of celebrity. Conan Doyle's earliest spiritualist experiences appear to have taken place in Southsea during his period as a practising physician. In a letter to the spiritualist periodical *Light* published on 2 July 1887, Conan Doyle writes that his most recent séance experience 'was the first time that I had ever had the opportunity of sitting with anyone who was not a novice and inquirer like myself'.[5] Houdini's early encounters with spiritualism came less than a decade later, and predictably took the form of a magical performance. Houdini's brother, the magician Hardeen (Theodore Weiss), notes in his own memoranda that Houdini had included a séance as part of his magical act soon after his marriage in July 1894. The playbill for this early Houdini performance proclaims a 'Grand, Brilliant, Bewildering and Startling Spiritualistic Séance, given by Prof. Harry Houdini The Great Mystifier assisted by Mlle. Beatrice Houdini the Celebrated Psycrometic [*sic*] Clairvoyant'.[6]

Conan Doyle's and Houdini's relationship always involved discussions of spiritualism. In his 1924 book *Houdini: A Magician Among the Spirits*, Houdini writes that 'The friendship of Sir Arthur and myself dates back to the time when I was playing the Brighton Hippodrome, Brighton, England [early in 1920]. We had been corresponding and had discussed through the medium of the mail, questions regarding spiritualism.'[7] A letter from Conan Doyle to Houdini, from 15 March 1920 confirms their correspondence by this date, although Conan Doyle's use of 'Mr. Houdini' (he later writes simply, 'Houdini') suggests that their friendship was still at an early stage.[8]

Conan Doyle's lecture tours to America in 1922 and 1923 cemented their relationship, although they also sowed the seeds of its breakdown. Despite Conan Doyle's promotion of spiritualism, and Houdini's growing scepticism, Houdini feted Conan Doyle when he first arrived on America's East Coast in 1922. As president of the Society for American Magicians, he invited Conan Doyle to attend, and to address, their annual meeting in New York on 2 June. In his travel narrative, *Our American Adventure* (1923), Conan Doyle records how he provided the magical community with some after dinner entertainment by showing them footage from the film of his novel, *The Lost World*, in which there was 'nothing occult … that it was psychic only as all human products are ultimately psychic, and that if it was preternatural it was only in the sense that it was not nature as we now could observe it'.[9] Clearly Conan Doyle drew from his lectures on spiritualism to introduce the film rushes, but it is interesting that he chose to offer the magicians something they could see rather than listen to, for observation of séance phenomena was to become central to Houdini and Conan Doyle's later disagreements. Houdini also performed at the meeting, offering his audience the celebrated 'Metamorphosis' box trick. Conan Doyle commented that 'Houdini is the greatest conjurer and this his greatest trick'.[10] He added:

> Houdini is not one of those shallow men who imagine they can explain away spiritual phenomena as parlour tricks, but that he retains an open – and ever, I think, a more receptive – mind towards mysteries which are beyond his art. He understands, I hope, that to get truth in the matter you have not to sit as a Sanhedrim of Judgment, like the Circle of Conjurers in London, since spiritual truth does not come as a culprit to a bar.[11]

In fact, Houdini was becoming ever more opposed to spiritualism's claims, which were further undermined, in his view, by Conan Doyle's apparent visual naivety which was increased by his obstinate belief in all spiritual activity. Joseph Rinn, a professional magician also present at the annual meeting, recalled later that Conan Doyle found it difficult to believe that 'the magicians ... were all posted on moving picture production' and so found his *The Lost World* film nothing out of the ordinary.[12] More telling, however, was Conan Doyle's inability to recognize that Houdini's 'Metamorphosis' was a simple illusion. 'All magicians knew in a general way ... how Houdini performed his feats', but Conan Doyle 'asserted that he performed his stunts through psychic power'.[13]

Following an unsuccessful séance with Conan Doyle's wife Jean, which took place two weeks after the Society for American Magicians meeting, Houdini's relationship with Conan Doyle began to deteriorate.[14] Houdini's growing belief in Conan Doyle's observational limitations is characterized in numerous letters, newspaper articles and published accounts from late 1922 onwards. In particular Houdini was astonished by Conan Doyle's failure to recognize that the physical manifestations of séance performances could easily be replicated by magicians. Even when Houdini explicitly pointed this out, Conan Doyle appeared to obfuscate: 'My dear chap', he wrote to Houdini, 'why do you go around the world seeking a demonstration of the occult when you are giving one all the time'.[15] Certainly, from the summer of 1922, until 1924 when Houdini and Conan Doyle appear to have stopped corresponding altogether, their letters to each other are increasingly fractious. In November 1922, Conan Doyle wrote in response to an article Houdini had written in the *New York Sun*: 'I have no fancy for sparring with a friend in public ... but none the less, I felt rather sore about it ... when you say that you have no evidence of survival, you say what I cannot reconcile with what I saw with my own eyes'.[16] By the beginning of 1923, Conan Doyle had accepted that Houdini now formed part of the anti-spiritualist movement. In another letter, he writes, 'I see that you are on the Scientific American Committee, but how can it be called an Impartial Committee ... It becomes biased at once'.[17] Houdini responded by focusing an entire chapter of *A Magician Among the Spirits* to Conan Doyle's spiritualist career.[18] Although careful not to injure Doyle personally, the chapter both attacks and jeers at Doyle's beliefs. It undoubtedly contributed to their disinclination to correspond thereafter.

Conan Doyle's and Houdini's relationship had a lasting effect on both men. Shortly before his death in 1930, and conscious of the little time he had left, Conan Doyle told an American acquaintance that 'there is just a chance that I may talk it [spiritualism] all over with Houdini himself before very long'.[19] Houdini went further. After his death in 1926 it became apparent that he had agreed with his wife, Bess, that she would conduct a séance annually, during which, if spirit return were possible, he would contact her. Bess conducted these séances for a decade, before concluding in 1936. This unsuccessful final séance took place on the roof of a skyscraper in Hollywood, California, on 31 October. The séance was overseen by Edward Saint, who noted in his introduction that, when he died, 'Houdini was working [on] a process that would permit him to take flash-light photographs in the dark without the flash being visible'.[20] The séance itself was brought to an end by Bess: 'I do not believe that Houdini can come back to me ... I do not believe that ghosts or spirits exist ... the Houdini shrine has burned for ten years ... I now reverently turn out the light'.[21] While Conan Doyle's last words restate his belief in the survival of the spirit, Houdini's final actions are a reminder of his fascination with seeing.

Why did the séance provoke Houdini's interest in modalities of vision? The comments of spiritualist participants in, and spectators of, séance performances are revealing of the complex visual dynamics that underpinned them. The séance demanded the active involvement of participants' full sensorium: sight, touch, hearing, and even smell and taste, were all brought to bear on the sometimes extraordinary events experienced. The first three senses were commonly employed in almost all séances, but with the emergence of ectoplasm and spirit manifestations from the 1870s, the olfactory and gustatory senses increasingly played their part. Despite this sensorial overload, vision continued to dominate; although séances often took place in darkened rooms that limited visual capacity, an optical consciousness still permeated participants' discussions and reflections. In the first of his three-volume *History of Spiritualism* (1926) Conan Doyle provides a series of instructive examples of this turn to visuality in the statements of witnesses to séance phenomena:

> We sat down at a moderately-sized table, the structure of which we were invited to examine. In a short time the table struggled, and a tremulous motion ran up all our arms ... The most unaccountable rappings were produced in various parts of the table, and the table actually rose from the ground when no hand was upon it. A small hand-bell was laid down with its mouth upon the carpet, and after lying for some time,

> it actually rang when nothing could have touched it ... Lights began to appear, and musical instruments floated and played above the heads of the company ... Loud and violent noises were heard and messages were spelt out ... Levitation quickly followed, and the boy was floated in the air above the heads of those in the room at a distance of nine feet from the floor.[22]

These comments are drawn from different séances performed in the 1860s by some of the period's best known mediums.[23] Each of the séances was performed in the dark, as is suggested by one witness noting that 'lights began to appear'. Yet it is a vocabulary of vision that dominates the register of the language. There exists, in fact, a subtle translation of the séance through a hierarchy of sense-experiences, each of which is ultimately subjugated to the perceived authority of visual evidence. The hand-bell lying on the carpet could only be heard, not seen, by the participants yet they perceive its untouchability as though they were looking at it. Similarly the boy's levitation is likely to have been experienced through touch – his shoes might graze the shoulder of one participant or land on the head of another – but the statements characterize this event as though it fell within participants' ocular horizon.

This pretence at seeing the phenomena of the séance, which might be read as either a desire to be visual or as a disregard for visual accuracy, betrays the key concern of spiritualists. To be taken seriously as a site of scientific inquiry the séance had to offer opportunities for intense, exact and repeatable observation. If, as spiritualists claimed, the séance made materially and visually accessible the existence of a natural world that expanded beyond the boundaries of human death, then this world should be available to the optical investigations of scientific practice. Confirmed spiritualists, like Conan Doyle, were enthusiastic that experiments to test this should be undertaken. Since, however, the majority of séance phenomena were produced when light (and therefore the ability to see with scientific precision) was partial or absent, the séance's role as a productive site of serious scientific scrutiny was never secure. However, spiritualism's significance was not absolutely determined by its status within science. Its popularity and importance as a 'new religion' that required belief rather than evidence also set it squarely against many of the principles of scientific practice. Its self-characterization as a series of phenomena that might offer truths that science could not determine called into question the authority of rigorous observation based in inductive and deductive methodologies. The séance, therefore, was an ambivalent site: at once determined to be accessible to scientific vision and yet simultaneously suggesting that vision (especially technologies of science that supported it) might fail to capture its faith-driven enchantments.

For Houdini, and many other professional magicians, there was rarely such doubt about séance performances. They were, quite simply, disguised acts of trickery and illusion that employed the art of the magician fraudulently. In the hope of revealing this to a public uncertain of spiritualism's provenance Houdini devised what became known as the 'bright light' séance: a magical performance of séance methods with a stage audience of blindfolded participants (recreating the darkened room of the séance) and an auditorium audience with unrestricted vision of the lit stage. The first recorded performance of this type was actually undertaken by the conjurer Joseph Rinn, on the suggestion of Houdini, on 12November 1911. Rinn later published a transcription of his blindfolded stage audience's comments:

> Dodson – A hand is feeling my ear while another is pulling my hair.
> Weston – The bell on the desk near me is ringing.
> Johnson – A tambourine is playing over my head – not far over ...
> Cross – The desk just bumped into me. It was levitated about three feet ...
> Johnson – Far away from me the lid of the desk is slapping up and down ...
> Rowley – The legs of the desk just floated past me and bumped my face ...
> Ellis – Rinn is near for I hear him crying stop-stop-don't-don't ...
> Chase – Rinn floated past me crying help-help ...
> Frank – I feel him standing on my knees ... He is being lifted upward – his shoes hit my ears. He is standing on my head – his weight hurts me. He has been carried off.[24]

Houdini's initial encouragement for this magical exposé, as Rinn records, stressed the importance of visual evidence: 'the public is occult-minded and but little impressed by the printed or spoken word describing trickery', Houdini says, 'People need to be educated by having ocular demonstrations'.[25] As the transcription reveals, Rinn's magical performance was similar in content to Conan Doyle's séances: the table moved, tambourines played and floated, bells rang and the medium levitated. Yet there is a decided difference in the descriptive language of the participants. Rinn's subjects make every effort to avoid a visual vocabulary, using tactile and aural words instead. Still, they find it difficult to avoid their sense-responses shifting into perception; when Cross states that the table 'was levitated about three feet' or Frank claims that Rinn 'has been carried off' they are making claims that cannot be verified without visual evidence.

However, the stage participants are not the real audience for Rinn's performance. It is, instead, the auditorium audience, with their clear view of the stage events, who are really the subjects of his ocular demonstration, and it is they who are receiving an optical education. This audience can see that the perceptual claims made by the stage participants are false. For example, and as Rinn makes clear, he has not been 'carried off' but has simply placed his shoes on his hands and then slowly raised them from Frank's knees to above his head.[26] However humorous this performance must have been for the auditorium audience, its effects are none the less significant. First, such a performance shows how the senses can be fooled by clever manipulation. It also reveals how the human sensorium works as a collective, and how easily things unseen can be translated into visual evidence that carries great authority. Finally, and perhaps most importantly, a performance like this not only uncovers the weaknesses inherent in human vision but also allows an audience to reflect upon and then celebrate both those weaknesses and the possibility of overcoming them with ocular education.[27] Unlike the spiritualist séance, the magical re-performance of séance phenomena does not suggest that vision might be unable to access truth. Yet it does, and again this is in opposition to the spiritualist séance, implicitly argue that scientific investigation does not necessarily hold a monopoly on observation and interpretation.

The relationships between spiritualists, scientists and magicians had been in development for some time before Conan Doyle's and Houdini's involvement. From the 1860s, when spiritualism became an influential phenomenon in both North America and Britain, magicians and scientists had taken a keen interest in the séance and its epistemological significance. They did so, however, from very different perspectives. Although it is not the purpose of this chapter to discuss exclusively scientific efforts to understand séance phenomena it is worth noting that in the 1860s and 1870s there was little correspondence or collaboration between the scientific and magical communities. The Dialectical Society's investigation of spiritualism from 1869, which concluded with an extensive published report in 1871, took no testimony from any practising magician nor included any magical expert on its extensive committee.[28] As a forerunner, as Roger Luckhurst argues, of the Society for Psychical Research the Dialectical Society's members might be regarded as the pre-eminent scientific investigators of spiritualism before the 1890s.[29] Similarly, William Crookes, whose research throughout the 1870s gave him the (dubious to some) position of leading scientific investigator of the séance, rarely offered any discussion of magicians' contributions to the examination of spiritualism. In his central work, published in 1874, Crookes does acknowledge that spiritualism 'is a subject which ... lends itself to trickery and deception', but does so not to introduce the work of magicians but rather the importance of scientific instruments utilized by the 'thorough scientific men'.[30] Even John Tyndall, as great a sceptic of spiritualism as Thomas Huxley, made no reference to magical performance in his short exposé of a séance published in 1872, despite implicitly arguing that the exhibition of table rapping he witnessed was nothing more than the subtle manipulations and misdirections of the medium.[31]

If magicians were ignored by scientists, then the reverse was also true, but to a lesser extent. Although it might easily have been argued that the magical community had some of the same aims as scientists – namely, the exposure of fraudulent mediums – they did not appear to recognize it. Indeed, those magicians who did investigate spiritualist phenomena considered it only from the perspective of performance, their own particular arena of knowledge. If they did invoke scientific investigations, it was largely to compare them unfavourably with the work of magicians. From one of the first magical texts on spiritualism, written by Herr Dobler in 1869, magicians' focused on its role in the varied and extensive exhibition culture of Britain and America. Dobler's *Exposé of the Davenport Brothers* signifies from its preface the author's interest in spiritualism as a form of popular show:

> The following pages are presented, not alone with the intention of exposing the means by which the celebrated Brothers Davenport succeeded in beguiling a large portion of the public with their spiritual pretensions, but also as a source, by the study of which, the amateur may possess himself of the most curious deceptions of the age.[32]

What follows is a work fixed firmly within the genre of the book of magical instruction, in which Dobler sets out the Davenport Brothers' techniques for deception, illustrated, as is the norm in such works, with numerous detailed pictures of rope ties and escape methods.[33]

In the many books, pamphlets and performances that followed Dobler's example, magicians continued to concentrate both on the relationship between spiritualism and other shows and exhibitions, and on exposing fraudulent spiritualists. J. N. Maskelyne, one of the most popular and successful Victorian magicians, led a campaign against spiritualism throughout the 1870s. At the Egyptian Hall, where he and his associate Cooke performed their magical act, Maskelyne introduced a mock-séance which replicated the tricks of famous mediums like the Davenport Brothers (just as Dobler had done). Maskelyne's aim was to reveal to the public that spiritualist manifestations were 'altogether mundane in their origin'. This was not a challenge to science – Maskelyne had 'no desire to enter the lists with the scientific investigators' – but simply a magical performance that revealed through reproduction that spiritualist phenomena were 'gross and harmful trickery and fraud'.[34] Maskelyne's aims met with a certain success: spiritualist séances began to be discussed in relation to other entertainments and exhibitions rather than as an unknown, perhaps supernatural phenomenon. Commentators began to note how the cabinets used by spiritualist performers were 'fitted up as the Proteus at the Polytechnic Institution' or like the stage effects used 'years ago by Professor Pepper'.[35] In an ironic circularity the spiritualist cabinet was even compared to 'the mysterious cabinet of Messrs. Maskelyne and Cooke'.[36]

More problematically for magicians, defenders of spiritualism began to respond in kind. George Sexton, editor of the leading spiritualist periodical, *The New Era*, revealed the secrets of Maskelyne's own illusions, comparing them unfavourably to séance phenomena. In a widely reported lecture at the Cavendish Rooms in London in June 1873, Sexton charged magicians, and Maskelyne in particular, with 'barefaced effrontery' for telling their audiences that their own tricks matched the character of the effects achieved by mediums, before proceeding to reveal how Maskelyne and Cooke performed some of their most successful tricks.[37] Sexton's lecture, which was later published as a popular pamphlet (see Figure 8.1) used as an epigraph a comment on vision. As Figure 8.1 shows, Sexton linked magicians' scepticism of spirit phenomena to the loss of vision one experiences when the eyes are closed. In the same year Maskelyne was involved in an epistolary disagreement with another defender of spiritualism, who used the pseudonym Iota. These letters were, like Sexton's lecture, published and widely discussed.[38] Iota's correspondence with Maskelyne revealed the magician's greater concern for performance fees than any altruistic motivation to oppose spiritualism. Counter-exposés such as these damaged the magical community's reputation, and, more pertinently, undermined their own entertainments by revealing how they were done. The satirical periodical *Punch* perhaps best captured the unedifying struggles between spiritualists and magicians by referring to the furore around the Davenport Brothers as the 'tie-fuss fever'.[39]

While the controversies over spiritualism had the effect of increasing audiences for magical performances of séance phenomena, magicians had little cause for concern. As Maskelyne had noted, when spiritualism's truths were called into question, as they routinely were, magicians found themselves playing 'to overflowing' auditoria.[40] However, as the spiritualist séance became of increasing interest to scientific investigators from various fields, it began to gain an authority and gravitas that magical performances could not match. Spiritualism's quest for scientific acceptance was lengthy and never fully successful. Yet it cannot simply be rejected as pseudo-science. As Richard Noakes has argued, there is 'a growing literature demonstrating the implausibility of such [pseudo-scientific] stories about spiritualism and a range of other fringe sciences'.[41] Certainly the evidence of spiritualism's evolution from the 1860s to the 1920s is one of increased scientific attention and experimentation. From the early work by the Dialectical Society, and by individual researchers such as Crookes, scientific engagement with the séance both increased and expanded across several disciplines. By the time of Houdini's and Conan Doyle's involvement in spiritualism the Society for Psychical Research had been undertaking experiments and investigations for forty years, spiritualism had been discussed at the British Association for the Advancement of Science (although this was not without opposition) and the new science of psychology had accepted the séance as a site suitable for its own investigations of mind.[42] Doyle himself drew on the language of psychology and medicine in mocking conjurers who would not accept that science had a serious purpose in investigating the séance: describing them as suffering from 'Conjuror's Complex' or the more serious 'Houdinitis'.[43]

No. 3.] DR. SEXTON'S ORATIONS.

SPIRIT-MEDIUMS

AND

CONJURERS.

AN ORATION

DELIVERED IN THE CAVENDISH ROOMS, LONDON, ON SUNDAY EVENING, JUNE 15TH, 1873,

BY

GEORGE SEXTON, M.A., LL.D., F.A.S., F.Z.S.

HONORARY MEMBER OF L'ACCADEMIA DEI QUIRITI, AT ROME; HONORARY MEMBER OF THE AMERICAN ANTHROPOLOGICAL ASSOCIATION; EDITOR OF "THE NEW ERA," ETC., ETC.

TO WHICH IS APPENDED RULES TO BE OBSERVED AT THE SPIRIT CIRCLE.

"Sey hinter ihm was will! Ich heb ihn auf."
DAS VERSCHLEIERTE BILD ZU SAIS.

"When people have resolved to shut their eyes, or to look only on one side, it is of little consequence how good their eyes may be."—ARCHBISHOP WHATELY.

LONDON:
J. BURNS, 15, SOUTHAMPTON ROW, HOLBORN, W.C.

Figure 8.1: Cover page of the pamphlet of George Sexton's anti-conjurer lecture of 1873, in which he characterized magicians as blind to spirit phenomena. Source: G. Sexton, *Spirit-Mediums and Conjurers: An Oration* (London: J. Burns, 1873), cover page; reproduced with permission of the Library of Congress.

While scientific practice became more common in the séance room, the problems of observation that had beset the earliest investigators remained. In the 1870s Crookes had claimed that his belief in spirit phenomena, which 'cannot be explained by any physical law at present known', was a result of 'exactness of observation' and 'careful investigation'.[44] However, he had quickly noted that 'no observations are of much use to the student of science unless they are truthful and made under test conditions; and here I find the great mass of spiritualistic evidence to fail'.[45] The French astronomer, Camille Flammarion, who undertook a series of spiritualist investigations in Paris in the early twentieth century, wrote with similar reticence in his 1907 book *Mysterious Psychic Forces*. Flammarion, who began by claiming that the medium Eusapio Palladino 'absolutely proved' the 'phenomenon of levitation'[46] then proceeded to highlight her fraudulence. 'Eusapia is a sly one. She is gifted with great sharpness of sight and has unusually sensitive ears', Flammarion noted, '[and] I have quite often been absolutely deceived. When I took the precautions that were necessary to put the medium beyond the possibilities of trickery, I obtained no result; if I pretended not to see anything I would perceive out of the corner of my eye attempts at deceit. And, in general, phenomena which took place happened only in the moments of distraction in which my attention was for an instant relaxed'.[47]

For the many investigators bringing to the séance the instruments and methods of accurate scientific observation, the séance proved a site of visual resistance which suggested an unstable ocularity. In this, the séance struck a decidedly prescient cultural note: visual authority was under threat in a number of ways in the second half of the nineteenth century, and even more disenfranchised from power in the early twentieth century, as the previous chapter argued. The séance's particular contribution to the ocular fragility expressed by Crookes and Flammarion was to reinforce an already existing concern with the limitations of human vision. The advances in physiological optics from the mid-century, accompanied by a marked increase in the practice of ophthalmology, brought a general understanding of the eye's frailty. As Ernst Fuchs had shown, the human eye was actually responsible for reducing vision; its various parts almost all contributing to make manageable the amount of light that reached the retina.[48] Helmholtz had previously revealed how certain parts of the eye actually worked to reject light, such as 'the black pigment of the choroid' that absorbed light, and even the retina itself, from which light was 'diffusely reflected and return[ed] through the pupil and out of the eye altogether'.[49] As John Plunkett has argued, the dissemination of this new physiological knowledge, 'challenged the positivism of Enlightenment conceptions of a stable, transparent, external world'.[50] Further advances in physiological understanding of the way the eye functioned also contributed to the destabilizing of the relationship between the world and human perception of it. Following earlier researchers, for example, David Brewster had shown how the colours of external objects were not inherent to those objects but determined by the eye's reception of reflected and refracted light.[51] The various illusions of vision that Brewster was investigating were also discussed by Helmholtz in his *Treatise on Physiological Optics*. Helmholtz claimed that even the healthy eye is never free from the spectres that Brewster had identified as examples of illusory vision: the eye, Helmholtz argued, is always subject to the 'luminous dust of the dark visual field' and therefore prone to errors of misperception.[52] Even when these arguments about the eye's potential limitations were precursors to conclusions about its tremendous adaptive power – and therefore its visual strength – it remained the case that discussions of the fragility of vision continued to undercut the more positive conclusions arrived at in physiological optics.

The séance, of course, did not invite perception based solely on the function of the eye, but on all of the senses. For scientific investigators, therefore, experimentation was a matter not only of seeing, but of managing, for example, auditory and tactile sensations so that they were subject to more accurate observations. There was an obvious reason for this: since vision was so limited, and the eye prone to error in such surroundings, the most effective way to see whether the medium touched an object, or made it move, was to delimit both the medium's range of movement and that of any object, and then concentrate all observation on those specific sites. Crookes attempted this with the accordion cage he constructed for an experiment with the medium D. D. Home.[53] Even these innovative experimental techniques, in the final instance, depended on sight. As Crookes attests, to be certain of the results an 'assistant went under the table' to report on the accordion's movements as he saw them.[54] Many experiments conducted by scientists actually depended on translating bodily sensations into visual evidence. The Scientific American Committee, on which Houdini worked, relied upon investigators' holding the medium's hands to negate the possibility of physical contact with objects that were supposed to be subject to psychical force alone.[55] Even Crookes's imaginative use of instruments was rare in comparison with the number of times that he asked the medium simply to 'take the hand of the person next to him'.[56]

The séance's demands on the full sensorium, coupled with investigators' failure to accomplish any adequate research, suggested that vision was not necessarily the pre-eminent sense that it had been believed to be in scientific methodology across the nineteenth century. In fact, the séance intimated that vision had no particular claim to its position as the most sophisticated of the human senses. Although Helmholtz had popularized the view expressed since the eighteenth century that the eye was the most sophisticated organ of human perception, there was, in the later nineteenth century, a return to a form of embodied vision that spiritualist phenomena appeared to endorse.[57] In the work of ophthalmologists the eye was increasingly being seen in relation to, and under the influence of, the body and its other senses. Jabez Hogg, an expert in diseases of the eye, showed as early as the 1850s that the health of the eye was closely related to the health of the body as a whole.[58] Later, in the 1870s, Edmund Landolt highlighted the extensive connections between the eye and the body, proving that 'all the tissues of the body are represented in the eye and its accessories'.[59] By the 1890s Landolt had shown through a series of detailed experiments that the nerves and muscles of the body were integral to the eye's ability to see correctly.[60] At the same time, Fuchs revealed that the pupil of the eye reacted to sensory stimuli on any part of the body, and that various 'disturbances of sensibility' had an adverse effect on vision.[61] The séance – as a site of potential new knowledge of the natural world – seemed to many to be an area of investigation that proved the unruliness of the eye and the embodiedness of perception. Even confirmed spiritualists, keen to accept the authority of scientific observation and subsequent ratification, vacillated between endorsement of the other senses and acceptance of visual weakness. Sexton, for example, noted in his lecture that 'darkness [in a séance] ... although highly unfavourable to seeing, is not at all so to feeling'[62] before acknowledging that the dark séance did undermine 'judging accurately' and anyone who rejected such an argument did so from 'ignorance of the laws that regulate the sense of sight, or rather from a misconception of the power of the eye in vision'.[63]

While séance phenomena and their investigation gave rise to serious concerns for the primacy of observation in scientific experiment and for the power of the eye as an instrument of perceptual accuracy, there remained a significant body of visual experts who were unconcerned by optical uncertainty. As Jutta Schickore has argued, many investigators of vision from the fields of physiological optics and ophthalmology were satisfied that any potential ocular ambiguity could be negotiated:

> Although the sense organ's activity was understood as effecting and occasionally impeding verdicial perception, it was assumed that the eye's impact on the outcome of acts of perception could be determined and truth and error in perception could be teased apart ... In fact, the experiments suggested that deceptions could be avoided – or at least uncovered – through correct interpretations of the perceptual situation in question.[64]

Certainly this had been the case at least since the work of Brewster, whose *Letters on Natural Magic* not only highlighted the problems for vision to be found in optical illusions but also performed the role of uncovering their deceptions and thereby annulling their power to undermine the truths of perception. In fact, many scientific demonstrations throughout the nineteenth century offered the same opportunities for controlling apparently chaotic visual phenomena.[65] Spiritualism also took this route: with the séance more likely to confuse the eye than offer enlightenment, spiritualists instructed observers in the meanings of the events they might be expected to (almost) witness. By setting instruction alongside its visually stimulating yet epistemologically ambiguous phenomena the séance became a key site of scientific spectacle.

The Séance and Spectacle

In cultural histories of popular entertainment spectacle has long been a key descriptive term. Although it has not been defined with any rigour, its extensive use in scholarship does suggest that historians implicitly accept certain definitional parameters. First, any form of entertainment, show or demonstration must be visually arresting. As Rebecca Stott has noted, the etymological roots of the word spectacle reside in descriptors of vision; either in looking, spying or watching.[66] One of the most important elements of this visuality was innovation. Spectators had to be able to see something that they could not see in the world around them everyday. Second, entertainments required a high standard of production and presentation; either artistically, in dioramas or panoramas for example, or mechanically, as was required in the many shows depicting train travel or in exhibitions of automata. Third, for entertainments to be regarded as spectacles they needed grandeur. This did not necessarily relate to size, although many entertainments were of enormous scale, as Wyld's Great Globe in central London testified. Rather, grandeur was to be found in the richness of a show's representation, its ability to evoke desired audience responses (often emotive), and to seem more consequential than the external world from which spectators had just come, and to which they would soon return.[67]

For historians of science, spectacle has more precise meanings. Scholarly investigations of scientific entertainments from the eighteenth century onwards, often regarded as 'popular' science,[68] all conclude that the categorization of science as spectacle is determined by the existence of the combined aims of instruction and amusement. Bernadette Bensaude-Vincent and Christine Blondel, in their consideration of scientific demonstrations, shows, and exhibitions prior to and during the Enlightenment, argue that the scientific spectacle was a 'scientific performance' that was 'a kind of tourism involving entertainment and learning'.[69] Bernard Lightman and Aileen Fyfe, who focus on science in the early nineteenth century, come to the conclusion that the sites of scientific shows to which they pay close attention 'were always playing with the balance between instruction and entertainment in their efforts to produce a form of rational recreation'.[70] Similarly, Ralph O'Connor posits that spectacular scientific entertainments were 'rooted in the human need for amusement' but also fed 'a demand for information'.[71] Iwan Morus, likewise, argues that scientific demonstrations 'were intended to combine entertainment and edification' and that '"amusement and instruction" were not opposite ends of the spectrum'.[72] Jon Agar takes this further in his analysis of twentieth century large-scale scientific projects. He argues that, unlike straightforward scientific displays which make no effort to influence audience response, the scientific spectacle aims specifically to control spectatorship. This results in a 'greater extent of organisation that creates, and attempts to guide, interpretation'.[73] The most common method for gaining pedagogic authority over an audience was the lecture or spoken narrative which accompanied the visual demonstration. These commentaries had been in use since the late eighteenth century and were aimed at 'stabilising [the] phenomena'[74] of the scientific spectacle and imbuing it with an 'intellectual purpose' beyond mere 'amusement'.[75] These narratives of instruction were also an attempt to hide the fact that the science being demonstrated often remained invisible. While the product of scientific knowledge might be displayed, the science itself was out of sight. This was certainly true for demonstrations of Pepper's Ghost, for example, which offered its audience an optical education, but kept the mechanism for the production of the ghost hidden from the audience's view.[76] Indeed Morus, in discussing Pepper's Ghost, concludes that 'the true object of exhibition was not always what seemed to be on display'.[77] Kevin Hetherington has made the same point about earlier optical entertainments, such as the phantasmagoria, where a 'principle of concealment was established by hiding the projectors from view'.[78] The scientific spectacle, therefore, appeared to show science but actually only made visible its proxy. The subject of science was actually revealed in narrative form, as lecture or explanation by the spectacle's demonstrator.

The spiritualist séance worked very much to this model of spectacle. Morus notes that Pepper's Ghost – as spectacle – had power as 'an antispiritualist instrument' because its rhetoric of 'authority over the Ghost' could be employed by 'antispiritualist campaigns' to undermine 'the claims of spiritualist fantasists'.[79] Yet the spiritualists themselves employed the rhetoric of spectacle to reinforce their own authority as demonstrators of science. John Tyndall's experience of a séance, although he was contemptuous of its claim to reveal new knowledge, shows how the instructive lecture was employed to guide the interpretation of spectators, before they were offered the opportunity to reflect for themselves on the phenomena of psychic force. Although 'facts were absent for a considerable time', Tyndall notes, 'a series of very wonderful narratives suppl[ied] their place'.[80] Tyndall then records that 'the duty of belief on testimony was frequently insisted on'.[81] In this instance the medium attempts both to instruct her audience and to point out the veracity of such rational 'testimony'. One witness of a séance revealed how mediums might shape interpretation at the conclusion of events. He reports how, as the phenomena came to a sudden end, 'the lady [medium] asked "what had happened," and appeared rather surprised, while the gentleman on my left appeared somewhat confused. The room was now lighted. The lady explained that the spirits sometimes were suddenly called away to other scenes and other séances.'[82]

These examples are, of course, drawn from private and domestic séances. Other séance performers employed lecturers to provide instructive narratives on their behalf at large-scale demonstrations. Maskelyne, in *Modern Spiritualism*, recalls that the séances of Annie Eva Fay, an American medium of repute, were directed by Henry Fay, 'whose lecture will never be forgotten by those who heard it'.[83] Sexton, who demonstrated séance phenomena and the conjuring tricks of magicians who opposed spiritualism, concluded his 'oration' with an instructive lecture on spiritualism's claims to scientific veracity:

> Spiritualism may now be reckoned, not by scores, or hundreds, or thousands, but by millions ... Then look at the character of men who have embraced Spiritualism, and judge whether they are people likely to be deceived and misled by rank imposters ... The benevolent old Robert Owen saw before his death how glorious was the truth of spirit-communion ... The late Robert Stephenson was a Spiritualist, so was Professor de Morgan, one of the most eminent mathematicians of his day – a man whose whole life therefore had been devoted to that branch of knowledge which allows nothing to be taken for granted, but demands demonstration at every step.[84]

The spiritualist journal, *Medium*, applauded Sexton's contribution, and recommended that 'committees could not promote the cause better than induce the Doctor [Sexton] to visit them, and deliver his lecture with the illustrative experiments'.[85] Whether or not Sexton did take his spectacular séance to spiritualist meetings around Britain, other spiritualists certainly did so. Conan Doyle's lecture tours of Britain and America from the early 1920s through until his death in 1930 are perhaps the most high-profile and spectacular. These tours not only included visits to, and reports on, specific mediums and séances, but also large-scale lectures on spiritualism with associated visual material: spirit photography, spirit hands in wax and images of ectoplasm.[86] While such lecture tours did not go unopposed – the *New York Times*, for example, ran significant campaigns against Conan Doyle during his visits in 1922 and 1923[87] – their structure and status as spectacle gave them authority by appealing to 'the more serious-minded' who responded positively to spiritualism's dissemination of knowledge rather than its inculcation that observers should show faith.[88]

Magicians' direction of ire at spiritualism is clearly an acknowledgement of the weaknesses of magical performance in the context of spectacle. While many magicians – Houdini foremost among them – believed that they were better placed to offer a credible critique of spirit phenomena than any scientist, they tried to do so by drawing on their own illusion practices. However, magical performance was not scientific spectacle: it did not offer any instruction for its audience but rather aimed to misdirect and mystify observers, to leave them in the dark rather than enlighten them. There appeared to be a central core of hypocrisy, then, in magicians' claims to reveal the truths of spirit phenomena when their own performances offered nothing more than obfuscation. Houdini's encounters with Harvard psychologist William McDougall, when both were invited to sit on the committee created to investigate spiritualism by the *Scientific American*, are exemplary of the disregard for magicians' arguments among certain scientific groups. At the same time as being invited to join the Scientific American Committee, Houdini argued that 'with my eyes trained by thirty years' experience in the realms of mystery and occultism it is not strange that I view these so-called phenomena from a different angle than the ordinary layman or even the expert investigator'.[89] McDougall responded coldly to Houdini's suggestion that he had greater powers of observation than the trained scientist: 'I do not require Houdini to teach me something about which I probably know more than he does'.[90] McDougall was supported, perhaps unsurprisingly, by Conan Doyle, who asked, in an incredulous tone, in a letter to the *Morning Post* in June 1926 whether '[Houdini's] manual dexterity ... enables its possessor to sit in judgement upon the Crookes, the Wallaces, and the Lombrosos?'[91]

Conscious of the limitations of magical performance in instructing observers, and of the position of dominance that this ceded to spiritualism, magicians devised a pastiche of the spiritualist séance that allowed them to offer rational entertainment. Maskelyne's and Cooke's reproduction of the Annie Eva Fay's séances at the Egyptian Hall in the 1870s is the first example of this form of opposition to spiritualist spectacle:

> At the same time that these manifestations [of Fay] were running at the Queen's Concert Rooms, my colleague, Mr. Cooke, was giving an exact reproduction of all Miss Fay's tricks in the light séance, to the most minute particulars, twice daily at the Egyptian Hall, where I had the pleasure of explaining the modus operandi of accomplishing the 'marvels' of the dark séance, which were too simple and absurd to bear any other treatment.[92]

While Maskelyne and Cooke both performed and explained, balancing entertainment with instruction, Houdini's 'bright light' séances focused more on visual performance, allowing spectators to see the methods employed by fraudulent spiritualists. In fact, J. Malcolm Bird, who attended séances with Houdini when they were both investigating the *Scientific American* medium 'Margery' claimed that Houdini took the performances of mediums directly into 'his own stage repertoire'.[93] In a lecture to the Faculty at Harvard in 1925 Houdini spoke at length about his 'bright light' séance and briefly performed a portion of it. The transcript of this lecture reveals the importance Houdini placed on vision, as is evident at the denouement of his performance when he addressed the audience as well as the committee on stage acting as witnesses:

> Mr. HOUDINI: (Raising the table cover) Look. The gentleman is quite right. See. (Mr. Houdini pulls his foot from the shoe and strikes the bell, reinserting it in the shoe being held by the foot of Mr. Sendo.) He couldn't understand why he could not feel my feet. Why, they were not there. They are never there.[94]

Houdini's insistent invocation that his audience 'look' and 'see' highlights the stress he placed on visual interpretation. Unlike the magical séances of Maskelyne and Cooke, which relied as much on narrative explanation as visual discovery, Houdini's spectacular séance relies on the dissemination of knowledge through the eye. Yet in providing a form of optical training for his spectators, his magical séances actually highlight the potential for observers to be fooled by a dexterous medium. This is, in turn, suggestive of the flaws in normal human vision, a suggestion that the spiritualist séance which Houdini opposed had already made. Ironically, in providing visual education, Houdini's magical séance engenders the same ocular anxiety as spiritualism.

Houdini also offered a visual education of séance phenomena in print. His pamphlet on the medium named Margery incorporated not only a commentary on her alleged trickery but an extensive series of illustrations depicting her methods (see figures 8.2 and 8.3). The pamphlet's cover (Figure 8.2) pictures the controlling experimental apparatus Houdini employed to reduce Margery's deceptions. This photographic illustration is intended to instruct the reader in the techniques required for proper vigilance of mediumistic fraud. Figure 8.3 illustrates a further technique Houdini used when vision was compromised. Houdini rolled up his trouser leg so as to heighten his sensitivity to Margery's movements; replacing sight with touch. The illustration itself, however, especially in its inset close view (bottom right) invites visual learning by the focused, attentive observation of expert practice.

While the spiritualist séance had insisted on the eye's deficiency, whether in scientific observation or in the visual perception of the spectator, Houdini's magical séance and his published work celebrated the eye's potential to learn. Underlying these divergent views are very different understandings of how vision functions. Spiritualists had highlighted vision's embodiedness, and implicitly attacked its status as the most sophisticated of the senses, by stressing its relationship to the body as a whole. Houdini distances the eye's knowledge from the body's knowledge and instead connects it to the intellectual facility of the observer. He claims that while vision may be subject to conscious misdirection it is also able to learn how to identify those moments of ocular chicanery – to 'look below the surface of any mystery' as he puts it – and to resist them.[95] In doing this, Houdini argues in *A Magician Among the Spirits*, the observer is restating the importance of intellectual engagement with what is seen at the expense of spiritualism's call to faith: rejecting 'those which [*sic*] were too actively and intensely willing to believe'.[96]

Figure 8.2: Cover of Houdini's pamphlet on the case of the medium Margery. This pamphlet offered readers an education via illustration on mediumistic fraud. Source: H. Houdini, *Houdini Exposes the Tricks Used by the Boston Medium 'Margery'* (New York: Adams Press, 1924), cover page; reproduced with permission of the Library of Congress.

FIG. 1.

Figure 8.3: An illustration from Houdini's pamphlet on the medium Margery, with Houdini on the right, revealing the techniques he employed to expose her fraudulence. Source: H. Houdini, *Houdini Exposes the Tricks Used by the Boston Medium 'Margery'* (New York: Adams Press, 1924), p. 4; reproduced with permission of the Library of Congress.

Houdini and other early twentieth-century magicians were beginning, therefore, to formulate a response to spiritualism that did not simply oppose the spectacle of the spiritualist séance with the spectacle of a magical séance, as their Victorian predecessors like Maskelyne and Cooke had done. It would have been remarkable if they had done so. The scientific spectacle's unselfconscious belief that it could both explain and entertain was under threat from new ideologies of vision that questioned whether any spectacle could be certain of its reception. Following Foucault's thesis on the nineteenth-century gaze, which articulated a disappearing social vision replaced by the disciplinary gaze of institutional authority,[97] Crary argues that, from the 1830s, and increasingly towards the end of the century, cultures of vision were not 'founded on the necessity of making a subject *see*, but rather on strategies in which individuals are isolated, separated, and *inhabit time* as disempowered'.[98] James Krasner argues that this disempowerment only fully came into being in the early years of the twentieth century, when the optical sciences concluded that recent research revealed 'a profoundly unstable interior world' of vision.[99] For critics such as Karen Jacobs, these 'new, distinctly modernist kinds of observers and visual relationships' suggested a 'diminished faith ... regarding the capacity of vision to deliver reliable knowledge'.[100]

In the performance strategies twentieth-century magicians conceived for their opposition to the spiritualist séance there emerges a growing confidence that illusion practice (the performance of magical tricks often based on a knowledge of optics) can offer a way of understanding vision that the spectacle cannot. In the séance illusion that Houdini performed in the final years of his career, entitled 'Do The Dead Come Back?' it is clear that scientific investigations of spiritualism form an important part of his background research. Houdini's inventory for the illusion, for example, lists a 'Margery bell box' and a 'Psychical Research book' as properties.[101] However, 'Do The Dead Come Back?' remained an illusion, not a lecture nor a spectacle with accompanying educational narrative. Oscar Teale, a leading theorist of magic in the early twentieth century, wrote an extensive essay on magic and spiritualism in 1925 (when Houdini was touring with 'Do The Dead Come Back?'). In it he articulates what Houdini was performing. 'The magician', Teale argued, 'because he knows forces and laws that have escaped your attention, because he can demonstrate how imperfect are your powers of observation, really does more toward exposing the fakers than any one else'.[102] The reason for this, Teale concluded, was because

> a real magician ... is of necessity a member of a distinct type of scientist, the nature of whose sphere is to discriminate ... to diagnose and dissect by scrupulous analysis; not content with superficial evidence when it comes to a matter of disintegration or segregation of the really phenomenal in nature.[103]

Frank Crane, another commentator on magic in the 1920s, wrote specifically of Houdini's séance illusions that 'he and other legitimate magicians are doing a real service to the community in demonstrating the failure of the [spiritualist] argument'.[104] In particular, Crane argued, 'the foundation for sound thinking is to believe there is a natural cause for every result. The magician helps to establish this foundation because he convinces us that the cause is there even when we cannot see it.'[105]

Wonder, Enchantment and Modernity

Teale's and Crane's conclusions go to the heart of magical performance's investigation of vision and spectacle. For magicians like Houdini performance was a 'distinct type' of scientific demonstration, an alternative route to truth through the concept of wonder. Magical performance was not science as explanation but rather as speculation and curiosity, spurs to understanding that opened a path from the phenomenal to the noumenal. Houdini's remark to the audience at the denouement of the majority of his performances highlights the importance to him of spectatorial inquisitiveness. Stepping forward to the footlights he would loudly call out a final question: 'do wonders never cease?'[106]

Wonder, as this book has consistently acknowledged, is a concept that has undergone a great deal of investigation in the history of science. It has been defined as a process of investigation; that moment when the mind is brought to a halt by a marvellous object or idea, and begins to speculate on its function. Wonder is intense, and often reckless, but its power to kick-start curiosity and lead the wonderer to understanding gives it its power and kept it at the forefront of scientific investigations before and after the scientific revolution. Most historians regard it as a component of science, and especially of scientific demonstration or exhibition, which survived only until the late eighteenth century, and was denuded of its power by the Enlightenment. This is particularly the view of Daston and Park, whose work has been most influential in this regard, who argue that wonder became disreputable during the Enlightenment due to a new sensibility that associated it with vulgarity.[107] Their view is supported by Barbara Stafford, whose investigation of vision in the eighteenth-century sciences revealed that the eye which might have observed the wonders of nature became increasingly regarded as misleading of rational judgement.[108] However, other scholars have challenged this sudden paradigm change as too reductive; claiming that wonder survived the Enlightenment and continued into the nineteenth century. Richard Holmes, for example, has argued that wonder united romantic scientists and artists during the early nineteenth century, while Stafford – productively complicating her own claims – revealed that wonder re-emerged in the 1830s as the imagination came to be seen as increasingly important to the intellect. Even Daston and Park do not entirely reject wonder's continuing influence, although they view it as remaining only as a tamed force in Victorian popular shows.[109]

Wonder is characterized as a 'cognitive passion' linked to speculation and curiosity, and rooted in the imagination.[110] Thomas Hankins and Robert Silverman argue that the role of the imagination as mediating 'between sense and reason' was an essential component of natural magic, those scientific phenomena most closely associated with the establishment of wonder as a part of scientific investigation.[111] To wonder, then, was to employ the imagination to take the evidence of the senses and turn them into reasoned hypotheses. At the same time the correspondence of wonder with curiosity and speculation made wondering a risky venture, a form of intellectual gamble, which Stott sees as symptomatic of all scientific hypothesis, but which Holmes has argued is also reckless, even perilous.[112] As wondering was instantiated by looking, the eye plays a key role in theories of wonder. In the process of wondering the eye is regarded as actively pursuing knowledge. Tony Bennett has argued that the connection between vision and reason for Enlightenment savants meant that the eye was characterized as a cognitive faculty, while M. Norton Wise has argued that visualization was argument rather than illustration.[113] Stafford, likewise, notes a significant body of work in the Romantic period that regarded the eye's mode of perception as a form of knowledge, so that to see was to begin to understand.[114] Stott, who takes the argument into the mid-nineteenth century, also notes the continuing relationship between speculation and vision, where understanding is formulated in 'the mind's eye' or imagination.[115] Nigel Thrift, whose non-representational theory focuses on contemporary visual culture, stresses the importance of continuing to pay attention to the 'pre-cognitive', which he defines as ways of looking that offer a route to knowledge by evoking 'a note of wonder'.[116]

Houdini's magical performances, his séance illusions included, exemplify the continued exhibition of wonder in the early twentieth century, upholding the tradition that had begun in the early modern period and which had remained a vital part of a popular scientific tradition through the Enlightenment. Stafford recognizes the importance of conjuring in the historical tradition of scientific exhibition when she argues that as the eye and the wonders that it saw began to be thought of as subject to trickery, it was the conjurer's tricks that 'shed light on physical and intellectual provinces that were believed vulnerable to adulteration'.[117] Similarly, Simon During, in his history of magical performance, claims that magic in the nineteenth and twentieth century should be 'regarded as the technology of a dynamic vision' that 'is organised and constructed in such a way as to induce experiences or sensations of amazement, wonder, and bewilderment'.[118] Houdini's call for his audience to search for knowledge in the wonder of his illusions was successful. The responses of spectators to his performances show that his presentation of wonder led them to 'speculation'; the most common word in the many remaining witness reports. One spectator noted of a show in 1906 that 'Houdini not only gives the spectators at Keith's [Theatre] something unusual to see, but material for comment and speculation long afterward'. Another reported that 'The spectators watched as intently as ever and at the end of the act the house hummed with the speculations of the audience. Never was a vaudeville turn the cause of more wonder.'[119]

Houdini himself also personified many of the characteristics of wonder that historians of science have seen as crucial to its continued existence. Houdini's escapology highlighted the risks of performance, even at times appearing perilous. His constant quest to perform ever more marvellous illusions – such as being thrown into a river while constrained by chains and inside a wooden box – may seem a reckless gamble. Yet Houdini always considered these wondrous illusions to have a purpose. One of his most revealing comments comes in a retort to William McDougall, who had opposed Houdini during the *Scientific American* investigations of Margery. When Houdini heard that McDougall believed scientific knowledge to far outweigh the practical skill of the conjurer when applied to spiritualism, Houdini responded 'I will wager him a sum equal to a year's salary that his knowledge of psychology will be useless to him if he lets me nail him into a heavily weighted packing-case and throw him into the Charles River'.[120] By challenging McDougall in this way, Houdini argues for the existence of knowledge outside the confines of the scientific establishment; knowledge which may, he claims, hold greater investigative power. For Houdini the performance of wonder offers a better opportunity to discover truth: it does not rely on the authority of observation, under threat from modernity's fragmented vision, but invites speculation on the processes of that fragmentation. Wonder thereby constructs a more flexible route to knowledge that may manage to resist the ocular deficiencies so damaging to scientific objectivity.

Any resistance to the disintegration of vision that developed in modernity is a political and cultural activity as much as it is a scientific one. As many scholars of early twentieth-century society have argued, the gradual unravelling of secure vision shifted control of knowledge from the individual to the state, and allowed its systems – capitalist and bureaucratic systems particularly – to grow more powerful.[121] How successful, then, was resistance to this growing power? Certainly magical performance claims to show that vision is not necessarily disempowered by its loss of objectivity. Indeed, to link sight to the speculative imagination and then to reason – the very process that wonder sets in motion – is to suggest that the reverse is true, or at least possible. Yet at the same time this is not the only route to resisting modernity's subjugation of vision to the powerful, destructive forces of Fordist capitalism that press the individual 'into myriad assemblages of work, communication, and consumption' and which create 'a patchwork of fluctuating effects in which individuals and groups [must] continually reconstitute themselves'.[122] The spiritualist séance, on the other hand, offers an alternative way out of this morass of multiple subjectivities and individual disenfranchisement. Spiritualism did not rely on vision's potential for resistance, but instead undermined its power by suggesting its inadequacy when faced with the extraordinary marvels of the séance. For spiritualism, resistance to modernity's bureaucratized world came through belief.

Spiritualism's emphasis on faith may provide an example of the enchantment that sociologists since Max Weber have understood as one of the key elements of resistance that modern rationalization of vision destroyed.[123] Weber's thesis claimed that the twentieth century's loss of faith led directly to the 'disenchantment of the world'[124] in the myriad processes of economic, political and scientific rationalization. These systems took particular control of knowledge, divesting the individual of any power to see (and therefore reason) for themselves: 'increasing intellectualization and rationalization do not, therefore, indicate an increased and general knowledge' but subjugate the individual's knowledge to that of institutions.[125] Indeed John Wallis has previously argued that spiritualism 'emerged as a reaction to a perceived disenchantment ... brought about by [the] transition to modernity'.[126] For Wallis, spiritualism aimed to '(re)discover a sense of wonder' by enchanting 'some of the very aspects of the new, modern world'.[127]

However, spiritualists' efforts to claim the séance as scientific spectacle, through which they hoped to gain cultural authority, undermined any attempt at enchantment through wonder. Since science was itself a cornerstone of modernity's rationalization, spiritualism's reliance on scientific practice was, as Wallis argues, 'a capitulation of the spiritual realm to materialism and – by extension – to the forces of disenchantment'.[128] Ultimately, spiritualists' decision to embrace the spectacle, as Conan Doyle did, made it unfit as a mode of resistance to modernity. As Guy Debord revealed, spectacle did not actually deliver enchantment, it simply functioned as a handmaiden to rationalization by obscuring and concealing the systems it apparently contradicted.[129] In Debord's political polemic 'The Society of the Spectacle' (1967) he argues that spectacles of capitalism (and as a totalizing power almost all spectacles are capitalist) distort the vision of the observer to such an extent that they can only see 'in an alienated way ... in which seeing and doing have become separate activities'.[130] By this process, observers become both passive recipients of rational systems and also 'susceptible to manipulation'.[131]

Houdini's séance illusions in particular, and the magical performances of early twentieth-century conjurors more generally, offer a different challenge to capitalist spectacle and its enclosure of vision within rationalized systems of power. Houdini's presentation of optical illusions as forms of wonder challenges the analysis of vision as disempowered, isolating and subject to the control of capitalist institutions. His engagement with an audience's vision explores the fragmentation and frailty of sense perception revealed by physiological optics and ophthalmology, but does not arrive at the fatalistic conclusions of spectacle's power over the observer. As During argues of magical performance in general, Houdini's illusion practice evoked 'a different set of cultural figurations and formations than those of a modernity conceived of as having lost its freedom, sense of connectedness, and spirit'.[132] While Houdini's performances echo Crary's view that the observer of spectacle is 'immersed in a collectivity and simultaneously separated in absorbed solitude'[133] they provide an empowering experience of this observational role. Houdini's illusions present the observer with an opportunity to become an individual creatively constituting their own visual knowledge. In doing this, they also reconfigure the observer as one part of a social body that finds pleasure in visual instability and responds not by withdrawing into a solipsistic inner space of subjective disenfranchisement but by discovering a renewed sense of community that together can speculate and wonder. One of the key features in the statements of spectators at Houdini's performances is how what they have seen engenders discussion that goes on long after the performance has ended.[134] The speculations in these statements, noted above, do not occur in silence; they happen as conversation and debate, they take place within, and serve to constitute, communities of observers rather than individual spectators.

It is particularly Houdini's insistence that wonder provides observers with a visual experience that can be translated into knowledge that gives magical performance a foundation from which to offer resistance to modernity's disenchantment. Eschewing rational science for the enchanted epistemology of wonder allows Houdini to exert pressure on capitalism's hegemony, and indeed reveal fractures in its apparently all-encompassing institutions of power. As sociologists have begun to argue, some of the knowledge structures that have previously been seen as underpinning disenchantment, are themselves subject to other forces. Science, so central to Weber's original thesis, may itself work through 'less "rational" processes than might be expected'.[135] As Stott has shown, one element of this irrationality is the employment of the imagination to speculate and hypothesize; rational science might also be imbued with wonder.[136] The power of spectacle as the vehicle for modernity's control of the individual subject's ocular horizon is also under threat. Hetherington argues that the spectacle's power may not be total; that there remains space for the 'active subject' who can act 'as a creative force in the world with a self-reflective consciousness and imagination'.[137]

Weber, too, suggested potential modes of resistance, although these have often been forgotten in contemporary scholarship. In his conclusion to 'Science as a Vocation' he advocated the 'brotherliness of direct and personal human relations' as an antidote to disenchantment, and closed by warning that 'nothing is gained by yearning and tarrying alone'.[138] It is Weber's recommendations that Houdini's magical performances most clearly articulate, but not those alone. His séance illusions, lectures and published works on spiritualism robustly defend wonder's role as the non-rational in science. His desire to fuel speculation draws on the observer's ability to connect imagination to their optical sensibility, and his invitation to an audience to wonder together reveals the potential for the individual subject to contribute to a cognitive community of their own making.

Conan Doyle's defence of spiritualism and Houdini's conflict with him sheds light on the role of vision both historically and in the present. To consider them comprehensively, and in a variety of contexts, across a key period of transition from the late nineteenth century into the twentieth century is to understand both the vitality of the debates around optical knowledge and the significance of them for the future of vision in the contemporary world. Whether in the sciences of physiological optics and ophthalmology, or broader debates on perception in psychology and sociology, or in the spiritualist and magical séances, it is clear that vision occupies a central position in how the external world – its science, politics and culture – can be understood. What remains apparent from the historical evidence is that vision retains social power of different kinds even when under pressure. Neither spiritualism's attempts to topple it from its place as the most sophisticated of the senses, nor modernity's challenge to its stable authority were able to negate its epistemological significance. The evidence also highlights vision's importance in maintaining a discrete sense of subjectivity: the modern world's efforts to control vision were aimed at controlling the individual, or at least turning the individual into a group or 'mass'. As Houdini's performance tactics have revealed, defending the active vision of the individual, and stressing the link between vision and the imagination, provided the individual subject with a method for resisting both assaults on their identity and their ability to determine for themselves the reality of the external world. That connection between vision, imagination and cognition remains essential today. Whether by speculating, hypothesizing or fictionalizing, imaginative vision still provides a path to truths that cannot only be determined by rationalized systems – political, scientific or economic. It is apposite, and not coincidental, that it was Herman von Helmholtz, the leading figure of physiological optics, who stated this more clearly than any other scientist of his or subsequent generations. In speaking of the poetry and prose of Goethe, and specifically of how Goethe's poetic intuition had contributed to his scientific work, Helmholtz concluded that 'art as well as science can represent and convey the truth'.[139]

AFTERWORD

The main route I have taken through this book has been the broad path where science and literature are knitted together. I have traced their individual paths, of course; and I imagine them topographically as running in parallel, one always visible to the other but kept apart by a thick vegetation between. More interesting territory, however, has been those moments when the two paths close up to form a wider route, broken still by a grassy central reservation but one which is worn by continual crossings over and is sometimes entirely eroded.

This analogy, I hope, registers the fact that science and literature remain different but sometimes come close to being the same. I have suggested, for example, that literature makes efforts to become science. This is not to say that the literary text is on a path that leads it to be, in the end, a work of science. In fact the opposite is true: in becoming science, the literary text enters into a series of correspondences with science which positions it in relation to science. The literary text does not imitate or resemble science, but by becoming science, as Gilles Deleuze and Felix Guattari argue, it 'form[s] a block that runs its own line "between" the terms in play and beneath assignable relations'.[1] Imaginative writing remains its own particular kind of discourse: its illustrations, observations, scholarship and consideration of science in the world might converge with science's texts and practices but it is not aiming for the same methods of seeking out truth (or knowledge), and nor does it succeed or fail on ontological certainty, as science does.

Nevertheless, there are significant correspondences between science and literature which have gone largely unnoticed since the twentieth century's intellectual communities accepted the 'two cultures' paradigm and entrenched themselves in disciplines. In 2010, however, James J. Bono introduced a Focus section of *Isis* on the convergences and divergences in history of science and literature and science.[2] Bono's view is that both these fields share a 'concern to understand the making of science', and he concludes that there is the 'possibility of a poetics of science: of understanding science as itself a form of *poiesis*, of making'.[3] This reminds me once again of O'Connor's important work on geology as a poetics of popular science. Bono's argument is reinforced by one of the essays he introduces: Henry S. Turner's contribution on form. For Turner, form is a method of 'thinking through making', or, as he puts it, a 'poetics – *poiesis*' which 'might suddenly catch the eye of the historian of science in new ways'.[4] Intriguingly Bono's introduction and several of the essays that follow it all begin with the clear instruction to consider the relationships between literary and history of science *scholarship* but they quickly forget their brief and turn to the relationships between literature and science *themselves* in their historical contexts. It seems to me that this signals a significant shift in the ways that the histories of literature and science are being considered by scholars: away from the study of language, further away from the influence of science on literature, and towards the comparable contributions both make to human knowledge. Certainly O'Connor's work suggests this, and, I hope, the present book might make its own contribution.

If it does, two important themes provide its foundations: wonder and the imagination. Interestingly, neither wonder nor the imagination were dealt with by Bono or the other *Isis* contributors. They are, for me, the lodestones of the book, and also of future work in the study of science in culture. To some extent wonder and the imagination are teleologically linked; either with wonder as the handmaiden to imagination's speculations, or with the imagination igniting the spark of wonder that leads on to more rational processes of discovery. I have dealt with these in visual culture because it is in observation that wondering and imagining begin. This is the central argument to emerge from this book's varied investigations: in the complex processes of observation cognitive wonder and controlled imagination acted as engines for a productive consideration of new phenomena. Whether in the sciences of medical microscopy, astronomy, archaeology or ophthalmology or in the genres of literature from travel writing and the Gothic to detective fiction, allowing what had been seen to be wondered over and imagined was a vital step in arriving at understanding. That this process was employed in both scientific research and creative writing or performance shows us how much these apparently disparate arenas were actually in sympathetic alignment.

Further than this, though, to recognize that both science and literature depended on wonder and the imagination as dynamic starting points for rational thought goes further in bringing them into alignment. It shows, first, how science in practice was never solely aiming for objectivity but made pragmatic use of subjective responses. It also illuminates how important it was for imaginative writers and performers to enhance their subjectivity with objective knowledge. Visual culture is a reminder of how far this alignment reaches. Since it is from observation that wonder and the imagination emerge, and observation takes place from within social, cultural and political as well as scientific contexts there are moments of considerable cross-over in what is seen and therefore also in what is wondered and imagined.

I have also come to a definite conclusion about aspects of the scholarship on the relationships between science and literature, and particularly on the position given to wonder and the imagination. It seems to me that Daston and Park are wrong to argue that wonder disappears at the end of the eighteenth century, just as Daston is too quick to argue that Victorian science rejected the imagination. Daston and Park's analysis of wonder as a concept is rich, provocative and hugely important. Yet they appear to me to fall rather suddenly under the spell of the two cultures paradigm in accepting that as soon as science began to fragment into disciplines it lost any connection to the subjective side of its character. Similarly, Daston's understanding of imagination as something loathsome to savants attempting to be more objective seems to ape once again the reductive alignment of science with rational objectivity and literature with subjective fancy that was a characterization of the debate on the two cultures. As I hope the previous chapters have shown, wonder did not cease (as Houdini said) nor did the imagination become feared and loathed. Indeed further investigation of their role in realigning science and literature needs to be undertaken, in the senses other than vision, for example, or in altogether new ways. This would be valuable work.

Equally, it would be just as valuable to extend the discussion into twentieth- and twenty-first century science and literature. Indeed it has been a central argument here that the ways of seeing I have investigated in the Victorian and Edwardian periods are those still in evidence now. Certainly, there was no distinct change in visual culture as it moved from the Victorian period into the early twentieth century, as is often argued, and I would go further, if tentatively, in claiming that the present visual cultures that combine science and creative arts develop but do not alter the observational paradigms of the nineteenth century. In popular science, for example, Rebecca Skloot's *The Immortal Life of Henrietta Lacks* (2010) makes visible hidden tissue samples by connecting them, through creative biography, to living humans.[5] In scientific demonstration, the British Science Museum's *Science on a Sphere* project consciously employs the wonder of large scale visual entertainment as a starting point for scientific education.[6] And in performance, the *Uncaged Monkeys* stage show delivers scientific content through the creative form of stand-up comedy for audience-observers across the nation.[7] I shall close, though, with an extended example of the continuing influence of Victorian visual culture in the contemporary world.

The example is primarily political; the extraordinary debacle of the (then) British Prime Minister Gordon Brown's letter of condolence to the mother of a British soldier killed on active service in Afghanistan (a theatre of war with which the Victorians were also familiar). The letter, which was reproduced several times in British newspapers on the 9th and 10th of November, 2009, was 'scrawled' in thick black ink, included a number of errors, and was, in parts, illegible.[8] The *Sun*, a newspaper opposed to Brown's government, had first revealed the letter, and its recipient's disgust at its insulting and disrespectful appearance, in order to attack Brown on his relationship with the British armed forces and the ongoing military engagement in Afghanistan. Their report led other national newspapers to claim that the letter contributed to, and was symptomatic of, a more general failure of Brown's leadership. Such opinion was not universal: the *Guardian* newspaper (supportive of the Labour government led by Brown) took the opportunity to attack the *Sun* rather than Brown, arguing that it was 'exploiting the grief of a mother whose son had been killed'.[9] Regardless of their political position, almost every newspaper took the opportunity to provide a visual representation of the letter itself, sometimes annotated with comments from handwriting experts, sometimes with its errors highlighted as though it were a piece of shoddy homework. The *Times* even compared it with a letter of condolence written in 1864: the well-known 'Bixby Letter' written by Abraham Lincoln to the mother of five young men killed in America's Civil War.[10] Despite Brown's letter's rather touching sentiment (in content not dissimilar to Lincoln's lauded effort) the visual representation of it was damning. Its handwriting and misspellings (5 in 118 words) suggested both linguistic weakness and a lack of attention to what was a serious responsibility. It was not what the letter said that was problematic but what it meant to look at it.

However, as both the *Independent* and *Guardian* noted, the episode of the badly-written letter was the culmination of a series of discussions centred not on Afghanistan, nor even on epistolary skill, but on Gordon Brown's eyesight.[11] For at least a month prior to the *Sun*'s reprinting of the letter of condolence the British and American media (both newspaper press and television news) had been speculating about the deteriorating vision of the Prime Minister, speculation set in motion by Brown's attendance at London's Moorfields Eye Hospital, founded in 1809 and the first hospital dedicated to diseases of the eye in Europe. Brown, as the British public was to learn in detail in October 2009, had lost the sight in his left eye in an accident at the age of sixteen, and suffered with weakened vision in his right eye from that time onwards, undergoing an operation to prevent blindness at age eighteen in 1970. The interest in Brown's vision rose when it became clear that his visit to Moorfields was due to 'damage to the retina in his one good eye', as the *Sunday Telegraph* reported.[12] Every national newspaper in Britain carried the story, with most providing an ophthalmological illustration of a damaged retina, of the type that had been common in guides to the diseases of the eye since the mid-nineteenth century. These illustrations – themselves visual representations of fragile vision – were accompanied by comments from leading ophthalmological surgeons, such as Hector Chawla, who had operated on Brown's eye in 1970, and Moorfields' consultant retinal surgeon, James Bainbridge.[13] Each of these surgeons responded to questions about the state of Brown's vision by denying that the type of retinal damage he had suffered was particularly threatening to health, yet the various reports continued to link his failing eyesight to his political fragility. The *Independent on Sunday*, for example, ran the story under the headline 'Damage to Brown's eye raises questions about his future'.[14] In addition several newspaper reports charged Brown with having hidden the problems with his vision from the British public, citing as evidence his denial of any deterioration in his eyesight when in conversation with the British journalist Andrew Marr.[15] However, reports also speculated that Brown might use the public knowledge of his failing vision as an opportunity to manage his own departure from national leadership, and as an escape from the expected controversy over parliamentary expenses.[16] The *Sunday Times*'s editor summed up the reports by asking: 'The One-Eyed Man May Be King, But For How Long?'[17]

While the understanding of vision here is, of course, specific to the context of Brown's political leadership and the social consequences of military conflict, some of the scopic relationships mirror those constructed and refined in the nineteenth century. The same visual materials were employed to provide a 'popular' scientific understanding of the effects of retinal weakness, and the same authoritative scientific voices were invited to contribute to the debate. Indeed, these voices do very much the same work as Victorian ophthalmologists, whose increasingly public discussions of eye disease suggested to the British public that their vision was closer than they imagined to breakdown. In several of the reports on Brown's retinal damage ophthalmologists pointed out that a significant percentage of the population suffered similar visual problems. Chawla, for instance, claimed that 'eight per cent of the population walk around with retinal tears, often with no idea they have got them'.[18] The effect of this was to promote the idea that approximately five million people in Britain lived with unseen yet weakened vision that had the potential for complete fragmentation. Brown's fragile eyesight becomes, in this example, a synecdoche of national vision; a metaphoric shift from the actual individual to the imagined collective that was characteristic of nineteenth-century visual culture.

Just as striking is the imbrication of scientific understandings of vision with wider social and political concerns. The failure of Brown's eyesight was read as a clear symbol of his political shortcomings, as though a lack of ocular clarity was symptomatic of an intellectual or professional blindness. The transference from optic instability to lack of knowledge marks out Victorian concerns about vision just as it does this contemporary moment. More than this, however, Brown's adumbrated vision becomes a marker of moral incapacity: his letter of condolence was not read as the material product of visual impairment but as the uncaring scribble of a man disrespectful of the armed forces and the sacrifices they make. The connection between vision and moral character does not emerge for the first time here. This, too, was a nineteenth-century phenomenon; when the invisible microbes of disease were regarded as signalling moral turpitude or ocular deficiency was suggestive of a failure to uphold appropriate levels of personal and national behaviour.

Implicitly, Brown's poor vision also revealed his problematic relationship to Britain and Britishness. Even before his retinal problems of 2009 Brown's Scottish heritage had made some sections of English society wary of his commitment to Britain as a united kingdom. This was often given as one of the reasons that he cared less for the 'English' armed forces (even though they were not solely English). The writing of the condolence letter appeared to confirm that, for Brown, England itself was a blind spot. His poor vision therefore became a shorthand for his Scottishness; monocular, fragmentary (especially of the united nation) and insular. The relationship between sight and national identity was also a hallmark of nineteenth-century visual culture: in medicine's concern to be able to identify the transmission of disease across continents, in Egyptological excavation, and in detecting the subversion of national science and culture by foreign infiltrators. It was certainly a key part of the increasingly over-determined symbolic significance of Brown's poor eyesight. The cartoonist Steve Bell captured this, and the accumulation of meanings that attached themselves to Brown's detached retina, in his satiric creation of an alternative headline for the *Sun* newspaper which had begun the debate on the condolence letter. Bell's cartoon shows a portrait of Brown (with suitably tiny eyes rendered as nothing more than slits) next to the headline 'Respect Our Cannon Fodder You Shameful Blundering Scotch Bastard'.[19] This, I must admit, is not something that the Victorians or Edwardians would have written. Yet it succinctly registers the ways in which Brown's weakened vision is put to metaphoric use to indicate something of his suspect nationality, immorality and political frailty. In using the word 'blundering', with its shades of Tennyson's 'Charge of the Light Brigade' and its comments on another military conflict, Bell also reveals how Brown's poor vision marks him out as a prime minister destabilized by his own ocular fragility. Yet at the same time Bell's sympathies are clearly with Brown. In highlighting his poor eyesight and its repercussions in such a performative and creative medium as the cartoon Bell actually makes an effort to empower Brown. Bell is more cynical and ironic than Houdini's illusions, certainly, but his aim is the same: satirize ocular weakness to illuminate ocular strength.[20]

NOTES

The following abbreviations have been used throughout the notes:

EEFA	Egypt Exploration Fund Archive
HHA	Harry Houdini Archive
M-YMC	McManus-Young Magical Collection
PLA	Percival Lowell Archive
PMA	Petrie Museum Archive
SA/LIS	Lister Institute Archive

Introduction

1. A. Conan Doyle, *The Stark Munro Letters* (London: Longmans, Green & Co., 1895), pp. 341–2.
2. This first sentence consciously parallels the opening sentence to Jonathan Crary's *Techniques of the Observer*, which reads 'This is a book about vision and its historical construction'. J. Crary, *Techniques of the Observer: On Vision and Modernity in the Nineteenth Century* (Cambridge, MA: MIT Press, 1992)
3. Ibid.
4. J. Krasner, *The Entangled Eye: Visual Perception and the Representation of Nature in Post-Darwinian Narrative* (New York: Oxford University Press, 1992).
5. C. T. Christ and J. O. Jordan (eds), *Victorian Literature and the Victorian Visual Imagination* (Berkeley, CA: University of California Press, 1995).
6. K. Z. Moore, 'Viewing the Victorians: Recent Research on Victorian Visuality', *Victorian Literature and Culture*, 25:2 (1997), pp. 367–85.
7. K. Flint, *The Victorians and the Visual Imagination* (Cambridge: Cambridge University Press, 2000); K. Flint, '"Seeing is Believing"? Visuality and Victorian Fiction', in F. O'Gorman (ed.), *A Concise Companion to the Victorian Novel* (Oxford: Blackwell, 2005); S. Smajić, *Ghost-Seers, Detectives, and Spiritualists: Theories of Vision in Victorian Literature and Science* (Cambridge: Cambridge University Press, 2010).
8. J. Green-Lewis, *Framing the Victorians: Photography and the Culture of Realism* (Ithaca, NY: Cornell University Press, 1996).
9. C. Otter, *The Victorian Eye: A Political History of Light and Vision, 1800–1910* (Chicago, IL: University of Chicago Press, 2008).
10. C. Yanni, *Nature's Museums: Victorian Science and the Architecture of Display* (New York: Princeton Architectural Press, 2005).

11. A. Jaffe, *Scenes of Sympathy: Identity and Representation in Victorian Fiction* (Ithaca, NY: Cornell University Press, 2000).
12. C. Otter, 'Victorian "Ways of Seeing"?', *Journal of Victorian Culture*, 14:1 (2009), pp. 95–102.
13. Smajić, *Ghost-Seers*, pp. 200–3.
14. L. Daston, 'Fear and Loathing of the Imagination in Science', in P. Galison, S. R. Graubard and E. Mendelsohn (eds), *Science in Culture* (New Brunswick, NJ: Transaction, 2001), pp. 73–95; L. Daston and K. Park, *Wonders and the Order of Nature, 1150–1750* (New York: Zone Books, 1998). Recent work on the imagination by J. H. Lindquist has begun to redress this: J. H. Lindquist, '"The Mightiest Instrument of the Physical Discoverer": The Visual "Imagination" and the Victorian Observer', *Journal of Victorian Culture*, 13:2 (2008), pp. 171–99.
15. G. Beer, *Darwin's Plots: Evolutionary Narrative in Darwin, George Eliot and Nineteenth-Century Fiction* (Cambridge: Cambridge University Press, 1983); G. Levine, *Darwin and the Novelists: Patterns of Science in Victorian Fiction* (Cambridge, MA: Harvard University Press, 1988).
16. R. O'Connor, *The Earth on Show: Fossils and the Poetics of Popular Science, 1802–1856* (Chicago, IL: University of Chicago Press, 2007).
17. In using the terms 'literary text' or 'imaginative writing' I refer not only to works of fiction, which I deal with extensively, but also other forms of narrative, such as the epistolary form or the travel narrative, that are part of a recognizable, if broader, definition of the literary in contemporary scholarship.

Part I: Small

1 Microscopy and Disease: Science, Imagination and the Phantasmagoria

1. G. Gooday, 'Placing or Replacing the Laboratory in the History of Science?', *Isis*, 99:4 (2008), pp. 783–95, on p. 792.
2. W. Henry, 'Report on the State of Our Knowledge of the Laws of Contagion', in *Report of the Fourth Meeting of the British Association for the Advancement of Science; held at Edinburgh in 1834* (London: BAAS, 1835), pp. 67–94, on pp. 67–8.
3. J. Quekett, *Lectures on Histology, Delivered at the Royal College of Surgeons of England, in the Session 1850–51* (London: Hippolyte Balliere, 1852), p. 1.
4. F.-J. O'Brien, 'The Diamond Lens', *Atlantic Monthly* (January 1858), pp. 354–67, on pp. 354, 356.
5. K. Alder, 'History's Greatest Forger: Science, Fiction, and Fraud along the Seine', *Critical Inquiry*, 30 (2004), pp. 702–16, on p. 712.
6. Daston and Park, *Wonders*, p. 136.
7. W. B. Carpenter, *The Microscope and its Revelations* (London: John Churchill, 1856), p. 37.
8. Daston and Park, *Wonders*, p. 136.
9. Anon., *The Wonders of the Microscope; or, An Explanation of the Wisdom of the Creator in Objects Comparatively Minute, Adapted to the Understanding of Young Persons* (London: Tabarat & Co., 1806), pp. 12–13.

10. J. Hogg, *The Microscope: Its History, Construction and Applications. Being a Familiar Introduction to the Use of the Instrument and the Study of Microscopical Science* (London: Herbert Ingram, 1856), p. 450.
11. G. Mantell, *Thoughts on Animalcules: or, A Glimpse of the Invisible World Revealed by the Microscope* (London: John Murray, 1846), p. 7.
12. T. Castle, 'Phantasmagoria: Spectral Technology and the Metaphorics of Modern Reverie', *Critical Inquiry*, 15 (1998), pp. 26–61.
13. M. Worboys, for example, argues that 'it was only after 1895 that bacteriologists played a major part' in defining the processes by which disease was transmitted (see M. Worboys, *Spreading Germs: Disease Theories and Medical Practice in Britain, 1865–1900* (Cambridge: Cambridge University Press, 2000), p. 193).
14. E. H. Ackerknecht, 'Anticontagionism Between 1821 and 1867', *Bulletin of the History of Medicine*, 22 (1948), pp. 562–93; P. Baldwin, *Contagion and the State in Britain, 1830–1930* (Cambridge: Cambridge University Press, 2000).
15. E. N. Bancroft, *An Essay on the Disease called Yellow Fever, with Observations Concerning Febrile Contagion, Typhus Fever, Dysentry, and the Plague, Partly Delivered as the Gulstonian Lectures, Before the College of Physicians, in the Years 1806 and 1807* (London: T. Cadell & W. Davies, 1811), p. 95.
16. T. S. Smith, *The Common Nature of Epidemics, and Their Relation to Climate and Quarantine. Also Remarks on Contagion and Quarantine* (London: John Trubner, 1866), p. 14.
17. F. Nightingale, 'Notes on Nursing for the Labouring Classes (1860)', in L. McDonald (ed.), *Florence Nightingale on Public Health* (Ontario: Wilfrid Laurier University Press, 2004), pp. 17–160, on pp. 34, 109.
18. Ibid., pp. 122–3.
19. J. Tyndall, 'On Haze and Dust', *Nature* (27 January 1870), pp. 339–42, on pp. 340–1.
20. J. K. Crellin, 'Airborne Particles and the Germ Theory, 1860–1880', *Annals of Science*, 22:1 (1966), pp. 49–60, on p. 51.
21. Nightingale, 'Letter to Dr Gillham Newlett, 27 July 1883', in McDonald (ed.), *Florence Nightingale*, pp. 567–8.
22. E. Lankester, *Half-Hours with the Microscope: Being a Popular Guide to the Use of the Microscope as a Means of Amusement and Instruction* (London: Robert Hardwicke, 1860), pp. 12, 79.
23. Ibid., p. 1.
24. S. La'mert, *Self-Preservation: A Medical Treatise on Nervous and Physical Debility, Spermatorrhea, Impotence, and Sterility with Practical Observations on the Use of the Microscope in the Treatment of the Diseases of the Generative System* (London: Samuel La'mert, 1859), pp. 145–6.
25. J. H. Bennett, *On the Employment of the Microscope in Medical Studies. A Lecture Introductory to a Course in Histology* (Edinburgh: Maclachlan, Stuart & Co., 1841), p. 11.
26. Ibid.
27. Quekett, *Lectures on Histology*, p. 3.
28. La'mert, *Self-Preservation*, p. 150.
29. Ibid., pp. 150–1.
30. L. Daston and P. Galison, 'The Image of Objectivity,' *Representations*, 40 (1992), pp. 81–128, on p. 119
31. I. Armstrong, *Victorian Glassworlds: Glass Culture and the Imagination, 1830–1880* (Oxford: Oxford University Press, 2008), p. 319.

32. L. Otis, *Membranes: Metaphors of Invasion in Nineteenth-Century Literature, Science, and Politics* (Baltimore, MD: Johns Hopkins University Press, 1999), p. 3.
33. J. Polidori, 'The Vampyre', in . A. O. Gladwell and J. Havoc (eds), *Blood and Roses: The Vampire in 19th Century Literature* (London: Creation Press, 1992), pp. 29–43.
34. Ibid., p. 35.
35. Ibid., p. 36.
36. Ibid., p. 42.
37. John Polidori's career at the Edinburgh Medical School ran from 1815–18, Southwood Smith's from 1816–19.
38. S. Le Fanu, 'Carmilla', in R. Tracy (ed.), *In a Glass Darkly* (Oxford: Oxford University Press, 1993), pp. 243–319, on p. 279.
39. Ibid., p. 274.
40. Ibid., p. 269.
41. Ibid., pp. 315–6.
42. Ibid., pp. 250–1.
43. Henry, 'Report on the State of Our Knowledge', p. 90.
44. B. Stoker, *Dracula*, ed. Maurice Hindle (1897; London: Penguin, 1993), p. 62.
45. Ibid., p. 66; Bancroft, *An Essay on the Disease called Yellow Fever*, p. 157.
46. Stoker, *Dracula*, pp. 186, 332.
47. Ibid., p. 269.
48. Ibid., p. 323.
49. Hogg, *The Microscope*, p. 450.
50. Mantell, *Thoughts on Animalcules*, p. 7.
51. T. Dick, *The Telescope and Microscope* (London: Religious Tract Society, 1851), p. 131.
52. K. Gelder, *Reading The Vampire* (London: Routledge, 1994); J. Halberstam, 'Technologies of Monstrosity: Bram Stoker's *Dracula*', in S. Ledger and S. McCracken (eds), *Cultural Politics at the Fin-de-Siècle* (Cambridge: Cambridge University Press, 1995), pp. 248–66; R. Mighall, '"A Pestilence which Walketh in Darkness": Diagnosing the Victorian Vampire', in G. Byron and D. Punter (eds), *Spectral Readings: Towards a Gothic Geography* (London: Macmillan, 1999), pp. 108–24.
53. W. Hughes, 'Fictional Vampires in the Nineteenth and Twentieth Centuries', in D. Punter (ed.), *A Companion to the Gothic* (Oxford: Blackwell, 2000), pp. 144–5.
54. A. Warwick, 'Vampires and the Empire: Fears and Fictions of the 1890s', in S. Ledger and S. McCracken (eds), *Cultural Politics at the Fin-de-Siècle* (Cambridge: Cambridge University Press, 1995), pp. 203–14, on pp. 203–4.
55. Le Fanu, 'Carmilla', p. 271.
56. Ibid., p. 310.
57. Stoker, *Dracula*, p. 147.
58. Ibid., p. 150,162.
59. Ibid., p. 162.
60. Ibid., p. 246.
61. Ibid.
62. Ibid.
63. Ibid., p. 249.
64. Ibid.
65. For various scientific, social and cultural assessments of the emergence of germ theory see Worboys, *Spreading Germs*; Ackerknecht, 'Anticontagonism'; W. F. Bynum, 'The Evolution of Germs and the Evolution of Disease: Some British Debates, 1870–1900', *History*

and Philosophy of the Life Sciences, 24 (2000), pp. 53–68; J. Waller, *The Discovery of the Germ: Twenty Years That Transformed the Way We Think About Disease* (Cambridge: Icon Books, 2002); N. Tomes, *The Gospel of Germs: Men, Women and the Microbe in American Life* (Cambridge, MA: Harvard University Press, 1998); and M. Pelling, 'The Meaning of Contagion: Reproduction, Medicine and Metaphor', in A. Bashford and C. Hooker (eds), *Contagion: Historical and Cultural Studies* (London: Routledge, 2001), pp. 15–38.

66. J. H. Bennett, 'The Atmospheric Germ Theory', *Edinburgh Medical Journal* (March 1868), pp. 810–34, on p. 812
67. Tomes, *The Gospel of Germs*, p. 7.
68. Quoted in Waller, *The Discovery of the Germ*, p. 187.
69. M. Willis, 'Unmasking Immorality: Popular Opposition to Laboratory Science in Late Victorian Britain', in D. Clifford, E. Wadge, A. Warwick and M. Willis (eds), *Repositioning Victorian Sciences: Shifting Centres in Nineteenth-Century Scientific Thinking* (London: Anthem Press, 2006), pp. 207–18.
70. Anon., 'Deputation to the Home Secretary: To protest against the building of the British Institute of Preventive Medicine at Chelsea', SA/LIS, 24 April 1894, p. 2.
71. Anon., 'Deputation to the Home Secretary', SA/LIS, p. 2.
72. Anon., 'Deputation to the Home Secretary', SA/LIS, p. 5.
73. Anon., 'Deputation to the Home Secretary', SA/LIS, p. 7.
74. Anon., 'Deputation to the Board of Trade', SA/LIS, 7 June 1894, p. 18.
75. Anon., 'Deputation to the Board of Trade', p. 22.
76. L. S. Jacyna, 'Moral Fibre: The Negotiation of Microscopic Facts in Victorian Britain', *Journal of the History of Biology*, 36 (2003), pp. 39–85, on p. 39.
77. Crary, *Techniques*, pp. 132–4; Castle, 'Phantasmagoria'.
78. Castle, 'Phantasmagoria', p. 27.
79. Crary, *Techniques*, p. 132; J. Crary, *Suspensions of Perception: Attention, Spectacle, and Modern Culture* (Cambridge, MA: MIT Press, 2001), p. 251.
80. Castle, 'Phantasmagoria', p. 30.
81. Ibid., pp. 31–43; Crary, *Techniques*, p. 132.
82. Daston and Park, *Wonders*, especially pp. 13–20. See also Daston, 'Fear and Loathing of the Imagination, pp. 73–95; and B. Stafford, *Artful Science: Enlightenment Entertainment and the Eclipse of Visual Education* (Cambridge, MA: MIT Press, 1996).
83. Crary, *Techniques*, p. 133.
84. Hogg, *The Microscope*, p. 6; Anon., *The Wonders*, p. 11.
85. J. H. Bennett, quoted in Waller, *The Discovery of the Germ*, p. 70.
86. See M. Cohen, 'Benjamin's Phantasmagoria: The Arcades Project', in D. S. Ferris (ed.), *The Cambridge Companion to Walter Benjamin* (Cambridge: Cambridge University Press, 2004), pp. 200–20; C. Britzolakis, 'Phantasmagoria: Walter Benjamin and the Poetics of Urban Modernism', in P. Buse and A. Stott (eds), *Ghosts: Deconstruction, Psychoanalysis, History* (London: Macmillan, 1999), pp. 72–90.
87. Britzolakis, 'Phantasmagoria', p. 73.
88. M. Weber, 'Science as a Vocation', in D. Owen and T. B. Strong (eds), *The Vocation Lectures*, trans. R. Livingstone (Indiana: Hackett, 2004), pp. 1–31.
89. Cohen, 'Benjamin's Phantasmagoria', p. 202.
90. G. Levine, *Darwin Loves You: Natural Selection and the Re-Enchantment of the World* (Princeton, NJ: Princeton University Press, 2006), p. xiv.

91. A. Hardy, *The Epidemic Streets: Infectious Disease and the Rise of Preventive Medicine, 1856–1900* (Oxford: Clarendon Press, 1993); F. Mort, *Dangerous Sexualities: Medico-Moral Politics in England Since 1830* (London: Routledge, 1987); K. Waddington, *The Bovine Scourge: Meat, Tuberculosis and Public Health, 1850–1914* (Woodbridge: Boydell & Brewer, 2006).
92. Waddington, *The Bovine Scourge*, p. 9.
93. Cohen, 'Benjamin's Phantasmagoria', p. 207.
94. *Westminster News*, 7 July 1896, Press Scrapbook, SA/LIS.
95. Anon., 'Deputation to the Home Secretary', SA/LIS, p. 5.
96. Anon., 'Deputation to the Home Secretary', SA/LIS, pp. 1,6.
97. *Westminster News*, 22 January 1898, Press Scrapbook, SA/LIS.
98. Anon., 'Report on the Parade and Mass Meeting at Pimlico', SA/LIS, p. 3; A. W. Whalley, 'Survey of Chelsea Residents', 11 April 1894, SA/LIS.
99. Britzolakis, 'Phantasmagoria', p. 76.
100. Le Fanu, 'Carmilla', p. 276.
101. Ibid., p. 274.
102. Hogg, *The Microscope*, pp. 1–12.
103. K. Maglen, 'Interpreting Infection: Quarantine, the Port Sanitary Authority and Immigration in Late Nineteenth and Early Twentieth Century Britain' (PhD dissertation, University of Glasgow, 2001). See also Mort, *Dangerous Sexualities*; Baldwin, *Contagion*; A. McClintock, *Imperial Leather: Race, Gender and Sexuality in the Colonial Contest* (London: Routledge, 1995). It is worth noting here that immigration was obviously an important factor in issues of quarantine, although it is not the focus of this chapter.
104. S. D. Arata, 'The Occidental Tourist: *Dracula* and the Anxiety of Reverse Colonization', *Victorian Studies*, 33:4 (1990), pp. 621–45.
105. Stoker, *Dracula*, p. 394.
106. Crary notes that the commodity is an 'image on which essence has been displaced by appearance', *Suspensions*, p. 279.

2 Microscopy and Disease: Place and Identity in Laboratory Science and Fiction

1. Those involved in lobbying for the support of Sir James Whitehead, Lord Mayor of London, included the zoologist and biologist Edwin Ray Lankester, the experimental surgeon Joseph Lister (for whom the Institute would later be named) and the evolutionary biologist T. H. Huxley. See SA/LIS.
2. Anon., 'Report of Parade and Mass Meeting at Pimlico', SA/LIS/E5, p. 1.
3. Ibid., p. 7.
4. Ibid., pp. 3, 17.
5. Ibid., p. 2.
6. Ibid.
7. See the various first-hand descriptions in E. Crawford (ed.), *The Women's Suffrage Movement: A Reference Guide, 1866–1928* (London: Routledge, 2001), pp. 149–51.
8. Mort, *Dangerous Sexualities*; Tomes, *The Gospel of Germs*.
9. Otis, *Membranes*.
10. Gooday, 'Placing or Replacing the Laboratory'.
11. R. E. Kohler, 'Lab History: Reflections', *Isis*, 99:4 (2008), pp. 761–8, on p. 764.

12. Public knowledge can also be found in many other types of written document, of course, but these are the documents under investigation in this chapter.
13. Kohler, 'Lab History', p. 765.
14. Anon., *How to Live in London: or, The Metropolitan Microscope and Stranger's Guide* (London: Joseph Smith, 1828), title page.
15. M. Rickards, *The Encyclopaedia of Ephemera: A Guide to the Fragmentary Documents of Everyday Life for the Collector, Curator, and Historian* (New York: Routledge, 2000), pp. 37–8.
16. Anon., 'On the Microscopical Character of Milk and the Use of the Microscope in the Choice of a Nurse', *Lancet*, 1 (1844), p. 531.
17. Mort, *Dangerous Sexualities*.
18. Smith, *The Common Nature of Epidemics*, p. 69.
19. E. A. Poe, 'The Masque of Red Death', in *Selected Tales*, ed. D. Van Leer (Oxford: Oxford University Press, 1998), p. 134.
20. Smith, *The Common Nature of Epidemics*, p. 12.
21. Nightingale, 'Letter to Henry Parker, 1 May 1867', in Macdonald (ed.), *Florence Nightingale on Public Health*, p. 533
22. Smith, *The Common Nature of Epidemics*, p. 14.
23. M. Gibson, *Dracula and the Eastern Question: British and French Vampire Narratives of the Nineteenth-Century Near East* (Basingstoke: Palgrave Macmillan, 2006).
24. Le Fanu, 'Carmilla', p. 244.
25. J. Robins, *Miasma: Epidemic and Panic in Nineteenth Century Ireland* (Dublin: Institute of Public Administration, 1995), pp. 66–8.
26. Le Fanu, 'Carmilla', pp. 266, 269.
27. Ibid., p. 270.
28. L. Brandeis, *Sanitarian*, 1873, SA/LIS/E4.
29. Anon., 'Report of Parade and Mass Meeting at Pimlico', SA/LIS/E5, p. 9.
30. Anon., 'Report of Parade and Mass Meeting at Pimlico', SA/LIS/E5, p. 9.
31. E. Carrington, 'Untitled', *Gloucester Chronicle*, 13 October 1894, SA/LIS/E8.
32. 'MORE SECRECY! District Councillors Kept Off Premises of Institute of Preventive Medicine', *Star*, 3 March 1896, SA/LIS/E8.
33. Kohler, 'Lab History', p. 762.
34. Gooday, 'Placing or Replacing the Laboratory', p. 786.
35. Arata, "The Occidental Tourist'; N. Auerbach, 'Dracula: A Vampire of Our Own', in G. Bryon (ed.), *Dracula: Contemporary Critical Essays* (London: Macmillan, 1999), pp. 145–72; G. Byron, 'Introduction', in Byron (ed.), *Dracula*, pp. 1–21; C. E. Prescott and G. A. Giorgio, 'Vampiric Affinities: Mina Harker and the Paradox of Femininity in Bram Stoker's *Dracula*', *Victorian Literature and Culture*, 33 (2005), pp. 487–515; and P. A. Roth, 'Suddenly Sexual Women', in Byron (ed.), *Dracula*, pp. 30–42 all argue that Lucy Westenra is most often assessed within the contexts of gender and sexuality.
36. Stoker, *Dracula*, p. 97.
37. Ibid., p. 186.
38. Lucy is described by Seward in this scene as speaking in 'a soft, voluptuous voice, such as I had never heard from her lips' (p. 208) when she calls for Arthur to kiss her. Seward's stressing of the voice which he has never before heard reinforces the point that Lucy's sexuality is altered by the onset of vampiric disease and is not the product of an already present sexuality.

39. W. Hughes, *Beyond Dracula: Bram Stoker's Fiction in Its Cultural Context* (New York: St Martin's Press, 2000), p. 141. There is an irony in the meaning of Lucy's name (light) that Hughes does not recognize: the importance of light to seeing clearly through the microscope. For microscopy light and vision are inextricably intertwined. Lucy, however, does not take any power from her name's association, rather she is subjected to the power of microscopic vision to foster a new way of considering her identity.
40. Stoker, *Dracula*, p. 119–22.
41. Ibid., p. 121.
42. Hogg, *The Microscope*, p. 450
43. Mort, *Dangerous Sexualities*, p. 68
44. R. Macleod, 'Introduction: Nature and Empire: Science and the Colonial Enterprise', *Osiris*, 15 (2000), pp. 1–13.
45. Ibid., p. 6.
46. M. L. Pratt, *Imperial Eyes: Travel Writing and Transculturation* (London: Routledge, 1992); B. Latour, *The Pasteurization of France*, trans. A. Sheridan and J. Law (Cambridge, MA: Harvard University Press, 1988); B. Latour, *Science in Action: How to Follow Scientists and Engineers Through Society* (Cambridge, MA: Harvard University Press, 1988).
47. Otis, *Membranes*, pp. 1–7.
48. Ibid., p. 5.
49. Henry, 'Report on the State of Our Knowledge', p. 83.
50. J. Haygarth, *Letter to Dr Percival*, quoted in Henry, 'Report on the State of Our Knowledge', p. 84; W. Heberden, *Comment*, again quoted in Henry, 'Report of the State of Our Knowledge', p. 84.
51. F. Fanon, *Black Skin White Masks* (New York: Grove Press, 1967), p. 111.
52. J. Killeen, 'Irish Gothic: A Theoretical Introduction', *Irish Journal of Gothic and Horror Studies*, 1 (2006), n.p.
53. R. F. Foster, *Paddy and Mr. Punch: Connections in Irish and English History* (London: Allen Lane, 1993), p. 220.
54. La'mert, *Self-Preservation*, pp. 145–6.
55. Anon., 'The Proposed British Institute of Preventive Medicine', SA/LIS/E8.
56. F. P. Cobbe, 'Vivisection and Its Two-Faced Advocates', in L. Otis (ed.), *Literature and Science in the Nineteenth Century: An Anthology* (Oxford: Oxford University Press, 2002), pp. 215–9.
57. J. Collinson, 'The Pasteur Institute', *Irish Figaro*, 11 May 1895, SA/LIS/E8. Collinson was a leading member of the Humanitarian League. His description of the escaping dog is reimagined by H. G. Wells in the Gothic novel, *The Island of Dr Moreau*, published a year later in 1896.
58. 'Mansion House Meeting', 1 June 1889, SA/LIS/M11.
59. 'Mansion House Meeting', 1 June 1889, SA/LIS/M11.
60. N. Rupke, 'Pro-Vivisection in England in the Early 1880s', in N. Rupke (ed.), *Vivisection in Historical Perspective* (London: Routledge, 1987), pp. 188–208, on p. 198.
61. 'Human Vivisection', *Daily Chronicle*, 5 June 1894, SA/LIS/E8.
62. *Field*, 6 November 1886, SA/LIS/E8.
63. G. Mitman, M. Murphy and C. Sellers, 'Introduction: A Cloud Over History', *Osiris*, 19 (2004), pp. 1–17, on p. 6.
64. Ibid., p. 8.
65. Worboys, *Spreading Germs*, pp. 6–12; D. Lee and T. Fulford, 'The Beast Within: The Imperial Legacy of Vaccination in History and Literature', *Literature and History*, 9:1

(2000), pp. 1–23; Maglen, 'Interpreting Infection', pp. 66–8; as well as Nightingale, 'Letter to Dr Gilham Newlett, 27 July 1883'. and Smith, *The Common Nature of Epidemics*, p. 12
66. See *Dracula* 7–8. Arata also argues this point in his discussion of the novel as an example of the fear of reverse colonization, 'The Occidental Tourist', pp. 634–7.
67. Stoker, *Dracula*, p. 32.
68. Ibid., p. 51.
69. Jabez Hogg speaks of how much more can be seen by 'the experienced eye' than by others. See Hogg, *The Microscope*, p. viii
70. See Daston and Galison, 'The Image of Objectivity', pp. 81–128.
71. Stoker, *Dracula*, p. 52.
72. Prescott and Giorgio have also noted Harker's transformation of his experiences in Dracula's castle into chivalric history, 'Vampiric Affinities', pp. 489–50.
73. Stoker, *Dracula*, p. 71.
74. Ibid., p. 35.
75. Lankester, *Half-Hours*, p. 1
76. Stoker, *Dracula*, p. 323.
77. Worboys, *Spreading Germs*, p. 5; N. J. Tomes and J. Harley, 'Introduction to Special Issue on Rethinking the Reception of the Germ Theory of Disease: Comparative Perspectives', *Journal of the History of Medicine and Allied Sciences*, 52:1 (1997), pp. 7–16, on p. 9.
78. Latour, *The Pasteurization*, pp. 149–50.
79. J. A. Mendelsohn, 'The Microscopist of Modern Life', *Osiris*, 18 (2003), pp. 150–70, on p. 158.
80. Mitman et al., 'Introduction', p. 15.
81. Ibid.

Part II: Large

3 Optical Shattering

1. P. Lowell, 'On Some Fallacies in Observing Mars', 1909, PLA, Ms–2U/19, pp. 1–2.
2. H. G. Wells, *The War of the Worlds* (1898; London, Penguin, 2005), p. 10.
3. These texts include various periodicals, such as *Nature*, which published articles on Lowell's work and Mars in this period, and to which Wells also contributed. See, for example, *Nature*, 50 (14 June 1894; 26 July 1894; 13 September 1894; and 27 September 1894).
4. J. F. W. Herschel, *Outlines of Astronomy* (London: Longman, Brown, Green & Longmans, 1849), p. 79. *Outlines* went through three editions in the year of its publication and was reprinted throughout the 1850s and 1860s, reaching its ninth edition by 1869. After Herschel's death in 1871 it continued to be successful, again going through numerous reprints in the 1870s and 1880s. For this reason it should be regarded as one of the most influential works of Victorian popular astronomy.
5. Ibid., pp. 81–2.
6. M. J. Crowe, *The Extraterrestrial Life Debate, 1750–1900: The Idea of a Plurality of Worlds from Kant to Lowell* (Cambridge: Cambridge University Press, 1986), p. 484.
7. S. J. Dick, *Plurality of Worlds: The Origins of the Extraterrestrial Life Debate from Democritus to Kant* (Cambridge: Cambridge University Press, 1982), p. 178.
8. W. Sheehan, *Planets and Perception: Telescopic Views and Interpretations, 1609–1909* (Tucson, AZ: University of Arizona Press, 1988); R. Markley, *Dying Planet: Mars in Science and the Imagination* (Durham, NC: Duke University Press, 2005).

9. L. Daston and P. Galison, *Objectivity* (New York: Zone Books, 2007).
10. Ibid. p. 139.
11. J. Tucker, *Nature Exposed: Photography as Witness in Victorian Science* (Baltimore, MD: Johns Hopkins University Press, 2005).
12. Ibid., p. 194.
13. Markley, *Dying*, p. 115.
14. A. Henchman, 'Hardy's Stargazers and the Astronomy of Other Minds', *Victorian Studies*, 51:1 (2008), 37–64, on p. 38.
15. Ibid., p. 43.
16. P. Gossin, *Thomas Hardy's Novel Universe: Astronomy, Cosmology and Gender in the Post-Darwinian World* (Aldershot: Ashgate, 2007), p. 80.
17. Ibid., p. 95.
18. Crary defines phantasmagoric vision as an ambiguous form of seeing in which it is difficult to know with any precision whether the observer is being deceived by fantastical or spectral images. See Crary, *Techniques*.
19. P. Lowell, *Mars* (1894; London: Longmans, Green & Co., 1896), p. v.
20. S. Newcomb, *Popular Astronomy* (New York: Harper, 1878), p. 146.
21. P. Lowell, 'Mars', 11 September 1894, PLA, Ms–2/2–1B, p. 1. This is the draft for a later article published in *Astronomy and Astrophysics*.
22. P. Lowell, *Annals of the Lowell Observatory: Volume 1 1894–5* (Boston, MA: Houghton & Mifflin, 1895), p. 250.
23. In many of his publications across his twenty-two-year career at Flagstaff, Lowell wrote of its excellent seeing conditions and its privileged position among observatory sites. He did so, for example, in the 1894 article 'Mars', listed above, and was still doing so in 1910, in the article 'Mars in 1909', PLA, Ms–6/4, pp. 1–7.
24. E. R. Burroughs, *A Princess of Mars* (1911; London: Penguin, 2007).
25. Ibid., p. 11.
26. Ibid., pp. 11–13.
27. Wells, *The War of the Worlds*, pp. 19, 24, 50–1.
28. Ibid., p. 51.
29. Lowell, *Annals*, pp. 192–3.
30. P. Lowell, 'Small Apertures Versus Large Ones in the Study of Planetology', PLA, Ms–2U/22, p. 5.
31. H. G. Wells, 'Introduction', in *The Scientific Romances* (London: Gollancz, 1930), p. viii. The novels collected under this title included *The War of the Worlds*.
32. B. Latour, 'On the Partial Existence of Existing *and* Nonexisting Objects', in L. Daston (ed.), *Biographies of Scientific Objects* (Chicago, IL: University of Chicago Press, 2000), pp. 247–69.
33. L. Daston, 'Introduction: The Coming into Being of Scientific Objects', in Daston (ed.), *Biographies of Scientific Objects*, pp. 9–24, on p. 13.
34. P. Lowell and V. M. Slipher, 'Air and Large Telescopes', 30 October 1908, PLA, Ms–5/13A, pp. 1, 5.
35. G. P. Serviss, *Edison's Conquest of Mars* (1898; Burlington: Apogee Books, n.d.), p. 36.
36. Lowell, *Annals*, p. 192.
37. Ibid., p. 437.
38. Serviss, *Edison's Conquest*, p. 145.
39. M. Wicks, *To Mars Via the Moon: An Astronomical Story* (1911; n.p.: Dodo Press, 2009), p. 107.

40. Wells, *The War of the Worlds*, pp. 88–9.
41. Ibid., p. 46.
42. Daston, *Biographies*, p. 13.
43. Ibid.
44. Daston and Galison, *Objectivity*, p. 139.
45. Ibid., p. 231.
46. S. Newcomb, *Astronomy for Everybody* (London: McClure, Phillips & Co.,1902), p. 183.
47. Ibid.
48. Lowell, 'Mars in 1909', PLA, Ms–6/4, p. 7.
49. J. Braesher, 'Letter to Percival Lowell', 19 July 1901, PLA, Corr–5/13A.
50. P. Lowell, 'Telegram to John Braesher', 14 August 1901, PLA, Corr–5/13A.
51. P. Lowell, 'Dawn', PLA, Ms–1U/1. This sonnet appears in five drafts before its final version, confirming the considerable time Lowell spent crafting it. Poetry ran in Lowell's family: his sister was the far more accomplished poet Amy Lowell.
52. P. Lowell, 'Untitled Ms', n.d., PLA, Ms–2U/12, pp. 4–5.
53. Lowell, 'Mars in 1909', PLA, Ms–6/4, p. 7.
54. Lowell, 'Small Apertures', pp. 7, 5.
55. P. Lowell, 'Means, Methods and Mistakes in the Study of Planetary Evolution', 13 April 1905, PLA, Ms–1U/12, pp. 25–6.
56. Daston and Galison, *Objectivity*, pp. 231, 55–113.
57. Ibid., p. 98.
58. S. Schaffer, 'Astronomers Mark Time: Discipline and the Personal Equation', *Science in Context*, 2 (1988), pp. 115–45.
59. C. Hoffmann, 'Constant Differences: Friedrich Wilhelm Bessel, the Concept of the Observer in Early Nineteenth-century Practical Astronomy and the History of the Personal Equation', *British Journal of the History of Science*, 40:3 (2007), pp. 333–65, on pp. 364, 365.
60. Burroughs, *A Princess of Mars*, p. 65.
61. Wicks, *To Mars*, pp. 78–9.
62. Wells, *The War of the Worlds*, p. 120.
63. Ibid., p. 124.
64. Ibid.
65. Lowell, *Mars*, pp. 158–9.
66. H. von Helmholtz, 'Concerning the Perceptions in General', in J. P. C. Southall (ed.), *Helmholtz's Treatise on Physiological Optics*, 3 vols (Bristol: Theommes Press, 2000), vol. 3, p. 10.
67. Lowell, *Mars*, p. 44.
68. Serviss, *Edison's Conquest*, p. 249.
69. See P. Lowell, 'Letter to E.W. Maunder', 28 November 1903, PLA, Corr–11/28/03.
70. P. Lowell, 'Areography', *Popular Science Monthly*, 61 (September 1902), p. 392.
71. Newcomb, *Astronomy*, p. 185.
72. Ibid., but only in the Lowell Observatory copy owned by Lowell (PLA, QB.44 NB 1908).
73. P. Lowell, 'On the Kind of Eye Needed for the Detection of Planetary Detail', PLA, Ms–4/11, n.p.
74. Ibid.
75. Lowell, 'Means, Methods and Mistakes', p. 11.
76. Ibid., p. 13.

77. Daston and Galison, *Objectivity*, p. 44.
78. Ibid., p. 48.
79. M. Wicks, 'Letter to Percival Lowell', PLA, Corr–9/24/07.
80. Ibid.
81. Wicks, *To Mars*, p. 207.
82. Ibid., p. 236.
83. Wells, *The War of the Worlds*, p. 177.
84. Ibid.
85. Ibid.
86. Ibid.
87. See W. G. Hoyt, *Lowell and Mars* (Tucson, AZ: University of Arizona Press, 1976).
88. J. Ritchie, Jnr, 'The Lowell Observatory, Arizona', *Nature*, 14 June 1894, p. 149.
89. Newcomb, *Astronomy*, p. 183; Anon., 'A Strange Light on Mars', *Nature*, 2 August 1894, p. 319.
90. P. de Bolla, *The Education of the Eye: Painting, Landscape, and Architecture in Eighteenth-Century Britain* (Stanford, CA: Stanford University Press, 2003), p. 16.
91. Ibid.
92. Daston and Galison, *Objectivity*, p. 324.
93. L. Daston, 'On Scientific Observation', *Isis*, 99 (2008), pp. 97–110, on pp. 100–6.
94. W. J. Lockyer, 'Mars as He Now Appears', *Nature*, 13 September 1894, pp. 476–8, on p. 476.
95. Ibid., p. 478.
96. Lowell, 'Means, Methods and Mistakes', p. 27.
97. Ibid., p. 33.
98. Ibid., p. 4.
99. E. H. Antoniadi, 'Letter to Percival Lowell', 9 September 1909, PLA, Corr–9/9/09.
100. P. Lowell, 'Letter to Antoniadi', 26 September 1909, PLA, Corr–09/26/09.
101. E. H. Antoniadi, 'Letter to Percival Lowell', 9 October 1909, PLA, Corr–10/09/09.
102. Daston and Galison, *Objectivity*, p. 311.
103. Antoniadi, 'Letter to Percival Lowell', 9 October 1909.
104. P. Lowell, 'Letter to Antoniadi', 2 November 1909, PLA, Corr–11/02/09.
105. Lowell, 'On Some Fallacies in Observing Mars', pp. 58, 59.
106. Daston, 'On Scientific Observation', p. 110.
107. Ibid., p. 107. Daston says little about seeing well, other than to suggest that defining it would be a useful project.
108. Wells, *The War of the Worlds*, pp. 10, 19.
109. Serviss, *Edison's Conquest*, p. 48.
110. Ibid., p. 3.
111. Wells, *The War of the Worlds*, p. 9.
112. Ibid., p. 11.
113. Ibid., p. 127. The writer is actually Wells himself, who had written an essay on the future evolution of the body of man in 1893. See G. S. Haight, 'H. G. Wells's "The Man of the Year Million"', *Nineteenth-Century Fiction*, 12:4 (1958), pp. 323–6.
114. Wicks, *To Mars*, p. 13.
115. R. Mortier, 'Illustration of Science and Technology and Artistic Perception in Eighteenth Century France', in B. Castel, J. A. Leith and A. W. Riley (eds), *Muse and Reason: The Relation of Arts and Sciences 1650–1850* (Canada: Queen's Quarterly, 1994), pp. 27–42, on p. 31.

116. Ibid.
117. S. Shapin and S. Schaffer, *Leviathan and the Air-Pump: Hobbes, Boyle, and the Experimental Life* (Princeton, NJ: Princeton University Press, 1985), p. 60.
118. Ibid.
119. Ibid., p. 61.
120. Ibid., pp. 62–3.
121. Ibid., p. 63.
122. See R. Cunningham, 'Virtual Witnessing and the Role of the Reader', *Philosophy and Rhetoric* 34:3 (2001), pp. 207–24.
123. Ibid., p. 222.

4 Lowell's Minimum Visible

1. Lowell, 'Mars', *c.* August 1894, PLA, Ms–1U/5, ll. 24–29. The manuscript of this part of the poem is handwritten in a style that also suggests speed, and is neither revised nor altered.
2. Ibid., l.222.
3. M. E. Martin, *The Ways of the Planets* (London: Harper, 1912), p. 12.
4. Ibid.
5. Ibid., p. 13.
6. Ibid.
7. W. Wordsworth, 'Star-gazers', in *Selected Poetry*, ed. S. Gill and D. Wu (Oxford: Oxford University Press), ll. 1–2.
8. Ibid., l. 30.
9. Ibid., ll. 7, 32.
10. Ibid., ll. 25–7.
11. See, for example, R. Cooter's analysis of popular science's place in the history of science scholarship in *The Cultural Meaning of Popular Science: Phrenology and the Organization of Consent in Nineteenth-century Britain* (Cambridge: Cambridge University Press, 1984). See also J. Topham, 'Focus: Historicizing 'Popular Science': Introduction', *Isis*, 100 (2009), pp. 310–18.
12. J. A. Secord, 'Knowledge in Transit', *Isis*, 95 (2004), pp. 654–72, on p. 661.
13. Ibid.
14. O'Connor, *The Earth on Show*.
15. R. O'Connor, 'Reflections on Popular Science in Britain: Genres, Categories, and Historians,' *Isis*, 100 (2009), pp. 333–45, on p. 336.
16. A. W. Daum, 'Varieties of Popular Science and the Transformations of Public Knowledge: Some Historical Reflections,' *Isis*, 100 (2009), pp. 319–32, on p. 320.
17. Ibid., p. 323.
18. Ibid.
19. A. Guillemin, *The Heavens: An Illustrated Handbook of Popular Astronomy* (London: Richard Bentley, 1867), p. v.
20. Ibid., p. vi.
21. See Lockyer's own preface to the second edition of Guillemin, *The Heavens*.
22. R. S. Ball, *The Story of the Heavens* (London: Cassell, 1886), p. 1.
23. M. Proctor, *Romance of the Planets* (London: Harper, 1929).
24. P. Lowell, *Mars as the Abode of Life* (New York: Macmillan, 1908), p. 3.
25. O'Connor, 'Reflections', p. 339.

26. H. G. Wells, *The Time Machine* (London: Heinemann, 1895). This first novel dealing with science was conceived in a short story in 1888 and broadened in scope in a serial running throughout 1894 and early 1895.
27. H. G. Wells, 'Letter to Grant Richards, Nov. 6, 1895', in D. C. Smith (ed.), *The Correspondence of H. G. Wells: Volume 1 – 1880–1903* (London: Pickering & Chatto, 1998), p. 250.
28. H. G. Wells, 'Letter to Morris Colles, Jan. 26, 1896' and 'Letter to Morris Colles, Mar. 25, 1896', in Smith, *The Correspondence*, pp. 260, 261.
29. R. Crossley, 'Percival Lowell and the History of Mars', *Massachusetts Review*, 41:3 (2000), pp. 297–318, on p. 307.
30. H. G. Wells, 'Popularising Science', *Nature*, 26 July 1894, pp. 300–1.
31. Ibid., p. 300.
32. Ibid., p. 301.
33. Ibid.
34. Ibid.
35. Ibid.
36. Ibid.
37. Tucker, *Nature Exposed*, p. 228.
38. J. S. L. Strachey, 'Review of *The War of the Worlds*, Spectator, 29 January 1898', in P. Parrinder (ed.), *H. G. Wells: The Critical Heritage* (London: Routledge & Kegan Paul, 1972), pp. 63–6, on p. 65.
39. Anon., 'Review of *The War of the Worlds*, Academy, 29 January 1898', in Parrinder (ed.), *H. G. Wells*, pp. 70–4, on pp. 71–2.
40. Strachey, 'Review', p. 64.
41. Gregory was a fellow student of Wells's and a lifelong friend. He would most likely, therefore, have been familiar with Wells's work before reading *The War of the Worlds*. However, he typifies the scientific man whom the *Academy* reviewer claims would have been impressed by Wells's scientific knowledge.
42. R. Gregory, 'Review of The War of the Worlds, Nature, 10 February 1898', in Parrinder (ed.), *H. G. Wells*, pp. 74–7, on p. 74.
43. Ibid., p. 77.
44. Wicks, *To Mars*.
45. Ibid., dedication page.
46. B. Lightman, 'The Visual Theology of Victorian Popularizers of Science: From Reverent Eye to Chemical Retina', *Isis*, 91 (2000), pp. 651–80, on p. 653.
47. Wicks, *To Mars*, p. 74.
48. Ibid., preface.
49. Ibid.
50. Ibid.
51. Ibid.
52. Secord, 'Knowledge', pp. 655, 664 and Topham 'Focus', p. 318.
53. Lightman, 'The Visual Theology', pp. 652, 657.
54. Guillemin, *The Heavens*, p. viii
55. Ibid., p. 485.
56. Lightman, 'The Visual Theology', p. 652.
57. Ibid., p. 680.
58. Lowell, *Mars*, p. 137.
59. Daston and Park, *Wonders*, p. 14.

60. Ibid., p. 134.
61. Lightman, 'The Visual Theology', p. 652.
62. Lowell, *Annals*, p. 192.
63. Ibid., p. 193.
64. Ibid., p. 192.
65. Antoniadi, 'Letter to Percival Lowell', 9 October 1909, PLA, Corr–10/09/09.
66. P. Lowell, 'Lecture D: Mars and the Earth', 1916, PLA, Ms–3U/13, p. 1.
67. Lowell, 'Mars', Auust 1894, ll. 229–35.
68. J. Keats, 'On First Looking into Chapman's Homer', in *Selected Poetry*, ed. E Cook (Oxford: Oxford University Press, 2008), ll. 9–12.
69. Lowell, *Mars*, pp. 92–3.
70. Wells, *The War of the Worlds*, pp. 31, 131.
71. Ibid., p. 157.
72. Wicks, *To Mars*, p. 19.
73. Ibid., p. 211.
74. Guillemin, *The Heavens*, p. v.
75. Ibid., p. vi
76. W. Whitman, 'When I Heard the Learn'd Astronomer', in *Selected Poems and Prose*, ed. A. N. Jeffares (Oxford: Oxford University Press, 1996), ll. 2, 7–9.
77. Daston, 'Fear and Loathing of the Imagination in Science', p. 81.
78. J. Tyndall, *On the Scientific Use of the Imagination: A Discourse* (London: Longmans, Green & Co., 1870), p. 6.
79. P. Lowell, 'Great Discoveries and Their Reception', 1916, PLA, Ms–3U/11, p. 9.
80. Ibid.
81. Proctor, *Romance*, p. 84.
82. Tyndall, *On the Scientific*, p. 10.
83. For example in the popular article, 'On the Kind of Eye', PLA, Ms–4/11, and the professional paper titled 'Double Canals and Separative Power', PLA, Ms–4/6.
84. Henchman, 'Hardy's Stargazers', p. 53.
85. Gossin, *Thomas Hardy's Novel Universe*, p. 95.
86. T. L. Hankins and R. J. Silverman, *Instruments and the Imagination* (Princeton, NJ: Princeton University Press, 1995), p. 6.
87. Cited in Flint, *Victorians and the Visual Imagination*, p. 308.
88. Lowell, 'Means, Methods and Mistakes', p. 32.
89. Ibid.
90. Lowell, 'Mars', August 1984, l. 21.
91. Ibid., ll. 90–9.
92. Martin, *The Ways of the Planets*, p. 179.
93. Guillemin, *The Heavens*, p. vi.
94. Wells, *The War of the Worlds*, pp. 51–2.
95. Anon., 'Review', p. 70.
96. Ibid., p. 72.
97. Ibid..
98. Gregory, 'Review', pp. 75, 74.
99. Wicks, *To Mars*, preface.
100. A. Fyfe and B. Lightman, 'Science in the Marketplace: An Introduction', in A. Fyfe and B. Lightman (eds), *Science in the Marketplace: Nineteenth-Century Sites and Experiences* (Chicago, IL: University of Chicago Press, 2007), pp. 1–19, on p. 9.

101. Ibid.
102. Ibid., p. 11.
103. Wells, *The War of the Worlds*, pp. 7, 177.
104. Ibid., p. 179.
105. Wicks, *To Mars*, p. 250.
106. Guillemin, *The Heavens*, p. 3.
107. Ibid., p. 5.
108. G. Simmel, 'Sociology of the Senses' (1907/8), in D. Frisby and M. Featherstone (eds), *Simmel On Culture* (London: Sage, 1997), pp. 109–20, on p. 116.
109. See, for example, Lowell, *Mars*; Lowell, 'Means, Methods and Mistakes'; Lowell, *Annals*; and Lowell, 'On the Kind of Eye Needed'.
110. P. Lowell, 'A Standard Scale for Telescopic Observations', 1902, PLA, Ms–3/7.
111. Ibid.
112. Weber, 'Science as a Vocation', pp. 1–31.
113. G. Simmel, 'The Metropolis and Mental Life' (1903), in Frisby and Featherstone (eds), *Simmel*, pp. 174–85, on p. 175.
114. Antoniaidi, 'Letter to Percival Lowell', 9 October 1909, PLA, Corr–10/09/09; and Lowell, *Annals*, pp. 279, 206.
115. Simmel, 'The Metropolis', p. 176.
116. Ibid., p. 177.
117. Ibid., p. 176.
118. Daum, 'Varieties', pp. 331–2.
119. E. Blackwelder, 'Our Friends, The Enemy: Mars as the Abode of Life', *Science*, 29:747 (23 April 1909), p. 661.
120. Anon., [no title], *Coconino Sun*, 15 November 1894, PLA, Scrapbooks, Clipping Books.
121. Marginalia, *Coconino Sun*, 15 November 1894.
122. Anon., 'Communication from Mars', *c.* 1894, PLA, Scrapbooks, Clipping Books.
123. M. Kendall, 'The Fatal Advertisements', in *Songs From Dreamland* (London: Longmans, 1894), ll. 5–28.
124. Ibid., l. 44.
125. Ibid., l. 41.
126. Ibid., ll. 57–64.
127. Daum, 'Varieties', p. 323.

Part III: Past

5 Looking as Tourists and Scientists

1. Latour, *Science in Action*; see ch. 6, 'Centres of Calculation', pp. 215–57.
2. A. B. Edwards, *A Thousand Miles Up the Nile* (1876; London: George Routledge & Sons, 1890).
3. 'Contact zone', a term conceived and defined by Pratt in her landmark work *Imperial Eyes*, is one of the key terms in theories of travel writing.
4. The scholarship on travel writing offers overlapping definitions for 'traveller' and 'tourist' along a continuum that ends with 'explorer'. Both the traveller and the tourist investigate the already known, while the explorer aims to discover the unknown. To some extent the traveller is more independent than the tourist, achieves greater self-discovery by being closer to the unknown, and must show greater self-reliance than the tourist who depends on others for the organization of their travel experience. See T. Fulford, D. Lee and P. J.

Kitson, *Literature, Science and Exploration in the Romantic Era* (Cambridge: Cambridge University Press, 2004).

5. See 'Thomas Cook Letter', EEFA, Box X/d/42. For prices of Thomas Cook's tours in Egypt in the nineteenth century see D. M. Reid, *Whose Pharaohs? Archaeology, Museums, and Egyptian National Identity from Napoleon to World War 1* (Berkeley, CA: University of California Press, 2002), pp. 64–92.
6. 'Thomas Cook Letter', 'Thomas Cook Invoice', EEFA, Box X/d/71.
7. W. M. F. Petrie, '1884 Diary', PMA, W. M. Flinders Petrie Diary, WFP1/9/3/1883–4, 7 January. Petrie, '7 November 1883–23 June 1884, Wady Tumilat, Tanis', in W. M. Flinders Petrie Journal, PMA, WFP1/16/1/1, p. 58.
8. Petrie, '30 November 1880 – 22 June 1881 – Gizeh', in W. M. Flinders Petrie Journal, PMA, WFP1/16/1/1, p. 32.
9. Petrie, '30 November 1880 – 22 June 1881 – Gizeh', in W. M. Flinders Petrie Journal, PMA, WFP1/16/1/1, p. 66.
10. E. Naville, 'The Route of the Exodus: Being the Annual Address of the Victoria Institute 1891', EEFA, Pamphlets NA-NEL, p. 1.
11. W. M. F. Petrie, 'The Grand Tour – Three Thousand Years Ago', *Harper's New Monthly Magazine*, 77 (June – November 1888), pp. 297–306, on p. 297, also held at the EEFA, Pamphlets PET.
12. Edwards, *A Thousand Miles*, pp. 325–53.
13. Petrie, '30 November 1880 – 22 June 1881 – Gizeh', pp. 12–30, in W. M. Flinders Petrie Journal, PMA, WFP1/16/1/1.
14. N. Leask, *Curiosity and the Aesthetics of Travel Writing, 1770–1840: 'From an Antique Land'* (Oxford: Oxford University Press, 2002), p. 1.
15. E. W. Lane, *Description of Egypt*, ed. J. Thompson (Cairo: American University in Cairo Press, 2000).
16. Ibid., p. 493.
17. Ibid., p. 495.
18. Ibid., p. 497.
19. Ibid.
20. W. H. Bartlett, *The Nile Boat; or Glimpses of the Land of Egypt* (London: Henry G. Bohn, 1862), p. 15.
21. Ibid., p. 23.
22. A. Butler, *Court Life in Egypt*, quoted in C. Pick, *Egypt: A Traveller's Anthology* (London: John Murray, 1991), p. 99.
23. Edwards, *A Thousand Miles*, p. 285.
24. S. Haight, *Letters from an Old World*, quoted in Pick, *Egypt*, pp. 101–4.
25. Haight, quoted in Pick, *Egypt*, p. 103.
26. Haight, quoted in Pick, *Egypt*, p. 104.
27. N. Lorimer, *By the Waters of Egypt* (London: Methuen, 1909), p. 198.
28. Ibid., pp. 198–9.
29. Edwards, *A Thousand Miles*, p. 129.
30. Virginia Zimmerman would argue that it is essential to all types of archaeology. See V. Zimmerman, *Excavating Victorians* (Albany, NY: SUNY Press, 2008), p. 14.
31. F. L. Griffiths, 'The Study of Egyptology. Inaugural Lecture Delivered in the Ashmolean Museum on May 8, 1901', p. 6, EEFA, Pamphlets G.
32. Ibid., pp. 6–7.
33. W. M. F. Petrie, 'The Discovery of Naukratis', EEFA, Pamphlets PETRIE-PIE.

34. Ibid., p. 4.
35. A. B. Edwards, 'Bubastis: An Historical Study', *Century Magazine*, 39:3 (January 1890), pp. 323–49, on p. 331.
36. Ibid., p. 332.
37. W. M. F. Petrie, C. Smith, E. Gardner & B. V. Head, *Naukratis, Part I. Excavation Report* (London: Egypt Exploration Fund, 1885), p. 25.
38. Daston, 'Fear and Loathing of the Imagination, p. 77.
39. Kate Bradbury was Amelia Edwards's travelling companion, her legatee, and later the wife of Egyptologist F. L. Griffiths.
40. 'W. M. Flinders Petrie to Kate Bradbury', in M. Dower (ed.), *Letters From the Desert: The Correspondence of Flinders and Hilda Petrie* (Oxford: Avis & Phillips, 2004), p. 76.
41. Edwards, *A Thousand Miles*, p. 13.
42. Ibid.
43. Lorimer, *By the Waters*, p. vii.
44. Ibid., p. vii.
45. E. A. W. Budge, *Cook's Handbook for Egypt and the Egyptian Sudan: With Chapters on Egyptian Archaeology* (London: Thomas Cook & Son, 1921). Budge's original work for Thomas Cook had been titled *The Nile: Notes for Travellers*, and was published in 1888. See also Reid, *Whose Pharaohs?*, pp. 64–92.
46. E. A. W. Budge, *By Nile and Tigris: A Narrative of Journeys in Egypt and Mesopotamia on Behalf of the British Museum Between the Years 1886 and 1913*, 2 vols (London: John Murray, 1920)
47. Petrie, 'Diary, 8 December 1897', in Dower (ed.), *Letters*, p. 119.
48. Petrie, 'Diary, 22 December 1897', in Dower (ed.), *Letters*, p. 126.
49. Petrie, 'Letter, 22 March 1886', in Dower (ed.), *Letters*, pp. 62, 69. See the following chapter for a discussion of fantasy in the context of seeing the archaeological object as a 'thing'.
50. Petrie, 'Excerpt from A Digger's Life', in Dower (ed.), *Letters*, p. 9.
51. John Payne's nine-volume translations appeared in 1882 and Richard Burton's ten volumes in 1885. These were printed privately and distributed via subscription but did instigate renewed interest in the tales. See D. Heller-Roazen, M. Mahdi and H. Haddawy (eds), *The Arabian Nights. Norton Critical Edition* (New York: W. W. Norton, 2008), pp. ix–xxiv.
52. Petrie, '7 November 1883–23 June 1884 – Wady Tumilat, Tanis', p. 127, in W. M. Flinders Petrie Journal, PMA, WFP1/16/1/1(2).
53. W. M. F. Petrie (ed.), *Egyptian Tales Translated From the Papyri: First Series. IVth to XIIth Dynasty* (London: Methuen & Co, 1895).
54. Ibid., p. 3.
55. A. B. Edwards, *Pharaohs, Fellahs and Explorers* (London: James R. Osgood, McIlvaine & Co, 1891), p. 3.
56. Ibid., p. 21.
57. Edwards, 'Bubastis', p. 323.
58. F. Nightingale, 'November 19th, 1849', in A. Sattin (ed.), *Letters From Egypt: A Journey on the Nile, 1849–50* (New York: Grove Press, 1987), p. 23.
59. C. D. Bell, *A Winter on the Nile, in Egypt and in Nubia* (London: Hodder & Stoughton, 1888), p. 23.
60. Budge, *Cook's Handbook*, p. 110.
61. Fulford et al., *Literature, Science and Exploration*, p. 12.

62. Latour, *Science in Action*, ch. 6.
63. L. Schiebinger, 'Forum Introduction: The European Colonial Science Complex', *Isis*, 96:1 (2005), pp. 52–5, on pp. 52, 54.
64. Zimmerman, *Excavating Victorians*, p. 8.
65. Ibid., p. 10.
66. Petrie, 'Wady Tumilat, Tanis'. See also W. M. F. Petrie, '1883–4 Diary', PMA, WFP1/9/3/1883–4.
67. Petrie, 'Wady Tumilat, Tanis', p. 30.
68. Ibid.
69. Ibid., p. 127.
70. Ibid., p. 147.
71. Petrie also finds evidence of Rameses II's destruction or reuse of artefacts at Guroh, Kahun and Illahun in 1888 and 1889. See W. M. F. Petrie, '24 Oct 1888 –23 May 1889 Guroh, Kahun and Illahun', in W. M. Flinders Petrie Journal, PMA, WFP1/16/1/2(5).
72. Daston, 'Introduction', in Daston (ed.), *Biographies*, p. 1.
73. Lorimer, *By the Waters*, pp. 277–8.
74. R. T. Kelly, *Egypt Painted and Described* (London: Adam & Charles Black, 1903), pp. 71, 60.
75. Nightingale, 'Ipsamboul, Jan 17th, 1850', in Sattin (ed.), *Letters*, p. 99.
76. Ibid.
77. Bartlett, *The Nile Boat*, p. 17.
78. Lorimer, *By the Waters*, p. 200.
79. Griffiths, 'The Study of Egyptology', pp. 10–11, EEFA, Pamphlets G.
80. Ibid., p. 11.
81. Ibid., p. 10.
82. Edwards, 'Bubastis', p. 334.
83. Ibid., p. 335.
84. Leask, *Curiosity*, p. 1.
85. Ibid., pp. 8–17.
86. Ibid., pp. 6–7.
87. Edwards, *A Thousand Miles*, p. xi.
88. Lorimer, *By the Waters*, p. vii.
89. Figures 5.3 and 5.4 reproduce pages 340 and 338, respectively, of Edward's *A Thousand Miles*.
90. Budge, *Cook's Handbook*, p. 678.
91. Edwards, *Pharaohs*, p. 12.
92. Ibid., p. 13.
93. Ibid., p. 27.
94. Pratt, *Imperial Eyes*, p. 9.
95. Leask, *Curiosity*, p. 7.
96. Griffiths, 'The Study of Egyptology', p. 6, EEFA, Pamphlets G. See also pp. 10–11.
97. Petrie, '30 November 1880 – 22 June 1881 – Gizeh', p. 55, in W. M. Flinders Petrie Journal, PMA, WFP1/16/1/1.
98. Petrie kept three forms of written record of his excavations: journals, notebooks and diaries. The diaries, really just tiny calendars, contain only a line or two on each day. The notebooks are taken up with on-site calculations and measurements, as well as small sketches. The journals are the most reflective written record, generally written at the close of each day after some time has been spent considering the day's excavatory work.

99. C. M. Hinsley, 'Revising and Revisioning the History of Archaeology: Reflections on Region and Context', in A. L. Christenson (ed.), *Tracing Archaeology's Past: The Historiography of Archaeology* (Carbondale, IL: Southern Illinois University Press, 2001), p. 80.
100. See Daston and Galison, *Objectivity*, pp. 17–53.

6 Egyptian Archaeology and Fiction

1. W. M. F. Petrie, 'Letter to Kate Bradbury', EEFA, Box 1C, No.1.
2. W. M. F. Petrie, 'Letter to E. Maunde Thompson', EEFA, Box 1C, No. 2.
3. Ibid.
4. Howard Carter was, later in his career, to find one of the most famous of all archaeological things: the tomb of the pharaoh, Tutankhamun.
5. Daston, 'Introduction', in Daston (ed.), *Things That Talk*, pp. 9–24; K. Alder, 'Focus: Thick Things: Introduction', *Isis*, 98:1 (2007), pp. 80–3.
6. S. J. M. M. Alberti, 'Focus: Museums and the History of Science: Objects and the Museum', *Isis* 96:4 (2005), pp. 559–71; C. Gosden and Y. Marshall, 'The Cultural Biography of Objects', *World Archaeology*, 31:2 (1999), pp. 169–78.
7. W. J. T. Mitchell, 'Romanticism and the Life of Things: Fossils, Totems, and Images', *Critical Inquiry*, 28:1 (2001), pp. 167–84, on p. 179.
8. L. Meskell, *Object Worlds in Ancient Egypt: Material Biographies Past and Present* (Oxford: Berg, 2004), p. 14.
9. W. M. F. Petrie, 'History in Tools', in *Smithsonian Report for 1918* (Washington, DC: Government Printing Office, 1920), pp. 563–72, on p. 572, EEFA. Pamphlets: PET.
10. Edwards, *Pharaohs*, p. 36.
11. H. R. Haggard, *She* (1887; London: Penguin, 2004); B. Stoker, *The Jewel of Seven Stars* (1903; Stroud: Sutton Publishing, 1996).
12. Rider Haggard's subscription form was dated 31 December 1889, EEFA.
13. Haggard, *She*, pp. 8–9 and 40–53.
14. Stoker, *The Jewel*, p. 16.
15. Alder, 'Focus: Thick Things', p. 82.
16. Edwards, *A Thousand Miles Up*, p. 148.
17. Lorimer, *By the Waters*, p. 363.
18. W. M. F. Petrie, 'Notebook Extract', in Drower (ed.), *Letters from the Desert*, p. 69.
19. F. L. Griffiths, *The Study of Egyptology* (Oxford: Horace Hart, 1901), pp. 7–8, EEFA, Pamphlets – G.
20. Ibid., p. 7.
21. Stoker, *The Jewel*, p. 22.
22. Haggard, *She*, p. 132.
23. Daston, 'On Scientific Observation', pp. 104, 102.
24. B. Brown, 'Thing Theory', *Critical Inquiry*, 28:1 (2001), pp. 1–22.
25. Ibid., p. 3.
26. Ibid., p. 5.
27. Alder, 'Focus: Thick Things', p. 82.
28. Edwards, *A Thousand Miles*, p. 151.
29. Budge, *Cook's Handbook for Egypt*, pp. 387, 682.
30. Pratt, *Imperial Eyes*.
31. Haggard, *She*, p. 136; Stoker, *The Jewel*, p. 21.
32. W. M. F. Petrie, 'Extract of Letter to Hilda Petrie', in Drower (ed.), *Letters*, pp. 62–3.

33. W. M. F. Petrie, '7 Nov. 1883 – 23 June 1884 – Wady Tumilat, Tanis', in W. M. Flinders Petrie Journals, PMA, WFP1/16/1/1(2), pp. 162–3.
34. The complex knowledge required in the deciphering of Egyptian hieroglyphs should remind us that Egyptian pictorial representation was far from simple, and that Petrie would have been well aware of the significant expertise required to undertake such analysis. See R. Parkinson, *Cracking Codes: The Rosetta Stone and Decipherment* (Berkeley, CA: University of California Press, 1999).
35. Stoker, *The Jewel*, pp. 31, 57.
36. Haggard, *She*, p. 145.
37. Ibid., p. 278.
38. Mitchell, 'Romanticism', p. 171.
39. Brown, 'Thing Theory', p. 5.
40. Ibid.
41. W. M. F. Petrie, 'Letter to Hilda Petrie', in Drower (ed.), *Letters*, p. 11.
42. Petrie, '7 Nov 1883 – 23 June 1884', p. 24. Professor Amos is probably Sir Percy Amos, an inspector of the Egyptian Ministry of Justice and lecturer at the Khedival School of Law in Cairo from 1889 to 1915.
43. W. M. F. Petrie, '30 Nov 1880 – 22 June 1881 – Gizeh', in W. M. Flinders Petrie Journals, PMA, WFP1/16/1/1, p. 30.
44. W. M. F. Petrie, 'Journal Extract, 19 – 25 January 1897', in Drower (ed.), *Letters*, p. 109.
45. W. M. F. Petrie, 'Notebook 3 – Abydos, Royal Tombs', in W. M. Flinders Petrie Notebooks, PMA, WFP1/99/1/3.
46. Petrie, '7 Nov 1883 – 23 June 1884', in W. M. Flinders Petrie Journals, PMA, WFP1/16/1/1(2), p. 131.
47. Brown, 'Thing Theory', p. 5.
48. A. Warwick, '"The City of Resurrections": Arthur Machen and the Archaeological Imagination', in R. Pearson (ed.), *The Victorians and the Ancient World: Archaeology and Classicism in Nineteenth-Century Culture* (Cambridge: Cambridge Scholars Press, 2006), pp. 124–38.
49. See pioneering work in this field by A. Smith, *Gothic Radicalism: Literature, Philosophy and Psychoanalysis in the Nineteenth Century* (London: Palgrave Macmillan, 2000). For a broader view of the genre, see also Smith's introductory volume on the Gothic, *Gothic Literature* (Edinburgh: Edinburgh University Press, 2007).
50. Bartlett, *The Nile Boat*, p. 44.
51. Stoker, *The Jewel*, p. 1.
52. Budge, *By Nile and Tigris*, pp. 77–8.
53. Ibid., p. 78.
54. Lorimer, *By the Waters*, p. 199.
55. Ibid.
56. Haggard, *She*, p. 146.
57. Edwards, *Pharaohs*, p. 5.
58. Ibid.
59. Ibid., p. 12.
60. W. M. F. Petrie, '24 Oct 1888 – 23 May 1889 – Guroh, Kahan, and Illahan', in W. M. Flinders Petrie Journals, PMA, WFP1/16/1/2(5), p. 47.
61. Ibid., p. 46.
62. Smith, *Gothic Literature*, p. 2.
63. W. M. F. Petrie, 'Journal Extract 1889', in Drower (ed.), *Letters*, p. 70.

64. Gosden and Marshall, 'The Cultural Biography', p. 169.
65. Brown, 'Thing Theory', p. 4.
66. Budge, *By Nile and Tigris*, p. 381.
67. Smith, *Gothic Literature*, pp. 2–4.
68. Budge, *By Nile and Tigris*, p. 381.
69. Ibid., pp. 381–2.
70. J. Grave, 'On The Aesthetics of Scientific Objects. Three Case Studies', in S. Vackimes and K. Weltersbach (eds), *Wandering Seminar on Scientific Objects* (Berlin: Max Planck Institute for the History of Science, 2007), pp. 35–47, on p. 36.
71. Smith, *Gothic Literature*, p. 2.
72. S. Haight, 'Extract from The Old World (1840)', in Pick (ed.), *Egypt*, p. 103.
73. Edwards, *A Thousand Miles*, p. 285.
74. Ibid., p. 304.
75. W. M. F. Petrie, 'Extract from Letter to Flaxman Spurrell, 11 Feb 1881, Gizeh', in Drower (ed.), *Letters*, p. 23.
76. Petrie, 'The Grand Tour', p. 304, also held at the EEFA, Pamphlets PET.
77. Ibid., p. 304.
78. E. Naville, *Excavations at Abydos: From the Smithsonian Report for 1914* (Washington, DC: Government Printing Office, 1915), p. 579.
79. Ibid., p. 579.
80. Haggard, *She*, p. 292.
81. Stoker, *The Jewel*, p. 171.
82. Ibid., pp. 148–9.
83. Haggard, *She*, p. 169.
84. Ibid., p. 146.
85. Stoker, *The Jewel*, p. 171.
86. B. Latour, 'The Berlin Key or How to Do Words with Things', in P. M. Graves-Brown (ed.), *Matter, Materiality and Modern Culture* (London: Routledge, 2000), pp. 10–21, on p. 20.
87. Haggard, *She*, pp. 188–9.
88. Alder, 'Focus: Thick Things', p. 82.
89. The word 'ab-sense' is a linguistic pun that draws on the predicate 'ab' (without) and vision (sense) with the complete word absence (a missing element).
90. V. Carroll, 'Natural History on Display: The Collection of Charles Waterton', in Fyfe and Lightman (eds), *Science in the Marketplace*, pp. 271–300, on pp. 295–6.

Part IV: Future

7 Arthur Conan Doyle and Harry Houdini: Optics, Ophthalmology and Magical Performance

1. Helmholtz, *Helmholtz's Treatise on Physiological Optics*; E. Landolt, *A Manual of the Examination of the Eyes: A Course of Lectures Delivered at the École Pratique* (1879; New York: Robert E. Kreiger, 1979); J. Hutchinson, *Aids to Ophthalmic Medicine and Surgery* (London: Ballière & Co, 1889).
2. Landolt, *A Manual*, p. 9.
3. Helmholtz, *Treatise*, no. 26, p. 7.

4. There are numerous works on visual culture where this view is pervasive. See, for example, N. Mirzoeff, *An Introduction to Visual Culture* (London: Routledge, 1999) and N. Mirzoeff (ed.), *The Visual Culture Reader* (London: Routledge, 1998).
5. J. Hogg, *The Ophthalmoscope: Its Mode of Application Explained, and its Value Shown, in the Exploration of Internal Diseases Affecting the Eye* (London: John Churchill, 1858), p. 34.
6. See B. Lightman, *Victorian Popularizers of Science: Designing Nature for New Audiences* (Chicago, IL: University of Chicago Press, 2007); Fyfe and Lightman (eds), *Science in the Marketplace*; I. R. Morus, 'Seeing and Believing Science', *Isis*, 97 (2006), pp. 101–10.
7. A. A. Hopkins, *Magic: Stage Illusions, Special Effects and Trick Photography* (1897; New York: Dover, 1976), pp. 520–3.
8. H. Houdini, *Houdini on Magic*, ed. W. B. Gibson and M. N. Young (1930; New York: Dover, 1953), p. 205.
9. See M. Polidoro, *Final Séance: The Strange Friendship Between Houdini and Conan Doyle* (Amherst, MA: Prometheus Books, 2001); B. M. L. Ernst and H. Carrington, *Houdini and Conan Doyle: A Strange Friendship* (New York: Benjamin Bloom, 1972)
10. J. Lellenburg, D. Stashower and C. Foley (eds), *Arthur Conan Doyle: A Life in Letters* (London: Harper Press, 2007), p. 265.
11. Ibid.
12. See, for example, D. Stashower, *Teller of Tales: The Life of Arthur Conan Doyle* (New York: Henry Holt, 1999), p. 117. Stashower has since been instrumental in bringing Conan Doyle's letters into the public domain and therefore making accessible an alternative reading of Conan Doyle's ophthalmological career.
13. A. Conan Doyle, *Memories and Adventures* (London: Hodder & Stoughton, 1924).
14. Lellenburg et al., *Arthur Conan Doyle*, p. 286.
15. E. Fuchs, *Text-Book of Ophthalmology* (1892; New York: D. Appleton & Co., 1901).
16. Lellenburg et al., *Arthur Conan Doyle*, p. 290.
17. E. Landolt, *The Refraction and Accommodation of the Eye and Their Anomalies* (Edinburgh: Young J. Pentland, 1886).
18. Landolt, *A Manual*, p. 20.
19. Lellenburg et al., *Arthur Conan Doyle*, p. 292.
20. Conan Doyle had published two Sherlock Holmes novels prior to 1891, *A Study in Scarlet* (1887) and *The Sign of Four* (1890).
21. A. Conan Doyle, *The Penguin Complete Sherlock Holmes* (London: Penguin, 1981), p. 161. All further references to the Sherlock Holmes canon refer to this work and will be found in parenthesis in the text.
22. Many of the works from Houdini's library are now collected in the Harry Houdini Archive (HHA) and the McManus-Young Magical Collection (M-YMC) at the Library of Congress, Washington DC.
23. 'Popular science' is a phrase much debated in recent history of science scholarship. I use it here, conscious of its connotations, to articulate Houdini's enactment of scientific principles at sites generally recognized as developed for popular entertainment. See Fyfe and Lightman (eds), *Science in the Marketplace*, introduction.
24. Jonathan Crary has argued most distinctively for the historical specificity of perception. See Crary, *Techniques*; and Crary, *Suspensions*.
25. D. Brewster, *Letters on Natural Magic* (1832; New York: Harper & Brothers, 1843), pp. 53–4.

26. H. V. Helmholtz, 'The Facts in Perception (1878)', in R. S. Cohen and Y. Elkana (eds), *Hermann Von Helmholtz: Epistemological Writings* (Boston, MA: D. Reidel, 1977), pp. 114–85.
27. T. Lenoir, 'The Eye as Mathematician: Clinical Practice, Instrumentation, and Helmholtz's Construction of an Empiricist Theory of Vision', in D. Cahan (ed.), *Hermann Von Helmholtz and the Foundations of Nineteenth-Century Science* (Berkeley, CA: University of California Press, 1993), p. 120.
28. Knight's *How To Observe* series covered both scientific and social topics. The first volume is well-known to historians of geology: H. T. de la Beche, *How To Observe Geology* (London: Charles Knight, 1835). The second volume, on Manners, was written by Harriet Martineau.
29. H. Houdini, *The Right Way to Do Wrong* (Boston, MA: Harry Houdini, 1906). Rather ironically this book was distributed by the Gambler's Book Club, based in Las Vegas.
30. Ibid., p. 3.
31. Ibid.
32. Ibid., p. 47.
33. Ibid.
34. Ibid., p. 48.
35. Ibid., p. 37.
36. Fuchs, *Text-Book of Ophthalmology*, p. 1.
37. Ibid.
38. See, for example, Lionel Laurance's explanation of the 'reduced eye' in L. Laurance, *Visual Optics and Sight Testing* (London: School of Optics, 1920), p. 18.
39. Fuchs, *Text-Book of Ophthalmology*, p. 2.
40. Helmholtz, *Hermholtz's Treatise on Physiological Optics*, pp. 26, 9.
41. N. Maskelyne and D. Devant, *Our Magic: The Art in Magic, the Theory of Magic, the Practice of Magic* (New York: E. P. Dutton & Co., 1911), p. 181.
42. H. J. Moulton, *Houdini's History of Magic in Boston, 1792–1915* (Illinois: Meyer Books, 1983), p. 132.
43. Houdini, *Houdini on Magic*, p. 24.
44. Houdini, *The Right Way*, p. 95.
45. Ibid., p. 3.
46. H. Houdini, *Handcuff Secrets* (New York: Routledge, 1910), p. ix.
47. Houdini, *The Right Way*, p. 3.
48. Laurance, *Visual Optics*, p. 56.
49. Ibid., pp. 56–89.
50. C. Wheatstone, 'Contributions to the Physiology of Vision', *Journal of the Royal Institution of Great Britain*, 1 (October 1830), pp. 205–33.
51. Houdini, *Houdini on Magic*, pp. 17–8.
52. Houdini, *The Right Way*, pp. 92–5.
53. J. F. Kasson, *Houdini, Tarzan, and The Perfect Man: The White Male Body and the Challenge of Modernity in America* (New York: Hill & Wang, 2001), p. 104.
54. Moulton, *Houdini's History*, p. 134.
55. Maskelyne and Devant, *Our Magic*, p. 17.
56. R. Houdin, *Les Secrets de la Prestidigitation et de la Magic* (1868), translated by Professor Hoffman as *The Secrets of Conjuring and Magic, or How To Become a Wizard* (London: George Routledge, 1878), p. 96.
57. Maskelyne and Devant, *Our Magic*, p. 190.

58. H. Houdini, 'The Psychology of Prestidigitation' (1925), HHA, Box 26.
59. E. Landolt, *Defective Ocular Movements and Their Diagnosis*, trans. A. Roemmelle and E. W. Brewerton (London: Oxford University Press and Hodder & Stoughton, 1913), p. 40.
60. J. Brooker, 'The Polytechnic Ghost: Pepper's Ghost, Metempsychosis and the Magic Lantern at the Royal Polytechnic Institution', *Early Popular Visual Culture*, 5:2 (2007), pp. 189–206, on p. 196.
61. Maskelyne and Devant, *Our Magic*, p. 176.
62. H. Houdini, *Miracle Mongers and Their Methods: A Complete Exposé* (1920; Buffalo, NY: Prometheus Books, 1981).
63. Advertising playbill from Houdini's private collection, HHA, Box 16.
64. Note to Houdini from Houdini's private collection, HHA, Houdini Pamphlets, no. 6.
65. Houdini, *Houdini on Magic*, p. 107.
66. See W. B. Gibson, *Houdini's Escapes: Prepared from Houdini's Private Notebooks and Memoranda with the Assistance of Beatrice Houdini, Widow of Houdini, and Bernard M. L. Ernst, President of the Parent Assembly of the Society of American Magicians* (New York: Harcourt, Brace & Co., 1930).
67. Challenge leaflet from Houdini's private collection, HHA, Box 20.
68. Inventory Book for Houdini's stage performances, HHA, Box 49b.
69. Houdini, *Houdini on Magic*, p. 103.
70. Maskelyne and Devant, *Our Magic*, p. v.
71. Ibid., p. 175–6.
72. Ibid., p. 176.
73. A. Binet, 'La Psychologie de la Prestidigitation', *Revue des Deux Mondes*, 54: 125 (1894), pp. 903–23. For a detailed consideration of Binet's aims, and his employment of magicians, see S. Lachapelle, 'From the Stage to the Laboratory: Magicians, Psychologists, and the Science of Illusions', *Journal of the History of the Behavioral Sciences*, 44:4 (2008), pp. 319–34.
74. Houdini, 'The Psychology of Predestination' (1925), HHA, Box 26.
75. Ibid.
76. Brewster, *Letters*, pp. 21–3.
77. Houdini, 'The Psychology of Predestination' (1925), HHA, Box 26.
78. Ibid.
79. Ibid.
80. Helmholtz, 'The Facts in Perception', p. 12.
81. Houdini, *Houdini on Magic*, pp. 238–40.
82. Houdin, *Secrets*, pp. 76–7.
83. Brewster, *Letters*, pp. 43–54. See also Brewster's original paper, 'On the Optical Illusion of the Conversion of Cameos into Intaglios, and of Intaglios into Cameos, with an Account of Other Analogous Phenomena' (1826), in N. Wade (ed.), *Brewster and Wheatstone on Vision* (London: Academic Press, 1983), pp. 56–65.
84. Brewster, *Letters*, p. 100.
85. J. Schickore, 'Misperception, Illusion and Epistemological Optimism: Vision Studies in Early Nineteenth-Century Britain and Germany', *British Journal for the History of Science*, 39:3 (2006), pp. 383–405, on p. 393. See also an extended and contextualized analysis in J. Schickore, *The Microscope and the Eye: A History of Reflections, 1740–1870* (Chicago, IL: University of Chicago Press, 2007) and in Morus, 'Seeing and Believing Science', pp. 101–3.

86. See R. S. Turner, 'Consensus and Controversy: Helmholtz on the Visual Perceptions of Space', in Cahan, *Hermann Von Helmholtz*, pp. 154–204, on p. 185.
87. Ibid., p. 185.
88. Helmholtz, 'The Facts in Perception', p. 132.
89. Ibid., p. 121.
90. Maskelyne and Devant, *Our Magic*, p. 31.
91. Gibson, *Houdini's Escapes*, p. 10.
92. Schickore, 'Misperception', p. 393.
93. Hogg, *The Ophthalmoscope*, p. 36.
94. H. P. Lovecraft and H. Houdini, 'Imprisoned With the Pharaohs', *Weird Tales*, May 1924, available online at http://gaslight.mtroyal.ca/gaslight/imppharo.htm [accessed 4 August 2011].
95. Ibid.
96. Ibid.
97. Schickore, 'Misperceptions', p. 404.
98. Hogg, *The Ophthalmoscope*, p. 36.
99. See I. R. Morus, '"More the Aspect of Magic than Anything Natural": The Philosophy of Demonstration', in Fyfe and Lightman (eds), *Science in the Marketplace*, pp. 336–70, for an extended discussion of the interaction between knowledge and wonder that characterizes the public displays of science of the period.
100. H. Groth, 'Reading Victorian Illusions: Dickens's *Haunted Man* and Dr. Pepper's Ghost', *Victorian Studies*, Autumn (2007), pp. 43–65, on p. 60.
101. Ibid.
102. Ibid., p. 62.
103. D. Abbott, *The Marvellous Creations of Joseffy* (Chicago, IL: Open Court, 1908), p. 1.
104. Maskelyne and Devant, *Our Magic*, p. 189.
105. Moulton, *Houdini's History*, p. 115.
106. Ibid., p. 117.
107. Ibid., p. 136.
108. Turner, 'Consensus', p. 154.
109. Hogg, *The Ophthalmoscope*, p. 106.
110. Ibid.
111. Laurance, *Visual Optics*, p. 89.
112. H. Houdini, *Magical Rope Ties and Escapes* (London: Will Goldston, 1921), p. 7.
113. J. Sully, *Illusions: A Psychological Study* (London: Kegan Paul, Trench & Co., 1887), p. 3.

8 Arthur Conan Doyle and Harry Houdini: Sensation, Spectacle and Spiritualism

1. There is no consensus over the definitions of these terms across the different fields of inquiry that this chapter addresses. It is still widely accepted (although not without challenge) in literary studies that the final years of the nineteenth century mark a transition from the Victorian to the modern. However, in the history of science, especially histories of science with an interest in vision, modernity might begin much earlier. Crary, for example, argues that signs of visual modernity can be traced to the 1830s. As should be clear from the context, this chapter will follow the traditional literary studies application of these two terms.

2. M. McGrath, *Ghosts of Futures Past: Spiritualism and the Cultural Politics of Nineteenth-Century America* (Berkeley, CA: University of California Press, 2008), p. 3. The other key works on spiritualism include: J. Oppenheim, *The Other World: Spiritualism and Psychical Research in England, 1850–1914* (Cambridge: Cambridge University Press, 1985); A. Owen, *The Darkened Room: Women, Power, and Spiritualism in Late Victorian England* (Chicago, IL: University of Chicago Press, 1989); L. Barrow, *Independent Spirits: Spiritualism and English Plebeians, 1850–1910* (London: Routledge and Kegan Paul, 1986); and C. Ferguson, 'Eugenics and the Afterlife: Lombroso, Doyle, and the Spiritualist Purification of the Race', *Journal of Victorian Culture,* 12:1 (2007), pp. 64–85.
3. R. Noakes, '"The Bridge which is Between Physical and Psychical Research": William Fletcher Barrett, Sensitive Flames and Spiritualism', *History of Science,* 42 (2004), pp. 419–64; and R. Noakes, 'Cromwell Varley FRS, Electrical Discharge and Victorian Spiritualism', *Notes and Records of the Royal Society of London,* 61 (2007), pp. 5–22.
4. Oppenheim, *The Other World*, pp. 24–7
5. A. Conan Doyle, *Letters to the Press*, ed. J. M. Gibson and R. L. Green (Iowa City, IA: University of Iowa Press, 1986), p. 25.
6. Hardeen Memorandum, HHA, Box 26.
7. H. Houdini, *Houdini: A Magician Among the Spirits* (1924; New York: Arno Press, 1972), p. 139.
8. This letter is reproduced in Ernst and Carrington, *Houdini and Conan Doyle*.
9. A. Conan Doyle, *Our American Adventure* (London: Hodder & Stoughton, 1923), p. 160.
10. Ibid., p. 161
11. Ibid., p. 162
12. J. F. Rinn, *Sixty Years of Psychical Research: Houdini and I Among the Spirits* (New York: Truth Seeker Co., 1950), p. 405.
13. Ibid.
14. Only Conan Doyle, Jean and Houdini were present at this séance, which occurred in Atlantic City on 17 June 1922. Conan Doyle describes the séance in *Our American Adventure*, p. 180, while Houdini offers his view in *Houdini: A Magician*, pp. 150–6.
15. Houdini, *Houdini: A Magician*, p. 162.
16. The letter is dated 19 November 1922. See Houdini, *Houdini: A Magician*, p. 156.
17. 1 January 1923. See Houdini, *Houdini: A Magician*, p. 159
18. Ibid., pp. 138–65.
19. Stashower, *Teller of Tales*, p. 435.
20. 'The Final Houdini Séance', HHA, Box 15, Folder 1, p. 4.
21. 'The Final Houdini Séance', HHA, Box 15, Folder 1, n. p.
22. A. Conan Doyle, *History of Spiritualism*, 2 vols (London: Cassell, 1926), vol. 1, p. 57.
23. Conan Doyle draws these specific examples both from séances he attended himself, where the medium was Eusapio Palladino, and from the statements of attendees at others, performed by William and Ira Davenport.
24. Rinn, *Sixty Years*, pp. 299–300. The transcription was recorded by Rinn's assistant Mary Young.
25. Ibid., p. 296.
26. Ibid., p. 304.
27. Ibid. Rinn notes that the auditorium audience broke into 'tumultuous applause' at the conclusion of his performance and that more than five hundred of them afterwards

assured him 'that never again would they believe anything genuine that they saw produced by a psychic' (p. 301).

28. The published report listed all committee members, witnesses and investigators. See *Report on Spiritualism by a Committee of the Dialectical Society* (London: Longman, 1871).
29. R. Luckhurst, *The Invention of Telepathy* (Oxford: Oxford University Press, 2002).
30. W. Crookes, *Researches in the Phenomena of Spiritualism* (London: J. Burns, 1874), p. 5. For more extensive discussions of Crookes, science and instrumentation in spiritualist contexts, see R. Noakes, '"Instruments to Lay Hold of Spirits": Technologizing the Body of Victorian Spiritualism', in I. R. Morus (ed.), *Bodies/Machines* (Oxford: Berg, 2002), pp. 25–63.
31. J. Tyndall, 'Science and the Spirits', in *Fragments of Science for Unscientific People: A Series of Detached Essays, Lectures and Reviews* (New York: D. Appleton, 1872), pp. 402–9.
32. H. Dobler, *Exposé of the Davenport Brothers* (Belfast: D & J Allen, 1869), p. 3
33. See, for example, Dobler's account of spirit-tying in *Exposé*, pp. 20–3.
34. J. N. Maskelyne, *Modern Spiritualism: A Short Account of its Rise and Progress, With Some Exposures of So-called Spirit Media* (London: Frederick Warne, 1876), p. vi
35. Anon., 'At a Spiritualist Séance', M-YMC.
36. Ibid.
37. G. Sexton, *Spirit-Mediums and Conjurers: An Oration* (London: J. Burns, 1873), p. 6.
38. Iota, *Maskelyne and Cooke: An Expose of the Falseness of their Pretensions* (London: J. Burns, 1873).
39. Quoted in H. R. Evans, *Hours with the Ghosts, or Nineteenth Century Witchcraft: Illustrated Investigations into the Phenomena of Spiritualism and Theosophy* (Chicago, IL: Laird & Lee, 1897), p. 136.
40. Maskelyne, *Modern Spiritualism*, p. 104.
41. Noakes, '"Instruments"', p. 155.
42. The physicist William Barrett first discussed spiritualism at the BAAS annual meeting in 1876. Besides the scientists, sciences and organizations listed here, scientific periodicals, such as *Scientific American*, also conducted séance investigations.
43. In a letter to the *Morning Post* dated 14 June 1926. See Lellenburg et al., *Arthur Conan Doyle*, p. 317.
44. Crookes, *Researches in the Phenomena of Spiritualism*, p. 4
45. Ibid., p. 5
46. C. Flammarion, *Mysterious Psychic Forces: An Account of the Author's Investigations in Psychical Research, Together with Those of Other European Savants* (Boston, MA: Small, Maynard & Co., 1907), p. 7.
47. Ibid., pp. 196–201.
48. Fuchs, *Text-Book of Ophthalmology*, p. 274.
49. Helmholtz, *Helmholtz's Treatise on Physiological Optics*, p. 226.
50. J. Plunkett, 'Depth, Colour, Movement: Embodied Vision and the Stereoscope', in J. Lyons and J. Plunkett (eds), *Multimedia Histories: From the Magic Lantern to the Internet* (Exeter: University of Exeter Press, 2007), pp. 117–31, on p. 118.
51. Brewster, *Letters*, p. 105.
52. Helmholtz, *Helmholtz's Treatise on Physiological Optics*, p. 12.
53. Crookes, *Researches in the Phenomena of Spiritualism*, pp. 10–13.
54. Ibid., p. 12.

55. See H. Houdini, *Houdini Exposes the Tricks Used By the Boston Medium "Margery" to Win the $2500 Prize Offered By the Scientific American* (New York: Adams Press, 1924)
56. Crookes, *Researches in the Phenomena of Spiritualism*, p. 13
57. Helmholtz's empirist view of vision was set down in his 1878 lecture on perception. His new analyses overturned those of his mentor, Muller, who had championed the relationship between the eye and the body, and claimed, crucially, that perception took place in the eye, rather than, as Helmholtz argued, in the mind. The embodied vision under discussion here, however, is not a return to Muller's nativism but a consideration of the relationship and influence of the body on the eye itself.
58. Hogg, *The Ophthalmoscope*, p. 34.
59. Landolt, *A Manual*, p. 9.
60. Landolt, *Defective Ocular Movements*, p. 40.
61. Fuchs, *Text-Book of Ophthalmology*, p. 500
62. Sexton, *Spirit-Mediums*, p. 16
63. Ibid., p. 22.
64. Schickore, 'Misperception', p. 405.
65. Pepper's Ghost is perhaps the best known example of this type of demonstration. See Groth, 'Reading Victorian Illusions', and Brooker, 'The Polytechnic Ghost'.
66. R. Stott, *Darwin and the Barnacle: The Story of One Tiny Creature and History's Most Spectacular Scientific Breakthrough* (London: Faber & Faber, 2003), p. 138.
67. See the enormously rich work of R. Altick, *Shows of London* (Cambridge, MA: Harvard University Press, 1978) and J. Flanders, *Consuming Passions: Leisure and Pleasure in Victorian Britain* (London: Harper Press, 2006). Wyld's Great Globe is explored by R. Bellon, 'Science at the Crystal Focus of the World', in Fyfe and Lightman, *Science in the Marketplace*, pp. 301–35.
68. The term 'popular' in relation to science is under re-evaluation in history of science scholarship. Fyfe and Lightman argue that the increased attention on popular science has 'generated increasing scepticism about the very workability of the concept' (*Science in the Marketplace*, p. 2).
69. B. Bensaude-Vincent and C. Blondel (eds), *Science and Spectacle in the European Enlightenment* (Aldershot: Ashgate, 2008), p. 4.
70. Fyfe and Lightman, *Science in the Marketplace*, p. 11.
71. O'Connor, *The Earth on Show*, p. 263.
72. Morus, '"More the Aspect of Magic than Anything Natural"', p. 348. See also Morus, 'Seeing and Believing Science'.
73. J. Agar, *Science and Spectacle: The Work of Jodrell Bank in Post-War British Culture* (Amsterdam: Harwood Academic Press, 1998), p. xvi.
74. Bensaude-Vincent and Blondel, *Science and Spectacle*, p. 8.
75. J. Riskin, 'Amusing Physics', in Bensaude-Vincent and Blondel, *Science and Spectacle*, pp. 43–63, on p. 46.
76. See Flanders, *Consuming Passions*, p. 272. Pepper took this further than many other demonstrators by deciding, for some time, not to explain the appearance of the ghost. However, as Groth has argued, audiences were aware of the opportunities for deception afforded by optics and would have been likely to understand Pepper's demonstration as an aspect of that discipline's new knowledge.
77. Morus, '"More the Aspect of Magic than Anything Natural"', p. 363.
78. K. Hetherington, *Capitalism's Eye: Cultural Spaces of the Commodity* (London: Routledge, 2007), p. 62

79. Morus, '"More the Aspect of Magic than Anything Natural"', p. 363.
80. Tyndall, 'Science and the Spirits', p. 403.
81. Ibid.
82. Anon., 'At a Spiritualist Séance', M–YMC.
83. Maskelyne, *Modern Spiritualism*, p. 121.
84. Sexton, *Spirit-Mediums*, pp. 27–8.
85. Reprinted in the introduction to Sexton, *Spirit-Mediums*, p. 4.
86. H. Green, 'The Second Coming of Sir Arthur', *New York Times*, 8 April 1923, p. SM6.
87. See P. W. Wilson, 'Elucidating Conan Doyle', *New York Times*, 8 June 8 1922, pp. 1, 19, 24, and Green, 'Second Coming'.
88. See Flanders, *Consuming Passions*, p. 269.
89. Houdini, *Houdini: A Magician*, p. xiv.
90. H. Houdini, 'Houdini on Spiritualism', lecture given at the Symphony Hall, Boston, MA, 3 January 1925, HHA.
91. Letter dated 14 June 1926. Lellenburg et al., *Arthur Conan Doyle*, p. 317.
92. Maskelyne, *Modern Spiritualism*, p. 121.
93. J. M. Bird, *"Margery" The Medium* (Boston, MA: Small, Maynard & Co., 1925), p. 420.
94. Houdini, 'Houdini on Spiritualism', lecture given at the Symphony Hall, Boston, MA, 3 January 1925, HHA.
95. Houdini, *Houdini: A Magician*, p. xiv
96. Ibid., p. xix
97. M. Foucault, *Discipline and Punish: The Birth of the Prison*, trans. Alan Sheridan (London: Allen Lane, 1977).
98. Crary, *Suspensions*, p. 3.
99. Krasner, *The Entangled* Eye, pp. 31, 30.
100. K. Jacobs, *The Eye's Mind: Literary Modernism and Visual Culture* (Ithaca, NY: Cornell University Press, 2001), p. 3.
101. Houdini, Inventory Book for Houdini's stage performances, HHA, Box 49b
102. O. Teale, 'The Mystery of Spiritualism – Magic Versus Spiritualism', *Sphinx*, April 1925 – February 1926, in April 1925, p. 46.
103. Ibid.
104. Frank Crane, cited in Teale, 'Mystery of Spiritualism', April 1925, p. 46.
105. Crane, cited in Teale, 'Mystery of Spiritualism', April 1925, p. 46.
106. Gibson, *Houdini's Escapes*, p. 10.
107. Daston and Park, *Wonders*, pp. 14–5, 331.
108. Stafford, *Artful Science*, p. 3
109. R. Holmes, *The Age of Wonder: How The Romantic Generation Discovered the Beauty and Terror of Science* (London: Harper Press, 2008), p. xvi; Stafford, *Artful Science*, p. 70; Daston and Park, *Wonders*, p. 357. See also V. Hunt, 'Raising a Modern Ghost: The Magic Lantern and the Persistence of Wonder in the Victorian Education of the Senses', *Romanticism and Victorianism on the Net*, 52 (2008), pp. 1–20.
110. Daston and Park, *Wonders*, pp. 14–5. See also Stott, *Darwin and the Barnacle*, for a discussion of speculation.
111. Hankins and Silverman, *Instruments and the Imagination*, pp. 4, 10.
112. Stott, *Darwin and the Barnacle*, p. 136; Holmes, *The Age of Wonder*, p. xvi
113. T. Bennett, 'Pedagogic Objects, Clean Eyes, and Popular Instruction: On Sensory Regimes and Museum Didactics', *Configurations*, 6:3 (1998), pp. 345–71, on p. 347; M. N. Wise, 'Making Visible', *Isis*, 97 (2006), pp. 75–82, on p. 81.

114. Stafford, *Artful Science*, p. 302.
115. Stott, *Darwin and the Barnacle*, p. 138.
116. N. Thrift, *Non-Representational Theory: Space, Politics, Affect* (London: Routledge, 2007), pp. 6, 12.
117. Stafford, *Artful Science*, p. 88
118. S. During, *Modern Enchantments: The Cultural Power of Secular Magic* (Cambridge, MA: Harvard University Press, 2002), pp. 285–6.
119. Moulton, *Houdini's History*, p. 136.
120. Houdini, *Houdini Exposes*, p. 5.
121. This has been argued by sociologists since the work of the founding figure of sociology, Max Weber. See M. Weber, *Protestant Ethic and the Spirit of Capitalism* (London: Allen & Unwin, 1930); and his 1918 essay 'Science as a Vocation'. For a contemporary perspective see G. Ritzer, *Enchanting a Disenchanted World: Revolutionizing the Means of Consumption* (Thousand Oaks: Pine Forge Press, 1999).
122. Crary, *Suspensions*, p. 370.
123. See R. Jenkins, 'Disenchantment, Enchantment and Re-Enchantment: Max Weber at the Millennium', *Max Weber Studies*, 1:1 (2000), pp. 11–32. Jenkins argues that 'for Weber the disenchantment of the world lay right at the heart of modernity' (p. 12).
124. Weber, 'Science as a Vocation', p. 17.
125. Ibid., p. 7.
126. J. Wallis, 'Spiritualism and the (Re-)Enchantment of Modernity', in J. A. Beckford and J. Wallis (eds), *Theorising Religion: Classical and Contemporary Debates* (Aldershot: Ashgate, 2006), pp. 32–43, on p. 32.
127. Ibid.
128. Ibid., p. 41.
129. See Ritzer, *Enchanting*, p. 105.
130. Hetherington, *Capitalism's Eye*, p. 37.
131. Ibid.
132. During, *Modern Enchantments*, p. 285.
133. Crary, *Suspensions*, p. 370.
134. See Moulton, *Houdini's History*, p. 136.
135. Jenkins, 'Disenchantment', p. 17.
136. Stott, *Darwin and the Barnacle*, pp. 135–53.
137. Hetherington, *Capitalism's Eye*, pp. 49, 70.
138. Weber, 'Science as a Vocation', p. 17.
139. Quoted in G. Hatfield, 'Helmholtz and Classicism: The Science of Aesthetics and the Aesthetics of Science', in Cahan (ed.), *Hermann Von Helmholtz*, pp. 522–57, on p. 552.

Afterword

1. G. Deleuze and F. Guattari, *A Thousand Plateaus: Capitalism and Schizophrenia*, trans. B. Massumi (London: Continuum, 2004), p. 263.
2. J. J. Bono, 'Focus: History of Science and Literature and Science: Convergences and Divergences', *Isis*, 101:3 (2010), pp. 555–9.
3. Ibid., pp. 557, 559.
4. H. S. Turner, 'Lessons from Literature for the Historian of Science (and Vice Versa): Reflections on Form', *Isis*, 101:3 (2010), pp. 578–89, on p. 586.
5. R. Skloot, *The Immortal Life of Henrietta Lacks* (London: Macmillan, 2010).

6. See images of the *Science on a Sphere* project available online at http://sos.noaa.gov/ [accessed 3 August 2011].
7. Promotional material for the *Uncaged Monkeys* stage performances are available online at http://www.stereoboard.com/content/view/162849/9 [accessed 3 August 2011].
8. A. Grice, '"Mortified" PM Apologises over Condolence Note', *Independent*, 10 November 2009, p. 13; A. Laing, 'Brown's Hasty Letter Insults My Son, Says Grieving Mother', *Times*, 10 November 2009, p. 5; A. Sparrow and A. Gabbatt, 'Sun "Exploiting" Grief of Dead Soldier's Mother', *Guardian*, 10 November 2009, p. 7.
9. Sparrow and Gabbatt, 'Sun "Exploiting" Grief', p. 7.
10. 'Two Ways to Write a Condolence Letter', *Times*, 10 November 2009, p. 17.
11. S. Arnold, 'Not Spelling, but Sight', *Guardian*, 10 November 2009, p. 30; 'The PM's Eyes: An Insight', *Independent*, 10 November 2009, p. 13.
12. P. Hennessy and L. Donnelly, 'New Fears Over Brown's Eyesight', *Sunday Telegraph*, 11 October 2009, p. 1.
13. P. Hennessy and L. Donnelly, 'Regular Eye Checks Unusual, Say Surgeons', *Sunday Telegraph*, 11 October 2009, p. 4; J. Oliver and S-K. Templeton, 'PM Admits Eye Damage', *Sunday Times*, 11 October 2009, p. 2.
14. J. Merrick and N. Lakhani, 'Damage to Brown's Eye Raises Questions about His Future', *Independent on Sunday*, 11 October 2009, p. 2.
15. Ibid.; G. Owen, 'Fears for Brown as No 10 Reveals New Eye Damage', *Mail on Sunday*, 11 October 2009, p. 6.
16. Owen, 'Fears', p. 2; G. Hinsliff, 'Brown Treated in Hospital for Damage to His Right Eye', *Observer*, 11 October 2009, p. 1.
17. 'The One-Eyed Man May Be King, But For How Long?', *Sunday Times*, 11 October 2009, p. 20.
18. Hennessy and Donnelly, 'Regular Eye Checks Unusual', p. 4.
19. S. Bell, 'Cartoon 2925', *Guardian*, 10 November 2009, p. 23.
20. History can be cruel to scholarship. In the British General Election of 2010 Brown was removed from office and his Labour government was replaced by a Conservative–Liberal government under the leadership of David Cameron.

WORKS CITED

Archival Sources

Egypt Exploration Fund Archive, Egypt Exploration Society, London, United Kingdom.

Griffiths, F. L., 'The Study of Egyptology. Inaugural Lecture Delivered in the Ashmolean Museum on May 8, 1901', Pamphlets – G.

—, *The Study of Egyptology* (Oxford: Horace Hart, 1901), pp. 7–8, Pamphlets – G.

Naville, E., 'The Route of the Exodus: Being the Annual Address of the Victoria Institute 1891', Pamphlets NA-NEL.

Petrie, W. M. F., 'The Discovery of Naukratis', Pamphlets PETRIE-PIE.

—, 'Letter to Kate Bradbury', Box 1C, No.1.

—, 'Letter to E. Maunde Thompson', EEFA, Box 1C, No. 2.

—, 'History in Tools', in *Smithsonian Report for 1918* (Washington, DC: Government Printing Office, 1920), pp. 563–72, Pamphlets: PET.

'Thomas Cook Letter', Box X/d/42.

Harry Houdini Archive, Library of Congress, Washington DC, United States of America.

Advertising playbill from Houdini's private collection, Box 16.

Challenge leaflet from Houdini's private collection, Box 20.

Hardeen Memorandum, Box 26.

Houdini, H., 'The Psychology of Prestidigitation' (1925), Box 26.

—, 'Houdini on Spiritualism', lecture given at the Symphony Hall, Boston, MA, 3 January 1925.

—, Inventory Book for Houdini's stage performances, Box 49b.

—, Note to Houdini from Houdini's private collection, Houdini Pamphlets, no. 6.

'The Final Houdini Séance', Box 15, Folder 1.

Hendricks Centre for Planetary Research, Lowell Observatory, Flagstaff, Arizona, United States of America.

Percival Lowell Archive.

Anon., [no title], *Coconino Sun*, 15 November 1894, Scrapbooks, Clipping Books.

Anon., 'Communication from Mars', *c.* 1894, Scrapbooks, Clipping Books.

Antoniadi, E. H., 'Letter to Percival Lowell', 9 September 1909, Corr–9/9/09.

—, 'Letter to Percival Lowell', 9 October 1909, Corr–10/09/09.

Braesher, J., 'Letter to Percival Lowell', 19 July 1901, Corr–5/13A.

Lowell, P., 'Dawn', Ms–1U/1.

—, 'Double Canals and Separative Power', Ms–4/6.

—, 'On the Kind of Eye Needed for the Detection of Planetary Detail', Ms–4/11.

—, 'Small Apertures Versus Large Ones in the Study of Planetology', Ms–2U/22.

—, 'Untitled Ms', n.d., Ms–2U/12.

—, 'Mars', *c.* August 1894, Ms–1U/5.

—, 'Mars', 11 September 1894, Ms–2/2–1B.

—, 'Telegram to John Braesher', 14 August 1901, Corr–5/13A.

—, 'A Standard Scale for Telescopic Observations', 1902, Ms–3/7.

—, 'Letter to E.W. Maunder', 28 November 1903, Corr–11/28/03.

—, 'Means, Methods and Mistakes in the Study of Planetary Evolution', 13 April 1905, Ms–1U/12.

—, 'Mars in 1909', Ms–6/4.

—, 'Letter to Antoniadi', 26 September 1909, Corr–09/26/09.

—, 'Letter to Antoniadi', 2 November 1909, Corr–11/02/09.

—, 'On Some Fallacies in Observing Mars', 1909, Ms–2U/19.

—, 'Great Discoveries and Their Reception', 1916, Ms–3U/11.

—, 'Lecture D: Mars and the Earth', 1916, Ms–3U/13.

Lowell, P., and V. M. Slipher, 'Air and Large Telescopes', 30 October 1908, Ms–5/13A.

Wicks, M., 'Letter to Percival Lowell', Corr–9/24/07.

McManus-Young Magical Collection, Library of Congress, Washington DC, United States of America.

Anon., 'At a Spiritualist Séance'.

Petrie Museum Archive, Petrie Museum, London, United Kingdom.

Petrie, W. M. F., '30 Nov 1880 – 22 June 1881 – Gizeh', in W. M. Flinders Petrie Journals, WFP1/16/1/1.

—, '1883–4 Diary', in W. M. Flinders Petrie Diary, WFP1/9/3/1883–4.

—, '7 Nov 1883 – 23 June 1884 – Wady Tumilat, Tanis', in W. M. Flinders Petrie Journals, WFP1/16/1/1(2).

—, '24 Oct 1888 – 23 May 1889 Guroh, Kahun and Illahun', in W. M. Flinders Petrie Journal, WFP1/16/1/2(5).

—, 'Notebook 3 – Abydos, Royal Tombs', in W. M. Flinders Petrie Notebooks, PMA, WFP1/99/1/3.

—, W. M. Flinders Petrie Diary Journal WFP1/16/1/1.

Wellcome Library for the History and Understanding of Medicine, London, United Kingdom.

Lister Institute Archive.

Anon., 'Deputation to the Home Secretary: To protest against the building of the British Institute of Preventive Medicine at Chelsea', 24 April 1894.

Anon., 'Deputation to the Board of Trade', 7 June 1894,

Anon., 'Report on the Parade and Mass Meeting at Pimlico', SA/LIS/E5.

Anon., 'The Proposed British Institute of Preventive Medicine', SA/LIS/E8.

Brandeis, L., *Sanitarian*, 1873, SA/LIS/E4.

Carrington, E., 'Untitled', *Gloucester Chronicle*, 13 October 1894, SA/LIS/E8.

Collinson, J., 'The Pasteur Institute', *Irish Figaro*, 11 May 1895, SA/LIS/E8.

Field, 6 November 1886, SA/LIS/E8.

'Human Vivisection', *Daily Chronicle*, 5 June 1894, SA/LIS/E8.

'Mansion House Meeting', 1 June 1889, SA/LIS/M11.

'MORE SECRECY! District Councillors Kept Off Premises of Institute of Preventive Medicine', *Star*, 3 March 1896, SA/LIS/E8.

Westminster News, 7 July 1896, Press Scrapbook.

Westminster News, 22 January 1898, Press Scrapbook.

Whalley, A. W., 'Survey of Chelsea Residents', 11 April 1894.

Published Sources

Abbott, D., *The Marvellous Creations of Joseffy* (Chicago, IL: Open Court, 1908).

Ackerknecht, E. H., 'Anticontagionism Between 1821 and 1867', *Bulletin of the History of Medicine*, 22 (1948), pp. 562–93.

Agar, J., *Science and Spectacle: The Work of Jodrell Bank in Post-War British Culture* (Amsterdam: Harwood Academic Press, 1998).

Alberti, S. J. M. M., 'Focus: Museums and the History of Science: Objects and the Museum', *Isis*, 96:4 (2005), pp. 559–71.

Alder, K., 'History's Greatest Forger: Science, Fiction, and Fraud along the Seine', *Critical Inquiry*, 30 (2004), pp. 702–16.

—, 'Focus: Thick Things: Introduction', *Isis*, 98:1 (2007), pp. 80–3.

Altick, R., *Shows of London* (Cambridge, MA: Harvard University Press, 1978).

Anon., *The Wonders of the Microscope; or, An Explanation of the Wisdom of the Creator in Objects Comparatively Minute, Adapted to the Understanding of Young Persons* (London: Tabarat & Co., 1806).

—, *How to Live in London: or, The Metropolitan Microscope and Stranger's Guide* (London: Joseph Smith, 1828).

—, 'On the Microscopical Character of Milk and the Use of the Microscope in the Choice of a Nurse', *Lancet*, 1 (1844), p. 531.

—, 'A Strange Light on Mars', *Nature*, 2 August 1894, p. 319.

—, 'Review of *The War of the Worlds*, Academy, 29 January 1898', in Parrinder (ed.), *H. G. Wells*, pp. 70–4.

Arata, S. D., 'The Occidental Tourist: *Dracula* and the Anxiety of Reverse Colonization', *Victorian Studies*, 33:4 (1990), pp. 621–45.

Armstrong, I., *Victorian Glassworlds: Glass Culture and the Imagination, 1830–1880* (Oxford: Oxford University Press, 2008).

Arnold, S., 'Not Spelling, but Sight', *Guardian*, 10 November 2009, p. 30.

Auerbach, N., 'Dracula: A Vampire of Our Own', in G. Byron (ed.), *Dracula: Contemporary Critical Essays* (London: Macmillan, 1999), pp. 145–72.

Baldwin, P., *Contagion and the State in Britain, 1830–1930* (Cambridge: Cambridge University Press, 2000).

Ball, R. S., *The Story of the Heavens* (London: Cassell, 1886).

Bancroft, E. N., *An Essay on the Disease called Yellow Fever, with Observations Concerning Febrile Contagion, Typhus Fever, Dysentry, and the Plague, Partly Delivered as the Gulstonian Lectures, Before the College of Physicians, in the Years 1806 and 1807* (London: T. Cadell & W. Davies, 1811).

Barrow, L., *Independent Spirits: Spiritualism and English Plebeians, 1850–1910* (London: Routledge and Kegan Paul, 1986).

Bartlett, W. H., *The Nile Boat; or Glimpses of the Land of Egypt* (London: Henry G. Bohn, 1862).

Beer, G., *Darwin's Plots: Evolutionary Narrative in Darwin, George Eliot and Nineteenth-Century Fiction* (Cambridge: Cambridge University Press, 1983).

Bell, C. D., *A Winter on the Nile, in Egypt and in Nubia* (London: Hodder & Stoughton, 1888).

Bell, S., 'Cartoon 2925', *Guardian*, 10 November 2009, p. 23.

Bellon, R., 'Science at the Crystal Focus of the World', in A. Fyfe and B. Lightman (eds), *Science in the Marketplace: Nineteenth-Century Sites and Experiences* (Chicago, IL: University of Chicago Press, 2007), pp. 301–35.

Bennett, J. H., *On the Employment of the Microscope in Medical Studies. A Lecture Introductory to a Course in Histology* (Edinburgh: Maclachlan, Stuart & Co., 1841).

—, 'The Atmospheric Germ Theory', *Edinburgh Medical Journal* (March 1868), pp. 810–34.

Bennett, T., 'Pedagogic Objects, Clean Eyes, and Popular Instruction: On Sensory Regimes and Museum Didactics', *Configurations*, 6:3 (1998), pp. 345–71.

Bensaude-Vincent, B., and C. Blondel (eds), *Science and Spectacle in the European Enlightenment* (Aldershot: Ashgate, 2008).

Beche, H. T. de la, *How To Observe Geology* (London: Charles Knight, 1835).

Binet, A., 'La Psychologie de la Prestidigitation', *Revue des Deux Mondes*, 54:125 (1894), pp. 903–23.

Bird, J. M., *'Margery' The Medium* (Boston, MA: Small, Maynard & Co., 1925).

Blackwelder, E., 'Our Friends, The Enemy: Mars as the Abode of Life', *Science*, 29:747 (23 April 1909), p. 661.

Bolla, P. de, *The Education of the Eye: Painting, Landscape, and Architecture in Eighteenth-Century Britain* (Stanford, CA: Stanford University Press, 2003).

Bono, J. J., 'Focus: History of Science and Literature and Science: Convergences and Divergences', *Isis*, 101:3 (2010), pp. 555–9.

Brewster, D., 'On the Optical Illusion of the Conversion of Cameos into Intaglios, and of Intaglios into Cameos, with an Account of Other Analogous Phenomena' (1826), in N. Wade (ed.), *Brewster and Wheatstone on Vision* (London: Academic Press, 1983), pp. 56–65.

—, *Letters on Natural Magic* (1832; New York: Harper & Brothers, 1843).

Britzolakis, C., 'Phantasmagoria: Walter Benjamin and the Poetics of Urban Modernism', in P. Buse and A. Stott (eds), *Ghosts: Deconstruction, Psychoanalysis, History* (London: Macmillan, 1999), pp. 72–90.

Brooker, J., 'The Polytechnic Ghost: Pepper's Ghost, Metempsychosis and the Magic Lantern at the Royal Polytechnic Institution', *Early Popular Visual Culture*, 5:2 (2007), pp. 189–206.

Brown, B., 'Thing Theory', *Critical Inquiry*, 28:1 (2001), pp. 1–22.

Budge, E. A. W., *By Nile and Tigris: A Narrative of Journeys in Egypt and Mesopotamia on Behalf of the British Museum Between the Years 1886 and 1913*, 2 vols (London: John Murray, 1920).

—, *Cook's Handbook for Egypt and the Egyptian Sudan: With Chapters on Egyptian Archaeology* (London: Thomas Cook & Son, 1921).

Burroughs, E. R., *A Princess of Mars* (1911; London: Penguin, 2007).

Butler, A., 'Court Life in Egypt', in C. Pick (ed.), *Egypt: A Traveller's Anthology* (London: John Murray, 1991), p. 99.

Bynum, W. F., 'The Evolution of Germs and the Evolution of Disease: Some British Debates, 1870–1900', *History and Philosophy of the Life Sciences*, 24 (2000), pp. 53–68.

Byron, G., 'Introduction', in G. Byron (ed.), *Dracula: Contemporary Critical Essays* (London, Macmillan, 1999), pp. 1–21.

Carpenter, W. B., *The Microscope and its Revelations* (London: John Churchill, 1856).

Canales, J., 'Exit the Frog, Enter the Human: Physiology and Experimental Psychology in Nineteenth-Century Astronomy', *British Journal of the History of Science*, 34:2 (2001), pp. 173–97.

Carroll, V., 'Natural History on Display: The Collection of Charles Waterton', in E. Fyfe and B. Lightman (eds), *Science in the Marketplace: Nineteenth Century Sites and Experiences* (Chicago, IL: University of Chicago Press, 2007), pp. 271–300.

Castle, T., 'Phantasmagoria: Spectral Technology and the Metaphorics of Modern Reverie', *Critical Inquiry*, 15 (1998), pp. 26–61.

Christ, C. T., and J. O. Jordan (eds), *Victorian Literature and the Victorian Visual Imagination* (Berkeley, CA: University of California Press, 1995).

Cobbe, F. P., 'Vivisection and Its Two-Faced Advocates', in L. Otis (ed.), *Literature and Science in the Nineteenth Century: An Anthology* (Oxford: Oxford University Press, 2002), pp. 215–9.

Cohen, M., 'Benjamin's Phantasmagoria: The Arcades Project', in D. S. Ferris (ed.), *The Cambridge Companion to Walter Benjamin* (Cambridge: Cambridge University Press, 2004), pp. 200–20.

Connan Doyle, A., *The Stark Munro Letters* (London: Longmans, Green & Co., 1895).

—, *Memories and Adventures* (London: Hodder & Stoughton, 1924).

—, *Our American Adventure* (London: Hodder & Stoughton, 1923).

—, *History of Spiritualism*, 2 vols (London: Cassell, 1926), vol. 1.

—, *The Penguin Complete Sherlock Holmes* (London: Penguin, 1981).

—, *Letters to the Press*, ed. J. M. Gibson and R. L. Green (Iowa City, IA: University of Iowa Press, 1986).

Cooter, R., *The Cultural Meaning of Popular Science: Phrenology and the Organization of Consent in Nineteenth-Century Britain* (Cambridge: Cambridge University Press, 1984).

Crary, J., *Techniques of the Observer: On Vision and Modernity in the Nineteenth Century* (Cambridge, MA: MIT Press, 1992).

—, *Suspensions of Perception: Attention, Spectacle, and Modern Culture* (Cambridge, MA: MIT Press, 2001).

Crawford, E. (ed.), *The Women's Suffrage Movement: A Reference Guide, 1866–1928* (London: Routledge, 2001).

Crellin, J. K., 'Airborne Particles and the Germ Theory, 1860–1880', *Annals of Science*, 22:1 (1966), pp. 49–60.

Crookes, W., *Researches in the Phenomena of Spiritualism* (London: J. Burns, 1874).

Crossley, R., 'Percival Lowell and the History of Mars', *Massachusetts Review*, 41:3 (2000), pp. 297–318.

Crowe, M. J., *The Extraterrestrial Life Debate, 1750–1900: The Idea of a Plurality of Worlds from Kant to Lowell* (Cambridge: Cambridge University Press, 1986).

Cunningham, R., 'Virtual Witnessing and the Role of the Reader', *Philosophy and Rhetoric*, 34:3 (2001), pp. 207–24.

Daston, L., 'Introduction: Speechless', in L. Daston (ed.), *Things That Talk: Object Lessons from Art and Science* (New York: Zone Books, 2004), pp. 9–24.

—, 'On Scientific Observation', *Isis*, 99 (2008), pp. 97–110.

—, 'Introduction: The Coming into Being of Scientific Objects' in L. Daston (ed.), *Biographies of Scientific Objects* (Chicago, IL: University of Chicago Press, 2000), pp. 9–24.

—, 'Fear and Loathing of the Imagination in Science', in P. Galison, S. R. Graubard and E. Mendelsohn (eds), *Science in Culture* (New Brunswick, NJ: Transaction, 2001), pp. 73–95.

Daston, L., and K. Park, *Wonders and the Order of Nature, 1150–1750* (New York: Zone Books, 2001).

Daston, L., and P. Galison, 'The Image of Objectivity', *Representations*, 40 (1992), pp. 81–128.

—, *Objectivity* (New York: Zone Books, 2007).

Daum, A. W., 'Varieties of Popular Science and the Transformations of Public Knowledge: Some Historical Reflections', *Isis*, 100 (2009), pp. 319–32.

Deleuze, G., and F. Guattari, *A Thousand Plateaus: Capitalism and Schizophrenia*, trans. Brian Massumi (London: Continuum, 2004).

Dick, S. J., *Plurality of Worlds: The Origins of the Extraterrestrial Life Debate from Democritus to Kant* (Cambridge: Cambridge University Press, 1982).

Dick, T., *The Telescope and Microscope* (London: Religious Tract Society, 1851).

Dobler, H., *Exposé of the Davenport Brothers* (Belfast: D & J Allen, 1869).

Dower, M. (ed.), *Letters From the Desert: The Correspondence of Flinders and Hilda Petrie* (Oxford: Avis & Phillips, 2004)

During, S., *Modern Enchantments: The Cultural Power of Secular Magic* (Cambridge, MA: Harvard University Press, 2002).

Edwards, A. B., *A Thousand Miles Up the Nile* (1876; London: George Routledge & Sons, 1890).

—, 'Bubastis: An Historical Study', *Century Magazine*, 39:3 (January 1890), pp. 323–49.

—, *Pharaohs, Fellahs and Explorers* (London: James R. Osgood, McIlvaine & Co, 1891).

Ernst, B. M. L., and H. Carrington, *Houdini and Conan Doyle: A Strange Friendship* (New York: Benjamin Bloom, 1972).

Evans, H. R., *Hours with the Ghosts, or Nineteenth Century Witchcraft: Illustrated Investigations into the Phenomena of Spiritualism and Theosophy* (Chicago, IL: Laird & Lee, 1897).

Fanon, F., *Black Skin White Masks* (New York: Grove Press, 1967).

Fanu, S. Le, 'Carmilla', in R. Tracy (ed.), *In a Glass Darkly* (Oxford: Oxford University Press, 1993), pp. 243–319.

Ferguson, C., 'Eugenics and the Afterlife: Lombroso, Doyle, and the Spiritualist Purification of the Race', *Journal of Victorian Culture*, 12:1 (2007), pp. 64–85.

Flammarion, C., *Mysterious Psychic Forces: An Account of the Author's Investigations in Psychical Research, Together with Those of Other European Savants* (Boston, MA: Small, Maynard & Co., 1907).

Flanders, J., *Consuming Passions: Leisure and Pleasure in Victorian Britain* (London: Harper Press, 2006).

Flint, K., *Victorians and the Visual Imagination* (Cambridge: Cambridge University Press, 2000).

—, '"Seeing is Believing"? Visuality and Victorian Fiction', in F. O'Gorman (ed.), *A Concise Companion to the Victorian Novel* (Oxford: Blackwell, 2005).

Foster, R. F., *Paddy and Mr. Punch: Connections in Irish and English History* (London: Allen Lane, 1993).

Foucault, M., *Discipline and Punish: The Birth of the Prison*, trans. Alan Sheridan (London: Allen Lane, 1977).

Fuchs, E., *Text-Book of Ophthalmology* (1892; New York: D. Appleton & Co., 1901).

Fulford, T., D. Lee and P. J. Kitson, *Literature, Science and Exploration in the Romantic Era* (Cambridge: Cambridge University Press, 2004).

Fyfe, A., and B. Lightman, 'Science in the Marketplace: An Introduction', in A. Fyfe and B. Lightman (eds), *Science in the Marketplace: Nineteenth-Century Sites and Experiences* (Chicago, IL: University of Chicago Press, 2007), pp. 1–19.

Gelder, K., *Reading The Vampire* (London: Routledge, 1994).

Gibson, M., *Dracula and the Eastern Question: British and French Vampire Narratives of the Nineteenth-Century Near East* (Basingstoke: Palgrave Macmillan, 2006).

Gibson, W. G., *Houdini's Escapes: Prepared from Houdini's Private Notebooks and Memoranda with the Assistance of Beatrice Houdini, Widow of Houdini, and Bernard M. L. Ernst, President of the Parent Assembly of the Society of American Magicians* (New York: Harcourt, Brace & Co., 1930).

Gooday, G., 'Placing or Replacing the Laboratory in the History of Science?', *Isis*, 99:4 (2008), pp. 783–95.

Gosden, C., and Y. Marshall, 'The Cultural Biography of Objects', *World Archaeology*, 31:2 (1999), pp. 169–78.

Gossin, P., *Thomas Hardy's Novel Universe: Astronomy, Cosmology and Gender in the Post-Darwinian World* (Aldershot: Ashgate, 2007).

Grave, J., 'On The Aesthetics of Scientific Objects. Three Case Studies', in S. Vackimes and K. Weltersbach (eds), *Wandering Seminar on Scientific Objects* (Berlin: Max Planck Institute for the History of Science, 2007), pp. 35–47.

Green, H., 'The Second Coming of Sir Arthur', *New York Times*, 8 April 1923, p. SM6.

Green-Lewis, J., *Framing the Victorians: Photography and the Culture of Realism* (Ithaca, NY: Cornell University Press, 1996).

Gregory, R., 'Review of *The War of the Worlds*, Nature, 10 February 1898', in P. Parrinder (ed.), *H. G. Wells: The Critical Heritage* (London: Routledge & Kegan Paul, 1972), pp. 74–7.

Grice, A., '"Mortified" PM Apologises over Condolence Note', *Independent*, 10 November 2009, p. 13.

Griffiths, F. L., *The Study of Egyptology* (Oxford: Horace Hart, 1901).

Groth, H., 'Reading Victorian Illusions: Dickens's *Haunted Man* and Dr. Pepper's Ghost', *Victorian Studies*, Autumn (2007), pp. 43–65.

Guillemin, A., *The Heavens: An Illustrated Handbook of Popular Astronomy* (London: Richard Bentley, 1867).

Haggard, H. R., *She* (1887; London: Penguin, 2004).

Haight, G. S., 'H. G. Wells's "The Man of the Year Million"', *Nineteenth-Century Fiction*, 12:4 (1958), pp. 323–6.

Haight, S., 'Letters from an Old World', in C. Pick (ed.), *Egypt: A Traveller's Anthology* (London: John Murray, 1991), pp. 101–4.

Halberstam, J., 'Technologies of Monstrosity: Bram Stoker's *Dracula*', in S. Ledger and S. McCracken (eds), *Cultural Politics at the Fin-de-Siècle* (Cambridge: Cambridge University Press, 1995), pp. 248–66.

Hankins, T. L., and R. J. Silverman, *Instruments and the Imagination* (Princeton, NJ: Princeton University Press, 1995).

Hardy, A., *The Epidemic Streets: Infectious Disease and the Rise of Preventive Medicine, 1856–1900* (Oxford: Clarendon Press, 1993).

Hatfield, G., 'Helmholtz and Classicism: The Science of Aesthetics and the Aesthetics of Science', in D. Cahan (ed.), *Hermann Von Helmholtz and the Foundations of Nineteenth-Century Science* (Berkeley, CA: University of California Press, 1993), pp. 522–57.

Heller-Roazen, D., M. Mahdi and H. Haddawy (eds), *The Arabian Nights: Norton Critical Edition* (New York: W. W. Norton, 2008).

Helmholtz, H. V., 'Concerning the Perceptions in General' in J. P. C. Southall (ed.), *Helmholtz's Treatise on Physiological Optics*, 3 vols (1856–67; Bristol: Theommes Press, 2000), vol. 3, pp. 1–37.

—, *Helmholtz's Treatise on Physiological Optics*, J. P C. Southall (ed.), 3 vols (1856–67; Bristol: Theommes Press, 2000).

—, 'The Facts in Perception (1878)', in R. S. Cohen and Y. Elkana (eds), *Hermann Von Helmholtz: Epistemological Writings* (Boston, MA: D. Reidel, 1977), pp. 114–85.

Henchman, A., 'Hardy's Stargazers and the Astronomy of Other Minds', *Victorian Studies*, 51:1 (2008), pp. 37–64.

Hennessy, P., and L. Donnelly, 'New Fears Over Brown's Eyesight', *Sunday Telegraph*, 11 October 2009, p. 1.

—, 'Regular Eye Checks Unusual, Say Surgeons', *Sunday Telegraph*, 11 October 2009, p. 4.

Henry, W., 'Report on the State of Our Knowledge of the Laws of Contagion', *Report of the Fourth Meeting of the British Association for the Advancement of Science; held at Edinburgh in 1834* (London: BAAS, 1835), pp. 67–94.

Herschel, J. F. W., *Outlines of Astronomy* (London: Longman, Brown, Green & Longmans, 1849).

Hetherington, K., *Capitalism's Eye: Cultural Spaces of the Commodity* (London: Routledge, 2007).

Hinsley, C. M., 'Revising and Revisioning the History of Archaeology: Reflections on Region and Context', in A. L. Christenson (ed.), *Tracing Archaeology's Past: The Historiography of Archaeology* (Carbondale, IL: Southern Illinois University Press, 2001).

Hinsliff, G., 'Brown Treated in Hospital for Damage to His Right Eye', *Observer*, 11 October 2009, p.1.

Hoffmann, C., 'Constant Differences: Friedrich Wilhelm Bessel, the Concept of the Observer in Early Nineteenth-Century Practical Astronomy and the History Of the Personal Equation', *British Journal of the History of Science*, 40:3 (2007), pp. 333–65.

Hogg, J., *The Microscope: Its History, Construction and Applications. Being a Familiar Introduction to the Use of the Instrument and the Study of Microscopical Science*. (London: Herbert Ingram, 1856).

—, *The Ophthalmoscope: Its Mode of Application Explained, and its Value Shown, in the Exploration of Internal Diseases Affecting the Eye* (London: John Churchill, 1858).

Holmes, R., *The Age of Wonder: How The Romantic Generation Discovered the Beauty and Terror of Science* (London: Harper Press, 2008).

Hopkins, A. A., *Magic: Stage Illusions, Special Effects and Trick Photography* (1897; New York: Dover, 1976).

Houdin, R., *The Secrets of Conjuring and Magic, or How To Become a Wizard*, trans. Professor Hoffmann (London: George Routledge, 1878).

Houdini, H., *The Right Way to Do Wrong* (Boston, MA: Harry Houdini, 1906).

—, *Handcuff Secrets* (New York: Routledge, 1910).

—, *Miracle Mongers and Their Methods: A Complete Exposé* (1920; Buffalo, NY: Prometheus Books, 1981).

—, *Magical Rope Ties and Escapes* (London: Will Goldston, 1921).

—, *Houdini: A Magician Among the Spirits* (1924; New York: Arno Press, 1972).

—, *Houdini Exposes the Tricks Used By the Boston Medium "Margery" to Win the $2500 Prize Offered By the Scientific American* (New York: Adams Press, 1924).

—, *Houdini on Magic*, ed. W. B. Gibson and M. N. Young (1930; New York: Dover, 1953).

Hoyt, W. G., *Lowell and Mars* (Tucson, AZ: University of Arizona Press, 1976).

Hughes, W., 'Fictional Vampires in the Nineteenth and Twentieth Centuries', in D. Punter (ed.), *A Companion to the Gothic* (Oxford: Blackwell, 2000), pp. 144–5.

—, *Beyond Dracula: Bram Stoker's Fiction in Its Cultural Context* (New York: St Martin's Press, 2000).

Hunt, V., 'Raising a Modern Ghost: The Magic Lantern and the Persistence of Wonder in the Victorian Education of the Senses', *Romanticism and Victorianism on the Net*, 52 (2008), pp. 1–20.

Hutchinson, J., *Aids to Ophthalmic Medicine and Surgery* (London: Ballière & Co, 1889).

Iota, *Maskelyne and Cooke: An Expose of the Falseness of their Pretensions* (London: J. Burns, 1873).

Jacobs, K., *The Eye's Mind: Literary Modernism and Visual Culture* (Ithaca, NY: Cornell University Press, 2001).

Jacyna, L. S., 'Moral Fibre: The Negotiation of Microscopic Facts in Victorian Britain', *Journal of the History of Biology*, 36 (2003), pp. 39–85.

Jaffe, A., *Scenes of Sympathy: Identity and Representation in Victorian Fiction* (Ithaca, NY: Cornell University Press, 2000).

Jenkins, R., 'Disenchantment, Enchantment and Re-Enchantment: Max Weber at the Millennium', *Max Weber Studies*, 1:1 (2000), pp. 11–32.

Kasson, J. F., *Houdini, Tarzan, and The Perfect Man: The White Male Body and the Challenge of Modernity in America* (New York: Hill & Wang, 2001).

Keats, J., *Selected Poetry*, ed. E. Cook (Oxford: Oxford University Press, 2008).

Kelly, R. T., *Egypt Painted and Described* (London: Adam & Charles Black, 1903).

Kendall, M., *Songs From Dreamland* (London: Longmans, 1894).

Killeen, J., 'Irish Gothic: A Theoretical Introduction', *Irish Journal of Gothic and Horror Studies*, 1 (2006), n.p.

Kohler, R. E., 'Lab History: Reflections', *Isis*, 99:4 (2008), pp. 761–8.

Krasner, J., *The Entangled Eye: Visual Perception and the Representation of Nature in Post-Darwinian Narrative* (New York: Oxford University Press, 1992).

Lachapelle, S., 'From the Stage to the Laboratory: Magicians, Psychologists, and the Science of Illusions', *Journal of the History of the Behavioral Sciences*, 44:4 (2008), pp. 319–34.

Laing, A., 'Brown's Hasty Letter Insults My Son, Says Grieving Mother', *Times*, 10 November 2009, p. 5.

La'mert, S., *Self-Preservation: A Medical Treatise on Nervous and Physical Debility, Spermatorrhea, Impotence, and Sterility with Practical Observations on the Use of the Microscope in the Treatment of the Diseases of the Generative System* (London: Samuel La'mert, 1859).

Landolt, E., *A Manual of the Examination of the Eyes: A Course of Lectures Delivered at the École Pratique* (1879; New York: Robert E. Kreiger, 1979).

—, *The Refraction and Accommodation of the Eye and Their Anomalies* (Edinburgh: Young J. Pentland, 1886).

—, *Defective Ocular Movements and Their Diagnosis*, trans. A. Roemmelle and E. W. Brewerton (London: Oxford University Press and Hodder & Stoughton, 1913).

Lane, E. W., *Description of Egypt*, ed. J. Thompson (Cairo: American University in Cairo Press, 2000).

Lankester, E., *Half-Hours with the Microscope: Being a Popular Guide to the Use of the Microscope as a Means of Amusement and Instruction* (London: Robert Hardwicke, 1860).

Latour, B., *Science in Action: How to Follow Scientists and Engineers Through Society* (Cambridge, MA: Harvard University Press, 1988).

—, *The Pasteurization of France*, trans. A. Sheridan and J. Law (Cambridge, MA: Harvard University Press, 1988).

—, 'On the Partial Existence of Existing *and* Nonexisting Objects', in L. Daston (ed), *Biographies of Scientific Objects* (Chicago, IL: University Of Chicago Press, 2000), pp. 247–69.

—, 'The Berlin Key or How to Do Words with Things', in P. M. Graves-Brown (ed.), *Matter, Materiality and Modern Culture* (London: Routledge, 2000), pp. 10–21.

Laurance, L., *Visual Optics and Sight Testing* (London: School of Optics, 1920).

Leask, N., *Curiosity and the Aesthetics of Travel Writing, 1770–1840: 'From an Antique Land'* (Oxford: Oxford University Press, 2002).

Lee, D., and T. Fulford, 'The Beast Within: The Imperial Legacy of Vaccination in History and Literature', *Literature and History*, 9:1 (2000), pp. 1–23.

Lellenburg, J., D. Stashower and C. Foley (eds), *Arthur Conan Doyle: A Life in Letters* (London: Harper Press, 2007).

Lenoir, T., 'The Eye as Mathematician: Clinical Practice, Instrumentation, and Helmholtz's Construction of an Empiricist Theory of Vision', in D. Cahan (ed.), *Hermann Von Helmholtz and the Foundations of Nineteenth-Century Science* (Berkeley, CA: University of California Press, 1993).

Levine, G., *Darwin and the Novelists: Patterns of Science in Victorian Fiction* (Cambridge, MA: Harvard University Press, 1988).

—, *Darwin Loves You: Natural Selection and the Re-Enchantment of the World* (Princeton, NJ: Princeton University Press, 2006).

Lightman, B., 'The Visual Theology of Victorian Popularizers of Science: From Reverent Eye to Chemical Retina', *Isis*, 91 (2000), pp. 651–80.

—, *Victorian Popularizers of Science: Designing Nature for New Audiences* (Chicago, IL: University of Chicago Press, 2007).

Lindquist, J. H., '"The Mightiest Instrument of the Physical Discoverer": The Visual "Imagination" and the Victorian Observer', *Journal of Victorian Culture*, 13:2 (2008), pp. 171–99.

Lockyer, W. J., 'Mars as He Now Appears', *Nature*, 13 September 1894, pp. 476–8.

Lorimer, N., *By the Waters of Egypt* (London: Methuen, 1909).

Lovecraft, H. P., and H. Houdini, 'Imprisoned With the Pharaohs', *Weird Tales*, May 1924, available online at http://gaslight.mtroyal.ca/gaslight/imppharo.htm [accessed 4 August 2010].

Lowell, P., *Annals of the Lowell Observatory: Volume 1 1894–5* (Boston, MA: Houghton & Mifflin, 1895).

—, *Mars* (1894; London, Longmans, Green & Co., 1896).

—, 'Areography', *Popular Science Monthly*, 61 (September 1902), p. 392.

—, *Mars as the Abode of Life* (New York: Macmillan, 1908).

Luckhurst, R., *The Invention of Telepathy* (Oxford: Oxford University Press, 2002).

Macleod, R., 'Introduction: Nature and Empire: Science and the Colonial Enterprise', *Osiris*, 15 (2000), pp. 1–13.

Maglen, K., 'Interpreting Infection: Quarantine, the Port Sanitary Authority and Immigration in Late Nineteenth and Early Twentieth Century Britain' (PhD dissertation, University of Glasgow, 2001).

Mantell, G., *Thoughts on Animalcules: or, A Glimpse of the Invisible World Revealed by the Microscope* (London, John Murray, 1846).

Markley, R., *Dying Planet: Mars in Science and the Imagination* (Durham, NC: Duke University Press, 2005).

Maskelyne, J. N., *Modern Spiritualism: A Short Account of its Rise and Progress, With Some Exposures of So-Called Spirit Media* (London: Frederick Warne, 1876).

Maskelyne, N., and D. Devant, *Our Magic: The Art in Magic, the Theory of Magic, the Practice of Magic* (New York: E. P. Dutton & Co., 1911).

Martin, M. E., *The Ways of the Planets* (London: Harper, 1912).

McClintock, A., *Imperial Leather: Race, Gender and Sexuality in the Colonial Contest* (London: Routledge, 1995).

McGrath, M., *Ghosts of Futures Past: Spiritualism and the Cultural Politics of Nineteenth-Century America* (Berkeley, CA: University of California Press, 2008).

Mendelsohn, J. A., 'The Microscopist of Modern Life', *Osiris*, 18 (2003), pp. 150–70.

Merrick. J., and N. Lakhani, 'Damage to Brown's Eye Raises Questions about His Future', *Independent on Sunday*, 11 October 2009, p. 2.

Meskell, L., *Object Worlds in Ancient Egypt: Material Biographies Past and Present* (Oxford: Berg, 2004).

Mighall, R., '"A Pestilence which Walketh in Darkness": Diagnosing the Victorian Vampire', in G. Byron and D. Punter (eds), *Spectral Readings: Towards a Gothic Geography* (London: Macmillan, 1999), pp. 108–24.

Mirzoeff, N. (ed.), *The Visual Culture Reader* (London: Routledge, 1998).

—, *An Introduction to Visual Culture* (London: Routledge, 1999).

Mitchell, W. J. T., 'Romanticism and the Life of Things: Fossils, Totems, and Images', *Critical Inquiry*, 28:1 (2001), pp. 167–84.

Mitman, G., M. Murphy and C. Sellers, 'Introduction: A Cloud Over History', *Osiris*, 19 (2004), pp. 1–17.

Moore, K. Z., 'Viewing the Victorians: Recent Research on Victorian Visuality', *Victorian Literature and Culture*, 25:2 (1997), pp. 367–85.

Mort, F., *Dangerous Sexualities: Medico-Moral Politics in England Since 1830* (London: Routledge, 1987).

Mortier, R., 'Illustration of Science and Technology and Artistic Perception in Eighteenth Century France', in B Castel, J. A. Leith and A. W. Riley (eds), *Muse and Reason: The Relation of Arts and Sciences 1650–1850* (Canada: Queen's Quarterly, 1994), pp. 27–42.

Morus, I. R., 'Seeing and Believing Science', *Isis*, 97 (2006), pp. 101–10.

—, '"More the Aspect of Magic than Anything Natural": The Philosophy of Demonstration', in A. Fyfe and B. Lightman (eds), *Science in the Marketplace: Nineteenth-Century Sites and Experiences* (Chicago, IL: University of Chicago Press, 2007), pp. 336–70.

Moulton, H. J., *Houdini's History of Magic in Boston, 1792–1915* (Illinois: Meyer Books, 1983).

Naville, E., *Excavations at Abydos: From the Smithsonian Report for 1914* (Washington, DC: Government Printing Office, 1915).

Newcomb, S., *Popular Astronomy* (New York: Harper, 1878).

—, *Astronomy for Everybody* (London: McClure, Phillips & Co.,1902).

Nightingale, F., 'November 19th, 1849', in A. Sattin (ed.), *Letters From Egypt: A Journey on the Nile, 1849–50* (New York: Grove Press, 1987), p. 23.

—, 'Ipsamboul, Jan 17th, 1850', in Sattin (ed.), *Letters*, p. 99.

—, 'Letter to Dr Gillham Newlett, 27 July 1883', in L. McDonald (ed.), *Florence Nightingale on Public Health* (Ontario: Wilfrid Laurier University Press, 2004), pp. 567–8.

—, 'Notes on Nursing for the Labouring Classes (1860)', in McDonald (ed.), *Florence Nightingale*, pp. 17–160.

Noakes, R., '"Instruments to Lay Hold of Spirits": Technologizing the Body of Victorian Spiritualism', in I. R. Morus (ed.), *Bodies/Machines* (Oxford: Berg, 2002), pp. 125–63.

—, '"The Bridge which is Between Physical and Psychical Research": William Fletcher Barrett, Sensitive Flames and Spiritualism', *History of Science*, 42 (2004), pp. 419–64.

—, 'Cromwell Varley FRS, Electrical Discharge and Victorian Spiritualism', *Notes and Records of the Royal Society of London*, 61 (2007), pp. 5–22.

O'Brien, F.-J., 'The Diamond Lens', *Atlantic Monthly* (January 1858), pp. 354–67.

O'Connor, R., *The Earth on Show: Fossils and the Poetics of Popular Science, 1802–1856* (Chicago, IL: University of Chicago Press, 2007).

—, 'Reflections on Popular Science in Britain: Genres, Categories, and Historians', *Isis*, 100 (2009), pp. 333–45.

Oliver, J., and S-K. Templeton, 'PM Admits Eye Damage', *Sunday Times*, 11 October 2009, p. 2.

'The One-Eyed Man May Be King, But For How Long?', *Sunday Times*, 11 October 2009, p. 20.

Oppenheim, J., *The Other World: Spiritualism and Psychical Research in England, 1850–1914* (Cambridge: Cambridge University Press, 1985).

Otis, L., *Membranes: Metaphors of Invasion in Nineteenth-Century Literature, Science, and Politics* (Baltimore, MD: Johns Hopkins University Press, 1999).

Otter, C., *The Victorian Eye: A Political History of Light and Vision, 1800–1910* (Chicago, IL: University of Chicago Press, 2008).

—, 'Victorian "Ways of Seeing"?', *Journal of Victorian Culture*, 14:1 (2009), pp. 95–102.

Owen, A., *The Darkened Room: Women, Power, and Spiritualism in Late Victorian England* (Chicago, IL: University of Chicago Press, 1989).

Owen, G., 'Fears for Brown as No 10 Reveals New Eye Damage', *Mail on Sunday*, 11 October 2009, p. 6.

Parkinson, R., *Cracking Codes: The Rosetta Stone and Decipherment* (Berkeley, CA: University of California Press, 1999).

Pelling, M., 'The Meaning of Contagion: Reproduction, Medicine and Metaphor', in A. Bashford and C. Hooker (eds), *Contagion: Historical and Cultural Studies* (London: Routledge, 2001), pp. 15–38.

Petrie, W. M. F. (ed.), *Egyptian Tales Translated From the Papyri: First Series. IVth to XIIth Dynasty* (London: Methuen & Co, 1895).

—, 'History in Tools', in *Smithsonian Report for 1918* (Washington, DC: Government Printing Office, 1920), pp. 563–72.

—, 'The Grand Tour – Three Thousand Years Ago', *Harper's New Monthly Magazine*, 77 (June – November 1888), pp. 297–306.

—, 'Notebook Extract' in M. Drower (ed.), *Letters from the Desert: The Correspondence of Flinders and Hilda Petrie* (Oxford: Aris & Phillips, 2004), p. 69.

Petrie, W. M. F., C. Smith, E. Gardner and B. V. Head, *Naukratis, Part I. Excavation Report* (London: Egypt Exploration Fund, 1885).

Plunkett, J., 'Depth, Colour, Movement: Embodied Vision and the Stereoscope', in J. Lyons and J. Plunkett (eds), *Multimedia Histories: From the Magic Lantern to the Internet* (Exeter: University of Exeter Press, 2007), pp. 117–31.

'The PM's Eyes: An Insight', *Independent*, 10 November 2009, p. 13.

Poe, E. A., 'The Masque of Red Death', in *Selected Tales*, ed. D. Van Leer (Oxford: Oxford University Press, 1998), pp. 129–34.

Polidori, J., 'The Vampyre', in A. O. Gladwell and J. Havoc (eds), *Blood and Roses: The Vampire in 19th Century Literature* (London: Creation Press, 1992), pp. 29–43.

Polidoro, M., *Final Séance: The Strange Friendship Between Houdini and Conan Doyle* (Amherst, MA: Prometheus Books, 2001).

Pratt, M. L., *Imperial Eyes: Travel Writing and Transculturation* (London: Routledge, 1992).

Prescott, C. E., and G. A. Giorgio, 'Vampiric Affinities: Mina Harker and the Paradox of Femininity in Bram Stoker's *Dracula*', *Victorian Literature and Culture*, 33 (2005), pp. 487–515.

Proctor, M., *Romance of the Planets* (London: Harper, 1929).

Quekett, J., *Lectures on Histology, Delivered at the Royal College of Surgeons of England, in the Session 1850–51* (London: Hippolyte Balliere, 1852).

Reid, D. M., *Whose Pharaohs? Archaeology, Museums, and Egyptian National Identity from Napoleon to World War 1* (Berkeley, CA: University of California Press, 2002).

Report on Spiritualism by a Committee of the Dialectical Society (London: Longman, 1871).

'Review of *The War of the Worlds*, Academy, 29 January 1898', in P. Parrinder (ed.), *H. G. Wells: The Critical Heritage* (London: Routledge and Kegan Paul, 1972), pp. 70–4.

Rickards, M., *The Encyclopaedia of Ephemera: A Guide to the Fragmentary Documents of Everyday Life for the Collector, Curator, and Historian* (New York: Routledge, 2000).

Rinn, J. F., *Sixty Years of Psychical Research: Houdini and I Among the Spirits* (New York: Truth Seeker Co., 1950).

Riskin, J., 'Amusing Physics', in B. Bensaude-Vincent and C. Blondel (eds), *Science and Spectacle in the European Enlightenment* (Aldershot: Ashgate, 2008), pp. 43–63.

Ritchie, Jnr, J., 'The Lowell Observatory, Arizona', *Nature*, 14 June 1894, p. 149.

Ritzer, G., *Enchanting a Disenchanted World: Revolutionizing the Means of Consumption* (Thousand Oaks: Pine Forge Press, 1999).

Robins, J., *Miasma: Epidemic and Panic in Nineteenth Century Ireland* (Dublin: Institute of Public Administration, 1995).

Roth, P. A., 'Suddenly Sexual Women', in G. Byron (ed.), *Dracula: Contemporary Critical Essays* (London: Macmillan, 1999), pp. 30–42.

Rupke, N., 'Pro-Vivisection in England in the Early 1880s', in N. Rupke (ed.), *Vivisection in Historical Perspective* (London: Routledge, 1987), pp. 188–208.

Schaffer, S., 'Astronomers Mark Time: Discipline and the Personal Equation', *Science in Context*, 2 (1988), pp. 115–45.

Schickore, J., 'Misperception, Illusion and Epistemological Optimism: Vision Studies in Early Nineteenth-Century Britain and Germany', *British Journal for the History of Science*, 39:3 (2006), pp. 383–405.

—, *The Microscope and the Eye: A History of Reflections, 1740–1870* (Chicago, IL: University of Chicago Press, 2007).

Schiebinger, L., 'Forum Introduction: The European Colonial Science Complex', *Isis*, 96:1 (2005), pp. 52–5.

Secord, J. A., 'Knowledge in Transit', *Isis*, 95 (2004), pp. 654–72.

Serviss, G. P., *Edison's Conquest of Mars* (1898; Burlington: Apogee Books, n.d.).

Sexton, G., *Spirit-Mediums and Conjurers: An Oration* (London: J. Burns, 1873).

Shapin, S, and S. Schaffer, *Leviathan and the Air-Pump: Hobbes, Boyle, and the Experimental Life* (Princeton, NJ: Princeton University Press, 1985).

Sheehan, W., *Planets and Perception: Telescopic Views and Interpretations, 1609–1909* (Tucson, AZ: University of Arizona Press, 1988).

Simmel, G., 'Sociology of the Senses (1907/8)', in D. Frisby and M. Featherstone (eds), *Simmel On Culture* (London: Sage, 1997), pp. 109–20.

—, 'The Metropolis and Mental Life (1903)', in D. Frisby and M. Featherstone (eds), *Simmel On Culture* (London: Sage, 1997), pp. 174–85.

Skloot, R., *The Immortal Life of Henrietta Lacks* (London: Macmillan, 2010).

Smajić, S., *Ghost-Seers, Detectives, and Spiritualists: Theories of Vision in Victorian Literature and Science* (Cambridge: Cambridge University Press, 2010).

Smith, A., *Gothic Radicalism: Literature, Philosophy and Psychoanalysis in the Nineteenth Century* (London: Palgrave Macmillan, 2000).

—, *Gothic Literature* (Edinburgh: Edinburgh University Press, 2007).

Smith, T. S., *The Common Nature of Epidemics, and Their Relation to Climate and Quarantine. Also Remarks on Contagion and Quarantine* (London: John Trubner, 1866).

Sparrow, A., and A. Gabbatt, 'Sun "Exploiting" Grief of Dead Soldier's Mother', *Guardian*, 10 November 2009, p. 7.

Stafford, B., *Artful Science: Enlightenment Entertainment and the Eclipse of Visual Education* (Cambridge, MA: MIT Press, 1996).

Stashower, D., *Teller of Tales: The Life of Arthur Conan Doyle* (New York: Henry Holt, 1999).

Stoker, B., *Dracula*, ed. Maurice Hindle (1897; London: Penguin, 1993).

—, *The Jewel of Seven Stars* (1903; Stroud: Sutton Publishing, 1996).

Stott, R., *Darwin and the Barnacle: The Story of One Tiny Creature and History's Most Spectacular Scientific Breakthrough* (London: Faber & Faber, 2003).

Strachey, J. S. L., 'Review of The War of the Worlds, Spectator, 29 January 1898', in P. Parrinder (ed.), *H. G. Wells: The Critical Heritage* (London: Routledge & Kegan Paul, 1972), pp. 63–6.

Sully, J., *Illusions: A Psychological Study* (London: Kegan Paul, Trench & Co., 1887).

Teale, O., 'The Mystery of Spiritualism – Magic Versus Spiritualism', *Sphinx*, April 1925 –February 1926, n.p.

Thrift, N., *Non-Representational Theory: Space, Politics, Affect* (London: Routledge, 2007).

Tomes, N., *The Gospel of Germs: Men, Women and the Microbe in American Life* (Cambridge, MA: Harvard University Press, 1998).

Tomes, N. J., and J. Harley, 'Introduction to Special Issue on Rethinking the Reception of the Germ Theory of Disease: Comparative Perspectives', *Journal of the History of Medicine and Allied Sciences*, 52:1 (1997), pp. 7–16.

Topham, J., 'Focus: Historicizing 'Popular Science': Introduction', *Isis*, 100 (2009), pp. 310–18.

Tucker, J., *Nature Exposed: Photography as Witness in Victorian Science* (Baltimore, MD: Johns Hopkins University Press, 2005).

Turner, H. S., 'Lessons from Literature for the Historian of Science (and Vice Versa): Reflections on Form', *Isis*, 101:3 (2010), pp. 578–89.

Turner, R. S., 'Consensus and Controversy: Helmholtz on the Visual Perceptions of Space', in D. Cahan (ed.), *Hermann Von Helmholtz and the Foundations of Nineteenth-Century Science* (Berkeley, CA: University of California Press, 1993), pp. 154–204.

'Two Ways to Write a Condolence Letter', *Times*, 10 November 2009, p. 17.

Tyndall, J., 'On Haze and Dust', *Nature* (27 January 1870), pp. 339–42.

—, *On the Scientific Use of the Imagination: A Discourse* (London: Longmans, Green & Co., 1870).

—, 'Science and the Spirits', in *Fragments of Science for Unscientific People: A Series of Detached Essays, Lectures and Reviews* (New York: D. Appleton, 1872), pp. 402–9.

Waddington, K., *The Bovine Scourge: Meat, Tuberculosis and Public Health, 1850–1914* (Woodbridge: Boydell & Brewer, 2006).

Wade, N. (ed.), *Brewster and Wheatstone on Vision* (London: Academic Press, 1983).

Waller, J., *The Discovery of the Germ: Twenty Years That Transformed the Way We Think About Disease* (Cambridge: Icon Books, 2002).

Wallis, J., 'Spiritualism and the (Re-)Enchantment of Modernity'. in J. A. Beckford and J. Wallis (eds), *Theorising Religion: Classical and Contemporary Debates* (Aldershot: Ashgate, 2006), pp. 32–43.

Warwick, A., 'Vampires and the Empire: Fears and Fictions of the 1890s', in S. Ledger and S. McCracken (eds), *Cultural Politics at the Fin-de-Siècle* (Cambridge, Cambridge UP, 1995), pp. 203–14.

—, '"The City of Resurrections": Arthur Machen and the Archaeological Imagination', in R. Pearson (ed.), *The Victorians and the Ancient World: Archaeology and Classicism in Nineteenth-Century Culture* (Cambridge: Cambridge Scholars Press, 2006), pp. 124–38.

Weber, M., *Protestant Ethic and the Spirit of Capitalism* (London: Allen & Unwin, 1930).

—, 'Science as a Vocation', in D. Owen and T. B. Strong (eds), *The Vocation Lectures*, trans. R. Livingstone (Indiana: Hackett, 2004), pp. 1–31.

Wells, H. G., 'Popularising Science', *Nature*, 26 July 1894, pp. 300–1.

—, 'Letter to Grant Richards, Nov. 6, 1895', in D. C. Smith (ed.), *The Correspondence of H. G. Wells: Volume 1 – 1880–1903* (London: Pickering & Chatto, 1998), p. 250.

—, *The Time Machine* (London: Heinemann, 1895).

—, 'Letter to Morris Colles, Jan. 26, 1896' and 'Letter to Morris Colles, Mar. 25, 1896', in D. C. Smith (ed.), *The Correspondence of H. G. Wells: Volume 1 – 1880–1903* (London: Pickering & Chatto, 1998), pp. 260–1.

—, *The War of the Worlds* (1898; London: Penguin, 2005).

—, 'Introduction', in *The Scientific Romances* (London: Gollancz, 1930).

Wheatstone, C., 'Contributions to the Physiology of Vision', *Journal of the Royal Institution of Great Britain*, 1 (October 1830), pp. 205–33.

Whitman, W., *Selected Poems and Prose*, ed. A. N. Jeffares (Oxford: Oxford University Press, 1966).

Wicks, M., *To Mars Via the Moon: An Astronomical Story* (1911; n.p, Dodo Press, 2009).

Willis, M., 'Unmasking Immorality: Popular Opposition to Laboratory Science in Late Victorian Britain', in D. Clifford, E. Wadge, A. Warwick and M. Willis (eds), *Repositioning Victorian Sciences: Shifting Centres in Nineteenth-Century Scientific Thinking* (London: Anthem Press, 2006), pp. 207–18.

Wilson, P. W., 'Elucidating Conan Doyle', *New York Times*, 8 June 1922, pp. 1, 19, 24.

Wise, M. N., 'Making Visible', *Isis*, 97 (2006), pp. 75–82.

Worboys, M., *Spreading Germs: Disease Theories and Medical Practice in Britain, 1865–1900* (Cambridge: Cambridge University Press, 2000).

Wordsworth, W., *Selected Poetry*, ed. S. Gill and D. Wu (Oxford: Oxford University Press, 2008).

Yanni, C., *Nature's Museums: Victorian Science and the Architecture of Display* (New York: Princeton Architectural Press, 2005).

Zimmerman, V., *Excavating Victorians* (Albany, NY: SUNY Press, 2008).

INDEX